Eagle Nest Public Library
00005498

IMPERILED PLANET

IMPERILED PLANET

Restoring Our Endangered Ecosystems

Edward Goldsmith
Nicholas Hildyard
Patrick McCully
Peter Bunyard

The MIT Press
Cambridge, Massachusetts

Managing Editors : David Burnie and Linda Gamlin
Art Direction : Ruth Prentice
Designers : Karen Bowen and Andrew Green
Picture Editor : Shona Wood
Typesetting : Richard Gray
Colour Reproduction : F.E. Burman

First MIT Press edition, 1990

Printed and bound in Spain

Library of Congress Cataloging-in-Publication Data

Imperiled planet: restoring our endangered ecosystems / Edward Goldsmith ... [et al.]. — 1st MIT Press ed.
p. cm.
ISBN 0-262-07132-0
1. Environmental protection. 2. Pollution — Environmental aspects.
I. Goldsmith, Edward, 1928-
TD170.I48 1990
363.7—dc20 90-5981
CIP

CONTENTS

1 GAIA: THE LIVING PLANET *6*

2 A WORLD IN CRISIS
The Balance of Nature *26*
The Changing Atmosphere *40*
Forests *54*
Agricultural Lands *96*
Rangelands *116*
Rivers *128*
Groundwater *142*
Wetlands and Mangroves *150*
Coasts and Estuaries *164*
Seas and Oceans *172*
Coral Reefs *186*
Islands *194*
Mountains *200*
Deserts *208*
Antarctica *218*
The Arctic *226*

3 THE HUMAN DIMENSION
The Diminishing Quality of Life *240*
The Future in Prospect *252*
The Dynamics of Destruction *264*
Solutions for Survival *272*

Index *283*
Acknowledgments *287*

1

GAIA
The Living Planet

Overwhelming Gaia *10*

◄► Lightning flashes over the Kavir National Park in Iran, as a storm brings revitalizing rain to an arid landscape. For millions of years, life has successfully met the challenges posed by habitats as diverse as deserts and the depths of the oceans. However, as the present century nears its end, many species face an uncertain future, as a result of habitat destruction by mankind.

Overwhelming GAIA

When, in the 1960s, the first astronauts looked back at the Earth, they saw a planet manifestly alive. Sapphire-blue seas, brown land, green forests, ice-capped poles, white clouds, mountains, deserts, cities, and - passing over this astonishing world - the daily rhythm of day and night. It was a remarkable sight. Through the images transmitted back to Earth, it was also one shared by millions around the globe. To see the planet in its entirety was in itself an unprecedented experience: to be able to witness its grandeur, its beauty and its vitality posed a challenge to conventional perceptions as profound as that which followed the first circumnavigation of the globe. This was no ball of inert matter, colonized haphazardly by organisms. On the contrary, the Earth and its life forms took on a new unity. It seemed quite simply as if the whole planet were a single living organism.

To many of those who watched, enthralled, as the first images flickered back from space, the concept that the Earth could indeed be 'alive' might have seemed novel. In fact, it is an idea that is older than civilization. The ancient Greeks, for example, deified the Earth, revering her as Gaia, the Earth Goddess. She was responsible for the well-being of her domain and would reward care for her bounties - the forests, rivers and creatures that inhabited the world - with good harvests and game to eat. But abuse her, destroy her forests, erode her hills, kill her creatures or pollute her rivers and lakes, and she would avenge herself through the agency of natural catastrophes - through drought, famine, earthquakes, violent storms and disease.

◀ The mountainous Hawaiian Island chain, seen by a camera aboard the Space Shuttle *Discovery*. From space, it is clear that the part of our planet that supports life - the 'biosphere' - is no more than a thin skin bounded by the ocean floor and the tops of mountains.

For thousands of years, the vast majority of human societies observed the mysteries of Gaia, and many continue to do so, living well within the limits of their natural environment. Their success is a reflection, not only of their social and economic organization, but also of their view of the natural world. The reverence shown to Gaia by the ancient Greeks, long out of favour in Europe and North America, now finds expression in the views of such tribal groups as the Yacuna Indians of western Amazonia.

The Yacuna do not regard the forests as existing simply for the benefit of themselves or of human beings in general. On the contrary, embedded in their world-view is the notion that each living part of the rainforest must be given the opportunity to exist in order to sustain the integrity of the whole, and that without that wholeness will come disease, disaster and death. The Yacuna see nature and the whole of creation as participating in an intricate network of exchanges and believe that both animals and plants are similar to them. They thus respect the territories and ways of life of their fellow creatures; they know how each behaves and they fear the consequences of any abuse.

To those raised in today's industrial society, the Yacuna and other groups like them may appear as quaint throwbacks, societies that 'progress' has passed by, leaving them at the mercy of a hostile environment. Through inquisitiveness, ingenuity and the assiduous application of science and technology, modern man may appear to have gained increasing mastery over the natural world - to the extent that the majority of our citizens happily eat three meals a day without a thought to where their food comes from.

Gaia has in effect been defrocked, removed from her pedestal and cast in the role of an unruly servant. We have taken seriously the exhortation of the 17th-century philosopher Francis Bacon to "wrest nature's secrets from her breasts, womb and bowels", moulding the natural world to our purpose. The invention of the steam engine, and later the internal combustion engine, gave us the ability to extract resources and manufacture goods on a scale undreamed of by pre-industrial technologists. The advent of the petrochemical revolution in the 1940s brought even greater powers, enabling the mass production of chemicals unknown in nature, the pesticide DDT being just one example. Today, we have moved into an era of unprecedented intervention, genetic engineering giving us the ability to create entirely new life-forms. Such is our faith in science and technology that at times it has seemed that there are no limits to what we might achieve. In 1966, the then Vice-President of the United States, Hubert Humphrey, told an audience of young scientists: "This can indeed be the Age of Miracles. It will be your age." By the year 2000, he predicted, scientists would have achieved "the virtual elimination of bacterial and viral diseases", "the correction of hereditary defects", and "the landing of men on Mars". Today, some scientists have even predicted infinite human lifespans, effectively putting an end to death itself. Other predictions include making available abundant cheap energy through nuclear fusion, setting up space colonies and redesigning the human body.

THE PRICE OF 'PROGRESS'

Today, the young scientists who listened to Hubert Humphrey might be wondering what happened to the Utopia he predicted. Despite two decades of economic growth and increasing technological sophistication, his Age of Miracles has palpably failed to materialize. Indeed, the technologies that have seemingly delivered nature into our hands - and the social and economic systems these technologies support - have wrought such destruction that the very survival of much of life on Earth is now in question. To manufacture the material goods that we associate with wealth, we have set about systematically transforming nature, with disastrous consequences. Whole mountains have been torn apart, seas, rivers and oceans polluted, forests destroyed, farmlands degraded, communities uprooted, and even the very chemistry of the atmosphere disrupted.

Undoubtedly there have been benefits, but they will soon be dwarfed by the costs of the destruction. For the vast bulk of humanity, the future is bleak indeed. As a 1980 report to President Carter concluded: "If present trends continue, the world in 2000 will be more crowded, more polluted, less stable ecologically, and more vulnerable to disruption than the world we live in today... Despite greater material output, the world's population will be poorer in many ways than they are today. For hundreds of millions of the desperately poor, the outlook for food and other necessities of life will be no better. For many it will be worse. Barring revolutionary advances in technology, life for most people on Earth will be more precarious than it is now - unless the nations of the world act decisively to alter current trends."

▲ An Orang Asli village in the highlands of Malaysia. Like the Amazonian Indians, the indigenous Orang Asli live in harmony with their surroundings. They hunt and practice slash-and-burn agriculture, growing rice and tapioca on patches of ground that they abandon after two seasons, letting the forest recolonize the land.

A GLOBAL IMPACT

Ours is not the first civilization to have devastated its environment. The archeological record is littered with the ruins of past civilizations which exceeded their ecological limits, leaving behind them a legacy of degraded landscapes. Today, it is difficult to believe that the sands which now drift across the once-splendid city of Babylon in Iraq used to grow the grain that fed a great civilization, or that the barren hills of modern Lebanon supported vast cedar forests which furnished King Solomon with the wood to build the Great Temple in Jerusalem. But they did - and it was human hands which brought them to their present state. Stripped of protective tree cover, relentlessly overgrazed and exploited, these previously fertile lands fell easy prey to soil erosion and were quickly reduced to arid, unproductive scrub or desert.

There are important differences, however, between the destruction of the past and that being caused today. In the past, ecological damage was limited in scale and extent, both because of the technologies involved and because the activities only affected relatively small areas. By

"Under the philosophy that now seems to guide our destinies, nothing must get in the way of the man with the spray gun. The incidental victims of his crusade against insects count as nothing; if robins, pheasants, raccoons, cats, or even livestock happen to inhabit the same bit of earth as the target insects and to be hit by the rain of insect-killing poisons no one must protest."
Rachel Carson

contrast, the consumer culture that propels our own headlong dash to destruction is generalized throughout the globe. For our economic system to survive, it is essential that more and more people are drawn into the system.

In that sense, our civilization is unique, for its impact is now global, destroying or undermining the viability of ecosystems from pole to pole. Toxic wastes poured into holes into the ground or discharged directly into waterways have contaminated rivers and groundwaters throughout the world *(see pp.128-149)*, polluting water supplies and destroying fisheries. Overintensive agriculture is condemning 200,000 km^2 (77,000 square miles) of land to becoming desert or semi-desert every year, with almost a quarter of the Earth's land surface now affected. Tropical forests, temperate forests, mangroves, wetlands, rangelands, oceans and seas - all are now threatened through our activities. Even Antarctica, the pristine continent, is in jeopardy, its mineral and oil resources being sized up for commercial exploitation.

Underlying this global assault is a way of life that demands more from the environment than the environment can sustain. We have become dependent on products and processes that are inherently destructive, either through their manufacture, their use or their disposal - or all three. To get to work, for example, many must drive by car. Making a car requires steel, copper, aluminium, plastics, rubber, water and numerous chemicals. Obtaining the steel requires mining the iron-ore, smelting it, cooling it. Fuelling a car requires petroleum. Obtaining the petroleum requires oil wells and refineries. Driving a car requires roads. Building roads demands the sacrifice of land, much of it highly productive. Nor does the process stop there. Every time it is driven, a car emits a range of harmful chemicals, some detrimental to human health and some destructive of the environment. And when, finally, the car has reached the end of its useful life, it must be disposed of, creating yet more environmental problems.

The warning lights have been flashing for decades. In the late 1950s, the poisoning of local people at Minamata in Japan, after they had eaten fish contaminated with mercury, highlighted the danger of chemicals dumped in the environment, and their potential for reaching toxic levels as they move through the food chain. A few years later, the dangers of pesticides were eloquently portrayed by Rachel Carson in her now classic book, *Silent Spring.*

In the early 1970s, the dangers of acid rain, the potential threat to the ozone layer, and the imminent disruption of our climate were first brought to public attention. Before the decade was out, the seeping of chemicals from an old toxic waste dump at Love Canal near Niagara City in the United States, and the subsequent evacuation of nearby residents, had brought home to the public the legacy of pollution left by years of irresponsible waste disposal. By the end of the 1980s, we had seen radioactive fallout over whole regions of Europe and the United States, following the explosion of a nuclear reactor at Chernobyl in the USSR, and enough signs of change in the climate for even mainstream climatologists to warn that 'global warming' was already becoming a reality.

▶ A woman tending a contaminated field in the Ukraine following the Chernobyl disaster. Unlike some forms of pollution, radiation is invisible, and its effects only become apparent through the passage of time. Many radioactive isotopes remain dangerous for decades or centuries, so contamination cannot easily be dealt with. For this old lady, the need to grow food has overcome the fear of a baffling and unseen threat.

▶ The green waters of a polluted river in the Austrian Tyrol. As more and more synthetic chemicals are used by industry, the chances of accidental leakages like this are constantly increasing. Pollution is rarely localized, because water and air can carry contaminants far and wide, spreading them throughout the environment. As we approach the end of the twentieth century, we are only just beginning to appreciate pollution's potential effects.

The Solar System

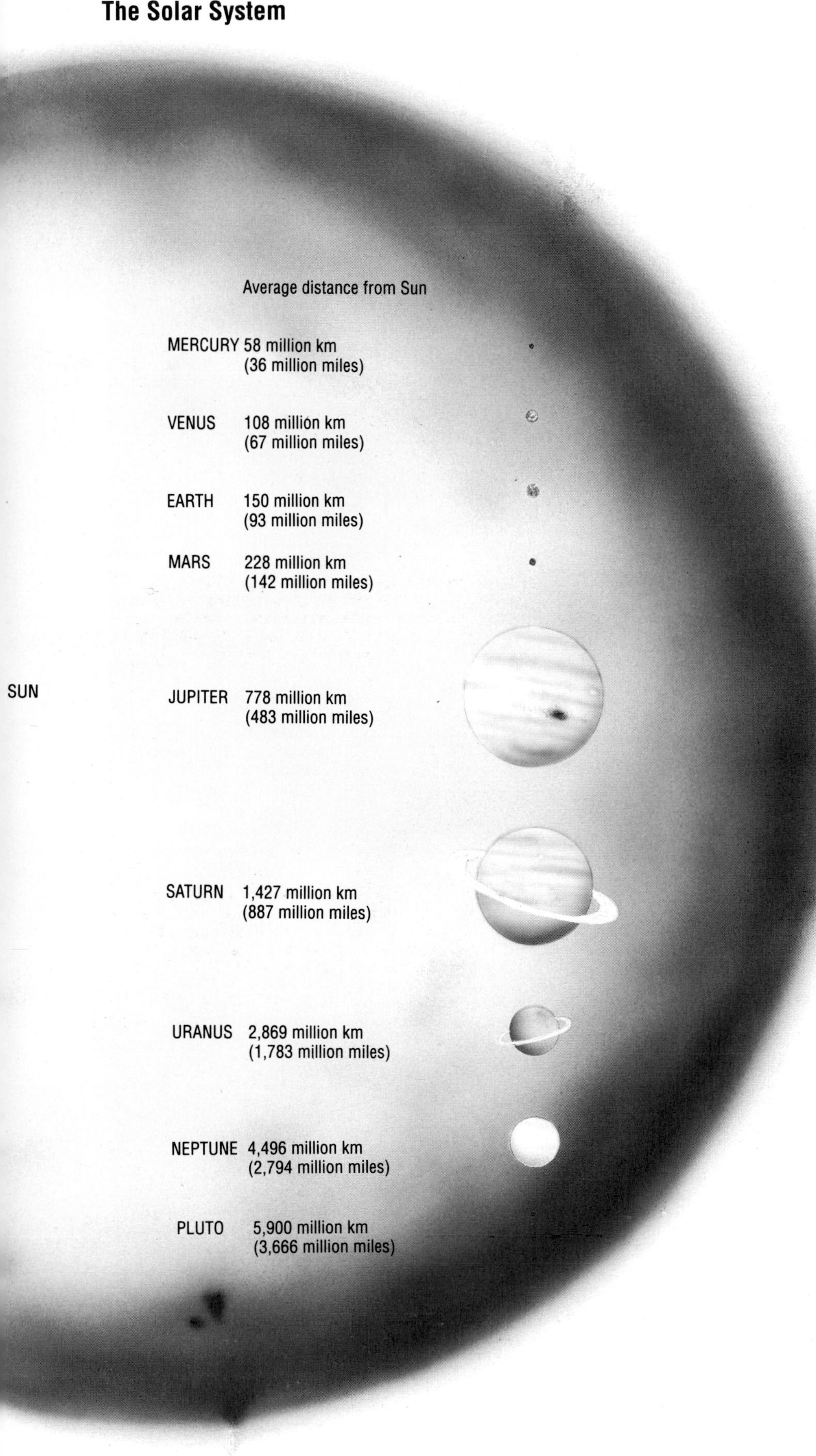

The danger is that we have gone beyond simply damaging ecosystems and are now disrupting the very processes that keep the Earth a fit place for higher forms of life. Indeed, the more we learn about life on Earth, the less fanciful the ancient idea of Gaia seems.

THE GAIA HYPOTHESIS

For life as we know it to continue, the balance of gases in the atmosphere must remain within certain limits. For the past 250 million years, oxygen levels in the atmosphere have remained remarkably constant. It is fortunate that they have done so. For should levels of oxygen fall to, say, 16 per cent, then fast-metabolizing animals such as humans would be left gasping for breath. Should levels rise to 25 per cent, the atmosphere would support combustion more easily. The surface of the Earth would be consumed by raging fires, even the wettest tropical forest catching alight with the first lightning strike.

We tend to take the Earth's atmosphere for granted. But we would be wrong to do so. As the atmospheric chemist James Lovelock points out, the chemical composition of the Earth's atmosphere is "highly improbable", given the expectations of orthodox chemistry. It contains gases which, left to their own devices, should react with each other at such a rate that only traces of the original gases remain.

For example, oxygen is found at the same time as methane, which can mop up oxygen. Certain derivatives of oxygen react with methane in the presence of sunlight to yield carbon dioxide and water. Without new sources of methane and oxygen to compensate for that used up, these gases would rapidly be exhausted.

This is indeed the case with Mars and Venus, the two flanking planets to Earth, both of which have atmospheres with minimal traces of oxygen and no methane. The oxygen content of the Earth's atmosphere, by contrast, is just under 21 per cent and levels of methane are 10^{35} (that is, 1,000 million trillion trillion) times greater than would be expected given such levels of oxygen. The explanation, argues Lovelock, is that both the methane and oxygen are being replenished by living organisms. In effect, it is life on Earth which keeps the Earth's atmosphere in the unique state that permits life to survive.

There is evidence, too, that life plays a vital role in regulating the carbon dioxide content of the atmosphere and thus, in part, the surface temperature of the Earth. Unlike the Earth, both Venus and Mars have atmospheres that are rich in carbon dioxide, with concentrations greater than 95 per cent. By contrast, the carbon dioxide content of the Earth's atmosphere is just 0.03 per cent. Carbon dioxide is a powerful greenhouse gas, meaning that it lets through the heat from the Sun but traps some of the heat that is radiated back. As a result of that heat-trapping blanket of carbon dioxide, surface temperatures on Venus, which is closest to the Sun, are a roasting 450°C (840°F), while Mars, furthest from the Sun, is considerably warmer than it would otherwise be at -53°C (-63°F).

GAIA AT WORK

The Earth had similar origins to Venus and Mars, and it seems likely that its atmosphere, some 4 billion years ago, was also rich in carbon dioxide. At that time, the Sun was about 25 per cent cooler than it is today, and only the high carbon dioxide content of the Earth's atmosphere would have kept our planet warm enough for life to begin. But as the Sun warmed up, that carbon dioxide blanket would have caused the Earth to become hotter and hotter.

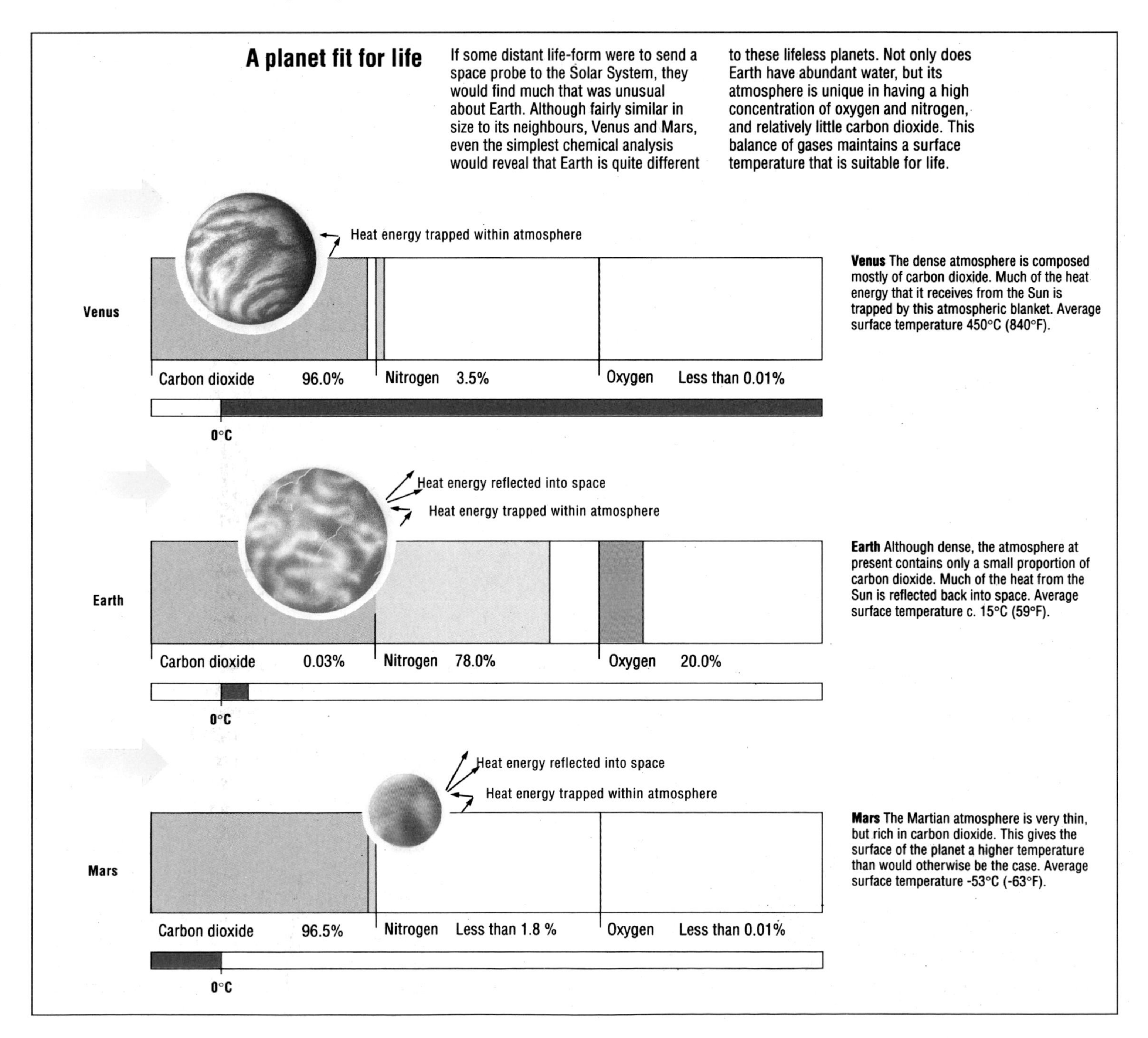

▲▲ Stromatolite mats at Shark Bay, Western Australia. Stromatolites are hard, cushion-like structures formed by cyanobacteria, shown immediately above. Stromatolites are the most ancient organic structures to be found on Earth.

How then did living organisms manage to reduce the carbon dioxide blanket at a rate which kept the average surface temperature of the Earth at an optimum for life? The answer lies in the ability of some organisms to make energy-rich substances, called carbohydrates, by combining carbon dioxide and water. They do this in a process known as photosynthesis, because it uses energy from sunlight to make the reaction work. As well as taking carbon dioxide out of the atmosphere and making carbohydrates, photosynthesis also generates oxygen.

When life first appeared on Earth, perhaps as long as 3.9 billion years ago, it would seem that photosynthesis was unnecessary because the seas abounded with energy-rich compounds. But these soon got used up and for life to continue it was vital that another source of food be generated. We now have evidence from fossilized remains in the Earth's oldest rocks that some of the first photosynthesizing organisms were cyanobacteria ('blue-green algae'), whose modern descendants are to be found in many different places including the stromatolite mats of some subtropical coasts.

◀ The history of the Earth's atmosphere, preserved in rock in the Badlands of North Dakota. The bands of colour that can be seen in the hills are evidence of an atmospheric 'switch-over' that occurred over 2,000 million years ago. It was at about this time that photosynthesizing organisms first began to release large amounts of oxygen into the atmosphere. The oxygen combined with minerals in the soil, producing a distinctive colour and thereby laying down evidence about out planet's past.

"No matter how we care to divide the phenomenon of life, regardless of the names we choose to give to species or the shapes we devise for family trees, the multifarious forms of life envelop our planet and, over aeons, gradually but profoundly change its surface. In a sense, life and Earth become a unity, each working changes on the other."

Lynn Margulis, Professor of Biology, Boston University

"To the centre of the world you have taken me and showed the goodness and beauty and strangeness of the greening Earth, the only mother."

Black Elk, holy man of the Sioux, 1931

▲▲ Chalk cliffs on the north coast of France - the result of millions of years of calcium carbonate deposition by marine animals. Today, the sea is reclaiming these cliffs, but yet more animals are laying down chalk on the sea floor.

▲ A nudibranch or sea-slug. Like most organisms, it is 'aerobic', needing oxygen to live. It absorbs this through its plume-like gills.

Those ancient, microscopic organisms not only released the first molecules of free oxygen, thereby paving the way for respiring organisms such as ourselves. They also began the process of depositing carbon dioxide as calcium carbonate ($CaCO_3$) or chalk. Since then, a host of different organisms, including single-celled marine plants and animals, molluscs and corals, have acquired the ability to lay down external chalky shells and skeletons. It is this process, Lovelock argues, that provided the mechanism whereby the Earth's carbon dioxide blanket was reduced. In effect, the white cliffs of Dover, the Great Barrier Reef and other limestone deposits may be seen as vast natural dumps of carbon dioxide.

Today, through the use of satellite images taken over the oceans, we can see enormous blooms of microscopic plants called coccolithophorid algae, sometimes more than 160 km (100 miles) long, which have turned the surface of the sea milky-white as a result of their chalky shells. Organisms such as these are still carrying on the Gaian task of countering the warming of the Earth's surface as the Sun gets hotter.

CRITICAL CYCLES

The laying down of limestone rock depends on a complicated cycle, one in which living organisms play a crucial role. Rain, falling through the air,

"Ecology has caught us by the throat."
Mikhail Gorbachev

picks up carbon dioxide gas to form a weak acid, carbonic acid (H_2CO_3). This acid causes the slow weathering of rock. If the rock is limestone, then the weathering process brings about the release of soluble bicarbonate ions (HCO_3-) which, when washed into rivers and the sea, provide the material from which marine organisms lay down their skeletal shells. Meanwhile, plants on the land accelerate the weathering process through releasing carbon dioxide, which can then form carbonic acid, into the soil around their roots. The plants benefit from the release of nutrients in the soil.

How much carbon dioxide remains in the atmosphere depends on many factors. For example, it is known that photosynthesizing plants tend to increase their growth as the carbon dioxide content of the atmosphere rises. This increased growth, by its very nature, tends to reduce the carbon dioxide levels. The cycle therefore regulates itself by a process known as feedback, which in this case controls the amount of carbon dioxide in the atmosphere.

Sulphur is another essential nutrient which, like the bicarbonate from the soil, tends to run off into the sea, entering a cycle in which living organisms are involved. Here, simple plants such as coccolithophorid algae have evolved an ingenious use of the sulphur, creating a special compound in their cells which prevents salts in the seawater from entering. When this compound eventually decomposes, after the death of the coccolithophorids, it produces the volatile substance, dimethysulphide. This quickly oxidizes in the atmosphere above the seas to give sulphur dioxide (SO_2). The sulphur dioxide has the effect of causing rainclouds to form over the oceans. These clouds then carry rain to the land, and with the rain comes sulphurous acid (H_2SO_3), produced by the combination of sulphur dioxide and water.

▼ **A forest killed by acid rain in Czechoslovakia. The exact way in which acid rain causes this sort of devastation is, as yet, not fully understood. All too often, the inability of scientists to fully explain such chains of cause and effect is used as an excuse for inaction. By the time corrective measures are taken, it may already be too late.**

OVERLOADING THE SYSTEM

Acid rain is thus a perfectly normal and essential phenomenon. But we, with our highly industrialized society, have added a great deal more sulphur dioxide to the natural output and begun to overload the natural system, overwhelming its controls. Acid rain has therefore become an acute problem in many parts of the world, turning lakes and rivers in Scandinavia, Scotland and North America into waters that can no longer support even the hardiest species of fish.

In relatively pollution-free areas, such as the more northerly parts of Canada, sulphur fallout is less than 10 kg/ha (9 lb/acre) per year. The average levels in areas afflicted by acid rain, such as southern Sweden, are two to three times higher. In Sweden, at least 18,000 out of a total of 20,000 lakes are now so acid that fish cannot live in them. Soils too are suffering, and some regions of southern Sweden are showing a tenfold increase in acidity. One of the consequences of this is that toxic minerals such as aluminium, cadmium and mercury have become soluble and therefore available to plants, with disastrous effects. Normally such toxic minerals remain harmlessly bound to the soil.

Worldwide, some 100 million tonnes of sulphur a year enters the atmosphere as a result of human activities, most of it coming from the burning of fossil fuels. Another source, often overlooked, is untreated sewage discharged into the sea. This helps stimulate the growth of those algae that generate dimethylsulphide.

Modern farming, with its heavy inputs of artificial fertilizers, also plays havoc with natural Gaian cycles. In the Netherlands, as much as 580 kg/ha (518 lb/acre) of nitrogen (in the form of nitrates or ammonium salts) are applied every year to farmland as fertilizer. At least 10 per cent of this evaporates directly into the atmosphere. In areas of northern Europe where fertilizer use is heavy, the fallout of nitrogenous compounds from the atmosphere has increased twenty-fold or even more. The rain falling through the forests of northern Europe is now so rich in nitrogen compounds that the trunks and branches of many trees have become covered in green slime, as microscopic, single-celled algae flourish on the fertilized bark.

INTO THE FUTURE

One of the major concerns arising out of the Gaia theory is that we are pushing natural processes beyond their capacity to maintain an atmosphere fit for higher forms of life. Beyond a certain point, the system may 'flip' to an entirely new state which could be extremely uncomfortable for life as we know it. At present we have no idea whether we are heading for a 'flip' but, once triggered, the change to the new state could occur with extreme rapidity - in perhaps as little as a few decades.

Can we survive such a 'flip' and - even more important - can we avert it? Whatever the answer, it would be foolhardy in the extreme to continue with our polluting ways in the hope that the changes we might be causing to the land, sea and air, are just transitory.

2

A World in CRISIS

The Balance of Nature *26*
The Changing Atmosphere *40*
Forests *54*
Agricultural Lands *96*
Rangelands *116*
Rivers *128*
Groundwater *142*
Wetlands and Mangroves *150*
Coasts and Estuaries *164*
Seas and Oceans *172*
Coral Reefs *186*
Islands *194*
Mountains *200*
Deserts *208*
Antarctica *218*
The Arctic *226*

◀▶ Rock Island Bend on the Franklin River, Tasmania. In the 1980s, this spectacular river was the scene of a major confrontation between the hydroelectric industry and conservationists. The region has since been declared a World Heritage Site.

The BALANCE of NATURE

In the 1960s, in an attempt to eradicate malaria, the World Health Organization embarked on a major campaign to rid the tropics of the mosquitoes that carry the disease. Borneo was to be one of the regions cleared, and a massive spraying campaign was initiated throughout the worst affected areas. The chosen pesticide was DDT, a highly toxic and cancer-causing chemical since banned in most western countries but still widely used in the Third World.

Initially the programme was successful and the mosquito population fell dramatically. But it was not only mosquitoes that died. Numerous other species were poisoned by the DDT, among them a minute wasp that preyed upon caterpillars living in the thatch of local houses. With the wasp gone, the caterpillar numbers increased to plague proportions, devouring the roofs of houses and causing them to collapse. Nonetheless, the spraying programme continued. The dead mosquitoes were eaten by gecko lizards which, as they became sick, proved easy prey for the local cats. As a result the cats accumulated large quantities of DDT, passed on from insect to lizard to cat. The cats began to die in their thousands - and the local population of rats exploded. The rats not only ate local crops but brought an even greater menace - bubonic plague. In desperation the Borneo government called for cats to be parachuted into the worst affected regions.

Today, the mosquitoes have returned to the sprayed areas and malaria is still rife. Many pesticides are now ineffective, the mosquitoes having developed resistance to them. But, as in other parts of the world, the spraying goes on unabated. And the subtle balance of nature continues to be disrupted.

◀ On a tropical island, palm forest stretches down to the sea, and coral reefs fringe the shore. When forests are clear-felled, the soil beneath is left unprotected and much of it is washed away by the heavy tropical rainfall. Soil carried into the sea by rivers then chokes the delicate corals. This sort of disregard for the interdependence of natural systems means that the whole planet is under threat from human activities.

ALL THINGS CONNECT

"No man is an island, entire of it self", wrote the Elizabethan poet John Donne. And so it is with the species of plants and animals that make up the natural world. None can survive in isolation. To eat, the lion must prey on the gazelle, the gazelle must graze on grass, and the grass must extract nutrients from the soil. And all are dependent on the Sun, without whose energy there would be little life on Earth.

The natural world is thus much more than a haphazard collection of plants and animals. The food chain that links one organism to another binds each into an interdependent community - or ecosystem - in which all living creatures, however small, have their place and function. At the bottom of the food chain are the green plants - flowering plants, ferns, mosses, seaweed and microscopic algae. These primary producers take carbon dioxide from the atmosphere and water from the soil and use the radiant energy from the Sun to produce energy-rich glucose, a type of sugar. In the process, they release oxygen into the atmosphere, without which plants and animals would be unable to survive.

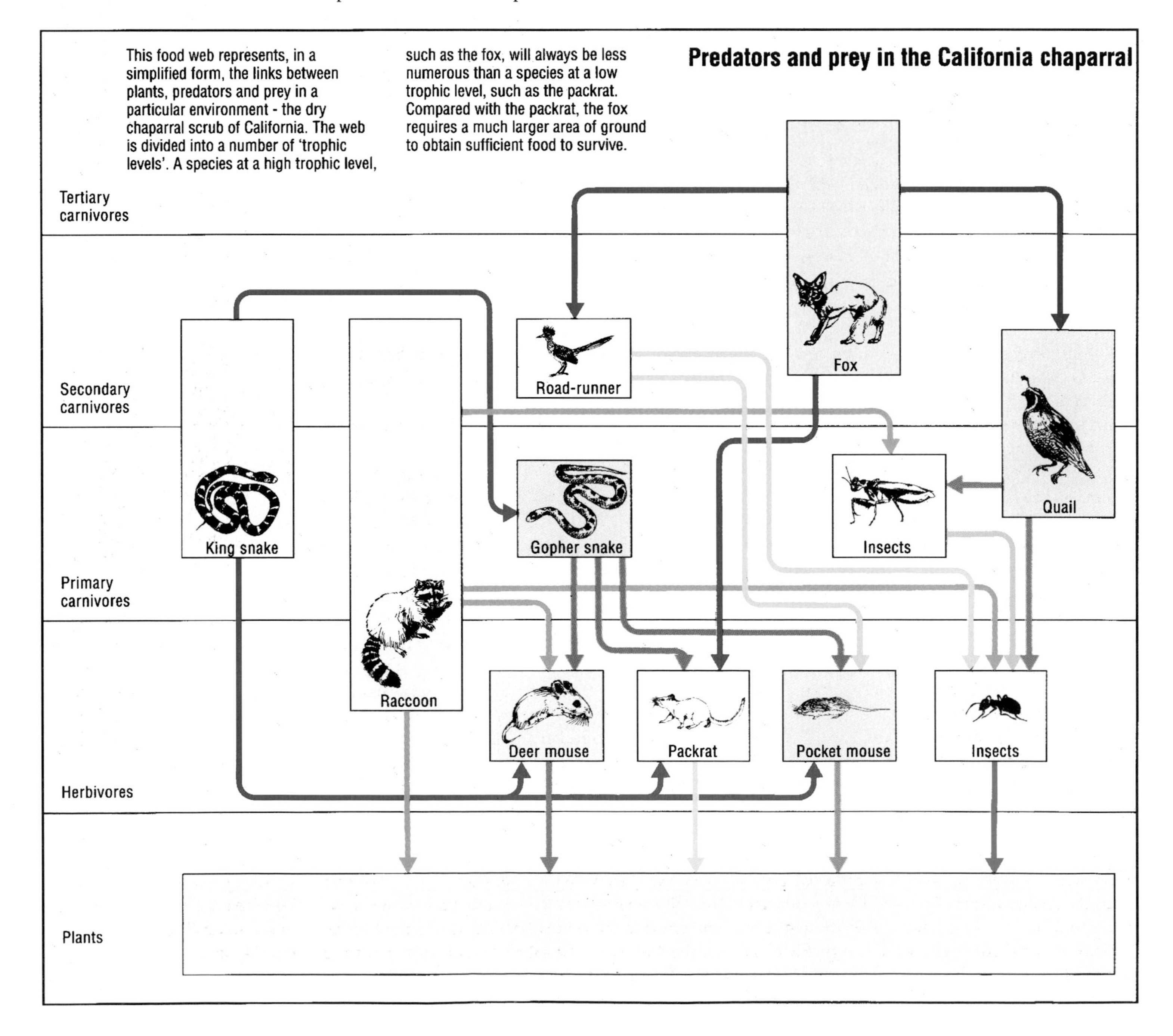

This food web represents, in a simplified form, the links between plants, predators and prey in a particular environment - the dry chaparral scrub of California. The web is divided into a number of 'trophic levels'. A species at a high trophic level, such as the fox, will always be less numerous than a species at a low trophic level, such as the packrat. Compared with the packrat, the fox requires a much larger area of ground to obtain sufficient food to survive.

Feeding on the primary producers are plant-eaters or herbivores - on land, deer, kangaroos, rabbits, caterpillars and the like, in the oceans, microscopic animals of the zooplankton, which graze on the tiny algae. These plant-eating primary consumers are preyed upon by meat-eating secondary consumers, who themselves are preyed upon by the larger meat-eaters such as the big cats, birds of prey and sharks at the top of the food chain. Meanwhile, decomposers, such as bacteria and fungi, break down dead matter, recycling its nutrients through the ecosystem. Bacteria and fungi also play a crucial role in dissolving minerals out of rocks, making the minerals available to plants and animals.

While green plants 'breathe in' carbon dioxide and release oxygen when they are photosynthesizing, primary and secondary consumers inhale oxygen and release carbon dioxide. This kind of give-and-take exchange between one living organism and another is a fundamental feature of the natural world. Deer, for instance, eat grasses and the leaves of trees but their droppings feed the beetles, fungi and bacteria which break down the dung, providing nutrients for the plants that keep the deer alive. Similar relationships, but at a microscopic level, go on between many trees and soil bacteria. Sugary substances, released from a tree's roots, encourage bacteria to proliferate. Living off the bacteria are tiny amoebae that excrete ammonia, which is then turned into nitrate by other bacteria and taken up by the trees for their own growth. By exuding sugars, a tree can increase the amount of ammonia available to its roots. Wherever we look, we see such mutually beneficial connections: we ourselves depend on a host of different bacteria, many of which inhabit our gut and without which our health would suffer.

THE WEB OF LIFE

The complex interrelationships between plants and animals in an ecosystem are critical to its stability. They ensure that the flow of energy through the system is kept constant, that nutrients are available, and that waste products are recycled. Although over time, individual species may change through evolution, with some disappearing altogether, while new species appear and proliferate, the overall system is kept in balance.

But that balance is easily upset. Over a century ago, one of the founding fathers of the modern environmental movement, George Perkins Marsh, warned of the ease with which ecosystems could be disrupted. An accomplished amateur botanist, Marsh - who had been America's first ambassador to the Kingdom of Italy - described how, by destroying the larvae of mosquitoes along the banks of rivers, man could have a fatal impact on salmon runs.

From his own patient observations, he had noted that, in wooded areas where mosquitoes and gnats abound, their larvae, which live in water, are the favourite food of the trout. The trout also feed on the larvae of mayflies, which eat the eggs of the salmon. "Hence, by a sort of house-that-Jack-built, the destruction of the mosquito, that feeds the trout that preys on the May fly that destroys the eggs that hatch the salmon that pampers the epicure, may occasion a scarcity of salmon in waters where it would otherwise be abundant," Marsh wrote.

Marsh was not the first to note that the natural world has a critical structure which, if disrupted too greatly, can spell unforeseen and disastrous consequences. Indeed the idea of 'the balance of nature', like that of Gaia, stretches back far into human prehistory. But, although Marsh's writings were widely read at the time, his message fell on deaf ears. His was the heyday of the Industrial Revolution and his views were out of tune with the spirit of his age. The natural world was seen as malleable, man's to shape as

▲ Charles Darwin was among the first scientists to comment on the complex interactions between living things which contribute to the balance of nature. In *The Origin of Species*, he reported that bumble bees were more common in districts with plenty of cats. Bumble bees nest underground, and mice dig up their nests to eat the eggs, pupae and honey, which are highly nutritious. Cats prey on mice and keep their numbers down, so cats indirectly benefit bumble bees.

▲ **The prickly pear - a plant that man has spread around the world. In Australia its numbers reached plague proportions, until it was eventually controlled by introducing a moth whose caterpillar eats the cactus. This was one of the earliest uses of modern 'biological control' methods.**

he pleased. Even today, despite the destruction we have clearly inflicted on the environment, the resilience of nature is all too often taken for granted.

Yet, the more we learn of the workings of the natural world, the clearer it becomes that there is a limit to the disruption that the environment can endure. Like the gossamer threads of a spider's web, an ecosystem can only take so much stress and only repair so much damage. Sever a vital link in the food chain and the ecosystem is thrown into disarray. Destroy an insect's natural predators and the insect's numbers will explode - often to pest proportions - before a balance is restored. Disrupt the recycling of nutrients and the whole ecosystem will begin to decay.

ECOLOGICAL INVASIONS

George Perkins Marsh highlighted just one of the ways in which the natural world can be disrupted. Just as damaging, potentially, as the elimination of a species from an ecosystem, is the introduction of an alien animal or plant.

At first, a newly introduced species may dramatically increase in numbers since there are no natural predators to keep it in check. For instance, the prickly pear cactus, introduced into gardens in Australia, soon spread into the wild. Without any natural controls, it rapidly colonized vast areas of the outback, its sharp spines making it inedible to the native wildlife. One reason for the cactus's success was the previous introduction into Australia of another alien species - the domestic sheep - which had grazed all other vegetation down to the bare ground. The sheep, however, could not eat the cactus, which they found too prickly.

In its native habitat in North and Central America, the cactus is controlled by the moth, *Cactoblastis cactorum*, whose caterpillars feed on the plant. But such moths are not found naturally in Australia. On this occasion, the ecological invasion was successfully controlled. The moth was brought over to Australia and let free, and in a matter of years the cactus population had been dramatically reduced.

Such problems are not always so easily solved. The native animals and plants of New Zealand, for example, have suffered particularly severely from ecological invasions. Rats, cats and - worst of all - ferrets, have had a devastating impact on several unique species of ground-nesting birds. Morning glory, gorse, honeysuckle, ragwort and South American pampas grass, brought over for hedging, have invaded vast areas where the native bush has been cleared. The Australian brush-tailed possum, whose population has now reached over 100 million, is defoliating local broadleaved trees, many of them found only in New Zealand.

CHEMICAL DISRUPTION

Recently, in an attempt to allay fears over chemical additives in food, a US chemical company ran a series of national newspaper advertisements with the reassuring by-line, "Life is just a bowl of chemicals". In a sense, it is. All living organisms are made up of chemical compounds and require certain chemicals to survive. But this does not mean that all the chemicals that occur in nature are harmless - for example, the venom

▲ Deer on a ranch in New Zealand. Such animals can escape all too easily, swelling the herds - from six different species - introduced at the turn of the century. Deer damage the vegetation and thus harm unique native birds. The rare and beautiful takahe ***(see p.196)*****, a bird that was thought to be extinct until rediscovered in 1948, is badly affected by the grazing of red deer on upland meadows.**

◀ The brush-tailed possum, introduced into New Zealand from Australia between 1837 and 1875, is now ubiquitous. It compounds the damage already inflicted on native vegetation, which has caused slopes and river banks to erode. In an effort to stabilize the soil, foreign trees such as willows and eucalyptus have been planted, but the leaf-eating possums have killed these, along with native trees.

◀ Cane toads, introduced into Australia from Hawaii to eat insect pests in sugar-cane plantations, have now reached plague proportions in Queensland.

produced by a snake - or that plants and animals can tolerate any chemical which man chooses to introduce into the environment.

Even those chemicals that living organisms need to keep healthy can cause ecological havoc when available in excess. Plants, for example, require nutrients such as phosphorus and nitrogen for their growth. If they receive too few nutrients, they die. But too rich a supply of nutrients can cause an explosive growth in plant populations. In aquatic ecosystems, such as seas, rivers, lakes and streams, this leads to a phenomenon known as eutrophication, the effects of which can be devastating.

One area which has been particularly badly affected is East Anglia, the grain belt of Britain. Seven hundred years ago, rising sea levels flooded the region, filling up old peat workings and forming a network of shallow lakes, marshes and meandering streams known as the Broads. All but a handful of the Broads have been drained for farmland and those that remain are now under dire threat from eutrophication. Fertilizer run-off from the surrounding fields is causing large quantities of nitrate and phosphate to enter the water of the Broads. Domestic sewage and silage effluent are adding further nutrients, stimulating carpets of algae to form on the surface, blocking sunlight and turning the water below into the aquatic equivalent of a desert. As the plants in the water die and rot, the organisms that decompose them use up the dissolved oxygen. The oxygen is further depleted by the growing numbers of bacteria feeding in the nutrient-rich waters. Fish, which obtain the oxygen they need directly from water, are suffocated, their bodies floating to the surface to join the putrefying mats of water weeds. Today, only four of the remaining 41 Broads remain healthy.

The nutrients that cause eutrophication are natural, indeed essential, components of the biosphere: the problem arises because of the great quantities which now enter rivers and other waterways. But the vast majority of the compounds now used in industry and agriculture, even when found in nature, have no role to play in the processes that maintain life. It is thus no coincidence that a large proportion of them are damaging to living organisms.

Some of these chemicals, such as cyanide, are acutely toxic, killing outright even in small doses. Others are chronically toxic, causing death or ill-health where living organisms are exposed to them repeatedly over a long period of time. The same chemical may be acutely toxic at high doses but chronically toxic at low doses. Where the chemical affects the processes that control cell division, it can result in cancer, the damaged cells multiplying out of control and invading other tissues, eventually causing death. Such chemicals are known as carcinogens. In many instances, carcinogens are also mutagens, so-called because they damage the genetic information, DNA, which determines the development of living organisms. Other chemicals affect the eggs, sperm or fetuses of an animal and result in birth defects. These are known as teratogens, the most infamous example being the drug Thalidomide, which was used to combat morning sickness during pregnancy.

Of the 103 chemical elements present on our planet, only 20 are actually made use of by animals and plants. Of the rest, most are damaging to living organisms. For example, the heavy metals lead and mercury are extremely poisonous to the nervous system. Cadmium, another heavy metal, is also dangerous. Nonetheless, all three are widely used in industry and in consumer products such as batteries. Cadmium is also a contaminant in cigarettes.

ALIEN TO LIFE

Since the dawn of the industrial age, some six million chemical compounds have been sythesised by man. Of these, at least 70,000 are now in common use, with over 1,000 new chemicals entering the market every year. The vast majority are xenobiotic, that is, 'alien to life'.

All organic compounds contain molecules consisting of different combinations of carbon and hydrogen atoms, but the number of combinations of these that occur naturally is only a fraction of those which are chemically possible. Science has enabled the synthesis of vast numbers of novel organic chemicals. By tacking chlorine atoms onto molecules made up of carbon and hydrogen, for example, chemists have produced chlorinated hydrocarbons such as trichloroethylene, a common industrial solvent, and the pesticides DDT, chlordane and pentachlorophenol (PCP). Further chemical manipulation has produced polychlorinated biphenyls (PCBs), used as fire retardants because of their resistance to heat.

Eventually, most of these substances end up in the environment. Some, like pesticides, are sprayed directly onto the land, while others, like chlorofluorocarbons (CFCs), enter the atmosphere. America alone generates 300 million tonnes of hazardous wastes a year - more than one tonne for every US citizen. And that figure does *not* include the vast volumes of waste spewed out into the atmosphere by industry or discharged into waterways and the seas.

Few of the chemicals in common use have been tested for their full range of effects on humans, let alone the environment. Nothing is known, for example, about the health effects of over a third of the pesticides in common use. Yet bitter experience has taught us that most are likely to have an adverse effect on living organisms and on ecosystems.

In particular, many synthetic organic chemicals are known to suppress or alter the biochemical processes which occur in nature and some cause cancer, birth defects or genetic damage in animals. The most damaging are those which are persistent in the environment, although even the most stable of them will eventually decompose. The danger with compounds such as

▲ Industrial pollution in Urumqui, in the northwest of China. The incidence of bronchitis, pneumonia and asthma is very high in many Chinese cities, due to air pollution. Ill-health caused by large industrial accidents, such as those at Bhopal in India or Seveso in Italy, receive ample publicity, but there is less awareness of the ill-effects of continuous low-level exposures to pollutants.

DDT, lindane and dieldrin lies in their tendency to collect in the fatty tissues of animals through the process of bioaccumulation. Any pollutants that are not excreted tend to build up within organisms, and are passed on through the food chain, becoming concentrated in the top predators. In the North Sea, for example, the levels of PCBs in seals are 80 million times higher than in the seawater in which they swim.

Bioaccumulation is not restricted to synthetic organic compounds. Heavy metals also accumulate as predators consume polluted prey. In predators at the top of the food chain, the levels of such pollutants may reach toxic concentrations, with deadly results. But equally serious are the sub-lethal effects of pollutants. Animals exposed to low levels of pollution over long periods of time may not die outright: instead, their health is slowly undermined to the point where their resistance to disease may become impaired. The epidemic which killed 18,000 North Sea seals during the summer of 1988, was a case in point. Although scientists eventually established that the actual cause of death was a virus, it is likely that pollutants such as PCBs undermined the seals' resistance to the virus. If that was the case, then the real cause of the seals' death was the ever-growing contamination of the North Sea and not the virus itself.

THE SCOURGE OF RADIATION

All living creatures are exposed to natural background radiation, some of it emanating from outer space and some from radioactive substances, such as uranium and thorium, which occur in rocks, soil and seawater.

Radiation is measured in becquerels, one becquerel being equivalent to one atomic disintegration per second. On average, a human being receives some 60,000 becquerels from natural

▲ Seals dying of a viral disease on the shores of Britain in 1988. Controversy still surrounds the cause of the epidemic, but many believe that pollutants in the North Sea weakened the seals' resistance to disease. Tumours and other pollution-related diseases have been seen in fish from the North Sea, and the level of pollutants in some fish make it unsafe for people to consume fish more than once a week.

▶ Many volcanic rocks, such as the granite in this landscape, are naturally radioactive. All radioactivity can harm the body's hereditary material, DNA, but living things constantly repair damaged DNA, and can cope with natural levels of radiation except where this is unusually high. When radiation is so great that it overwhelms the repair system, tumours and birth defects result from the damaged DNA.

▼ Rain leaching radioactive materials from the spoil of a uranium mine has killed vegetation nearby. Pollution and ill-health caused by the mining of uranium is one of the hidden costs of nuclear energy.

▶ A view into the core of a nuclear reactor, which is shielded by a deep tank of water.

“The rapidity of change and the speed with which new situations are created follow the impetuous and heedless pace of man rather than the deliberate pace of nature. Radiation is no longer merely the background radiation of rocks, the bombardment of cosmic rays, the ultra-violet of the sun, that have existed before there was any life on earth; radiation is now the unnatural creation of man's tampering with the atom.”
Rachel Carson

sources. The radiation comes from outside the body as well as from natural radioactive substances, such as radioactive potassium, taken in with food. Such exposure cannot be prevented, although it can sometimes be reduced.

Radiation damages the body by bombarding the cells with high-energy rays and particles, generated when an atom (the fundamental building block of all substances) disintegrates. Plants and animals have had little choice but to live with natural background radiation, and they have protective mechanisms to combat the damage it does. But if such radiation is unavoidably part of life, that does not signify that additional amounts from man-made sources are harmless. The evidence is that a small addition to the natural levels will cause an increase in disease - and that, in some areas, even background levels can cause cancers and genetic damage.

Yet, over the past 40 years, as a result of human activity, ecosystems have become increasingly polluted with radiation-emitting substances. After more than 1,000 nuclear weapons were tested by exploding them in the atmosphere, the fallout of radiation covered the Earth in a fine radioactive dust, containing highly carcinogenic plutonium, a few grains of which, taken into the lungs, can significantly increase the chances of lung cancer. Plutonium will remain an environmental hazard for tens of thousands of years. In 1963, as a result of increasing awareness of the risks to health, the Soviet Union and the US agreed to a Partial Test-Ban Treaty which outlawed atmospheric testing of nuclear weapons.

To fuel nuclear power stations, now the major source of electricity in some countries, notably France, uranium must be mined. Exposure to radiation may increase following mining operations which produce enormous piles of tailings that must be dumped on the surface. To obtain the uranium to run one nuclear reactor for one year requires 100,000 tonnes of rock to be brought up to the surface, most of which gets dumped as waste tailings. Nearly 90 per cent of the original radioactivity in the rock remains in the tailings. The number of reactors that need fuelling is increasing every year, and thus the mountains of tailings increase day by day.

Nuclear power plants, meanwhile, produce an extremely dangerous cocktail of radioactive substances. The splitting of the atom increases the original radioactivity in the uranium a million times or more. The high-level wastes that result are so radioactive that they must be isolated for

Pollution without frontiers

On April 26th 1986, one of the nuclear reactors at the Chernobyl nuclear power station in the Soviet Union exploded. Radioactive gases poured into the atmosphere, and were swept across the northern hemisphere, following a complex path as they were carried by winds of varying directions. Within days, the radioactive gases had travelled thousands of kilometres from their source, vividly demonstrating that pollution respects no frontiers.

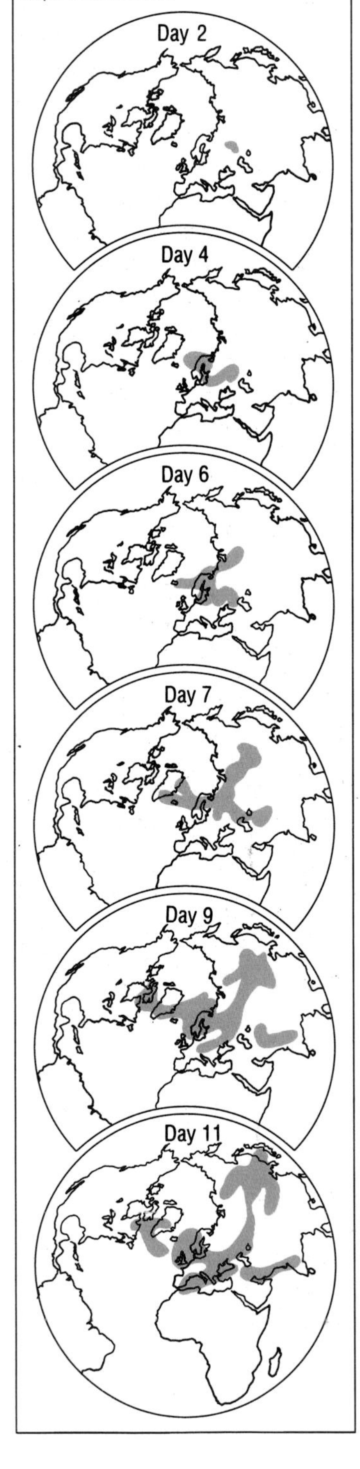

thousands of years from living things. To date, there is little agreement of the best way to dispose of such waste. Proposals vary from sending it by rocket out to space to burying it deep in the ground. The large quantities of low-level waste produced in operating nuclear installations are meanwhile discharged into the environment within limits set by governments - even though the evidence that they are damaging the environment is now overwhelming.

Accidents have occurred at nuclear installations with depressing regularity, frequently releasing radioactive substances into the environment. The most devastating accident to date took place in April 1986, when a reactor at Chernobyl in the Ukraine exploded. Hundreds of thousands of people had to be evacuated. Four years after the accident, people were still being moved. Levels of radiation in areas as far as 400 km (250 miles) from Chernobyl are still between 8 and 25 times more than the background average. Trees close to the reactor died, but further away, they began to grow abnormally, with strange shoots and deformed leaves. Even in Bavaria, in southern Germany, cows which had been allowed to graze outside on contaminated grass a month after the Chernobyl accident showed three times as many stillbirths as animals kept indoors. Fallout levels in parts of Britain were also high. One year after the Chernobyl accident, 500,000 sheep in Cumbria, North Wales, Yorkshire and Scotland had to be banned from slaughter because their flesh contained more than 1,000 becquerels per kg (2,200 becquerels per lb).

People living in the contaminated region of the Ukraine are reported to be suffering from various illnesses, and to have difficulty recovering because of damage to their immune systems. The radioactive iodine emitted at Chernobyl is particularly dangerous to children because it accumulates in the thyroid gland in the neck, and it is this gland which regulates growth. In Corsica, some 2,000 km (1,250 miles) from Chernobyl, the French medical authorities failed to warn that heavy fallout from Chernobyl had been received in the mountains. As a result of eating sheep and goats' cheeses, a number of children received radiation doses to their thyroids which were more than 100 times that from background radiation.

THE LOSS OF BIODIVERSITY

The Earth possesses a staggering diversity of plant and animal life. To date, biologists have classified almost 2 million species, but recent research suggests that the final count could top 40 million, insects alone accounting for 30 million species. But as we have set out to conquer nature, subjugating one ecosystem after another to our perceived needs, so species after species is being wiped off the face of the planet.

Some species, such as the African elephant *(see p.120)* and the black rhino, are being driven to extinction through overhunting, their tusks and horns being prized commodities. Many fear that neither will survive beyond this century. Leopards, tigers, ocelots and other big cats have also been hunted mercilessly to satisfy the demands of the fur trade. So, too, the capture of animals such as parrots, tortoises and terrapins for the pet trade, and the collection of rare plants for sale to nurseries, is condemning numerous species to extinction. Although an international treaty - the Convention on International Trade in Endangered Species (CITES, for short) - has been agreed to regulate the trade in threatened species, it is widely flouted. It is estimated that $1.5 billion worth of threatened plants and animals are traded every year on the international market.

But the gravest threat to species comes from the destruction and degradation of their habitats. As forests are cleared, marshlands drained, coral reefs mined, rangelands eroded and desertified, seas and oceans polluted, so the animals and plants which rely upon them are lost forever. Most species have specific ecological requirements and can only survive if those needs are satisfied. Some have a very limited distribution as well. The golden slipper orchid, for example, only reproduces in the wild on one limestone mountain in the Yunnan province of China. Destroy that habitat and the extinction of the species becomes inevitable.

ISLANDS OF LIFE

During the 1970s, the creation of nature reserves and national parks was an approach to conservation that was very much in vogue. Areas that were of special interest or rich in wildlife were preserved while the area all around was developed for agriculture or other uses. The logic behind such conservation practice was that a residual pool of wildlife would be left in these protected, pristine areas, so as to conserve it for posterity. The Amazon Basin could, by this logic, be developed over much of its extent as long as islands of forest were left behind in regions known to be especially rich in wildlife.

The threat from fragmentation

In Australia, much of the original forest has been reduced to scattered fragments of different sizes, surrounded by farmland. A study of some of these fragments in Western Australia showed that there were more forest-loving native mammals in the large fragments than in the smaller ones. This is common in habitats that have been broken up by man, or by natural forces. Position, as well as size, is also important. A fragment close to a large area of forest, for example, is richer than one that is isolated because species that die out by chance,as often occurs,can be replaced by immigrants. Sometimes a small fragment can have more species than a larger one. This happens if it is invaded by animals that are adapted to the land around it. However, these invaders are likely to be common, widespread species, and in time they may wipe out the rarer species that are confined to the fragment.

NATIVE MAMMALS IN FOREST FRAGMENTS, AUSTRALIA

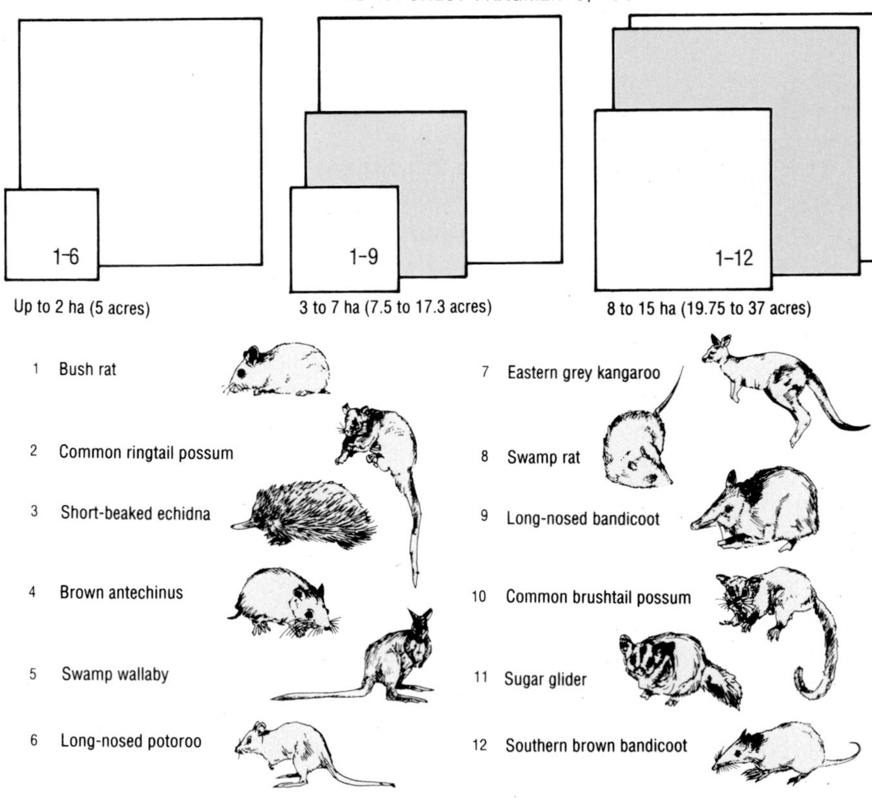

MIGRATORY BIRDS IN FOREST FRAGMENTS, EASTERN U.S.

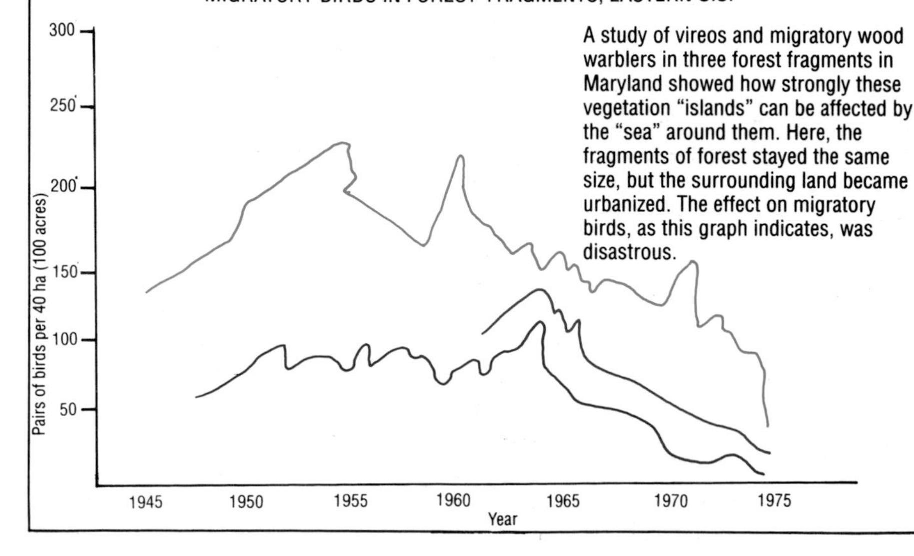

Given that the Brazilian Amazon encompasses 5 million km^2 (1.9 million square miles), 10 per cent of that would be almost the size of France, a seemingly generous area of land.

In recent years, the fundamental flaws in this approach have become all too obvious. The key to understanding the problem is to look at fragments of forest - or fragments of any habitat - as if they were islands. Studies of island natural history have shown, time and time again, that islands are not microcosms of life on the mainland. They are always poorer in species than mainland ecosystems of the same type, and the smaller the island the fewer species it has. This basic rule forms part of the theory of island biogeography, which has great relevance for wildlife conservation everywhere. The implications of the theory are that when an ecosystem is drastically reduced in size the number of species that can be sustained must also fall, perhaps catastrophically. Such extinctions clearly present a problem for conservation and the problem is compounded by our ignorance of all the interactions between plants and animals.

▲ **Black rhinoceroses have been hunted to the verge of extinction. There is now a total ban on the sale of rhino horn, but this action has probably come too late to save the species.**

▲ A hunter with birds of paradise. If New Guinea had not banned the export of these birds in the 1920s, most would now be extinct. Before the ban, as many as two million dead birds had been imported by European hatmakers.

Some fundamental rules have become clear from the study of islands and their man-made counterparts. Size is only one factor in deciding how many species are present. The distance of the island from the mainland (the nearest large body of pristine habitat in the case of habitat fragments) is also crucial. The further away, the less chance there is of colonists reaching the island and enriching the population - or replenishing it after natural losses. As they have studied fragments, biologists have realized that the corridors along which migrants can pass are crucial to maintaining their species richness. Small reserves surrounded by a vast 'sea' of agricultural land or concrete are the least viable fragments. For many species that 'sea' will prove impassable, and will spell terminal decline for the isolated population. It is also clear that one large reserve will do far more to preserve wildlife than several small reserves adding up to the same total area.

The World Wide Fund for Nature (WWF) and the National Institute of Amazonian Research in Manaus, Brazil, have set out to test what is the minimum critical size of ecosystems that will survive intact in the Amazon forest, given that the rest of the forest around has been cut down. In their study, inventories of species are being undertaken in isolated plots of forest ranging in size from a few hectares to some 10,000 ha (24,700 acres).

In small plots, of 1,000 ha (2,470 acres) or less, the number of animal species rapidly declines within a couple of years of the forest fragment becoming isolated. The effects on vegetation take a little longer. Yet within the first couple of years, the forest suffers wind-damage at the fringes, with trees blown over, and there are signs of drying out and water-stress. The leaf litter, being drier, does not rot down as quickly, but piles up beneath the trees. As well as increasing the risk of fire, this may result in nutrients becoming scarce, because they are locked up in the leaf-litter rather than being recycled.

Loss of insect species, and the knock-on effect of this on other species, may be less obvious than immediate changes in the vegetation, but is just as damaging in the long run. Euglossine bees play a vital role in pollinating the trees of the forest, and the study of forest fragments has revealed that some species are unable or unwilling to cross felled areas, even when these are relatively narrow. Many bees have specific relationships with particular plants, especially orchids, and without their pollinators some plant species will be doomed to extinction. Their loss may have unexpected consequences, because of the intricate relationships of rainforest species. The euglossine bees that pollinate the Brazil nut tree are females, but the male of the same species relies on a particular species of orchid for its courtship display. If that orchid dies out, so will the bees, leaving the Brazil nut trees unable to set seed and so produce nuts.

Less specialized pollinators can be equally at risk. Since no single species of tree flowers all the year round, animal pollinators - whether bee, bat or bird - need other sources of food at other times. In fact, it seems as if the flowering season of different forest plants is staggered throughout the year. Again, the conservation of a patch of rainforest could be undermined if a particular tree that bloomed at a time when the rest of the forest trees were not in flower was left out. Botanists are now discovering species of trees that survive as rare isolated individuals in the forest and which appear to fulfil the role of providing the pollinators with food when little else is available. The other tree species need the rare specimens of such trees for their own survival, and unless such links are taken into account when deciding on a national park, it may ultimately fail as a conservation area.

These studies of tropical rainforest are relevant to other habitats as well - to marshlands and prairies, to coniferous forests and scrubland. Wherever ecosystems are reduced to small scattered remnants, species are invariably lost.

Even where a species manages to survive the fragmentation and degradation of its habitat, its numbers may be so reduced that it is unable to maintain a breeding population, as happened with the Californian condor, and could easily happen with the giant panda. These belong to the growing army of species that biologists refer to as 'the living dead', because their extinction is inevitable, though it may take decades before the last individual disappears. Just as important, the decline in breeding populations whittles away the genetic diversity of the species, rendering it less and less able to adapt to long-term changes in the environment. The chances of survival are still further diminished by pollution and the overall impoverishment of the environment. Food chains are disrupted; the subtle links between species are shattered; and the vulnerability of ecosystems to disruption increased.

UNSEEN EXTINCTIONS

Although most attention has focused on the 'cuddlier' animals under threat - cuddly if only because they make good children's toys and excellent photographs - our fate, and that of millions of other species, is likely to depend more on the survival of insects and plants than pandas or leopards. For, important as pandas and leopards are, their loss would not have the same impact as the insect pollinators or decomposers. Without decomposers, for example, the nutrient cycles within ecosystems would be fatally disrupted, depriving the soil of its fertility. Inevitably, given the role that biological organisms play in regulating the chemical composition of the atmosphere, climatic patterns would also be affected.

As one species follows another into extinction, so a chain reaction is triggered. With 50 species currently being lost every day, some warn that we are on the brink of mass extinction *(see p.259)*. Indeed, if current rates of habitat destruction continue unabated, then several hundred species could soon be lost every day. In that respect, the destruction of the tropical forests - home to more than 50 per cent of the species on Earth - threatens a biological holocaust which, because of the forests' role in regulating climate, could have dire repercussions throughout the globe.

The resilience of nature is being stretched to the limit. Wherever we care to look we can see the devastation caused by our activities and the disruption of one ecosystem after another. The greenhouse effect, the death of marine mammals, the die-back of trees throughout Europe and North America, the disruption to global climate all provide clear warnings that the natural world is sick. It may take very little more before the life-support systems on which we humans depend are finally overwhelmed.

▼ A common fungus of soil and water, seen under the electron microscope. The largely invisible army of decomposers, breaking down the remains of living things, is a vital part of all ecosystems.

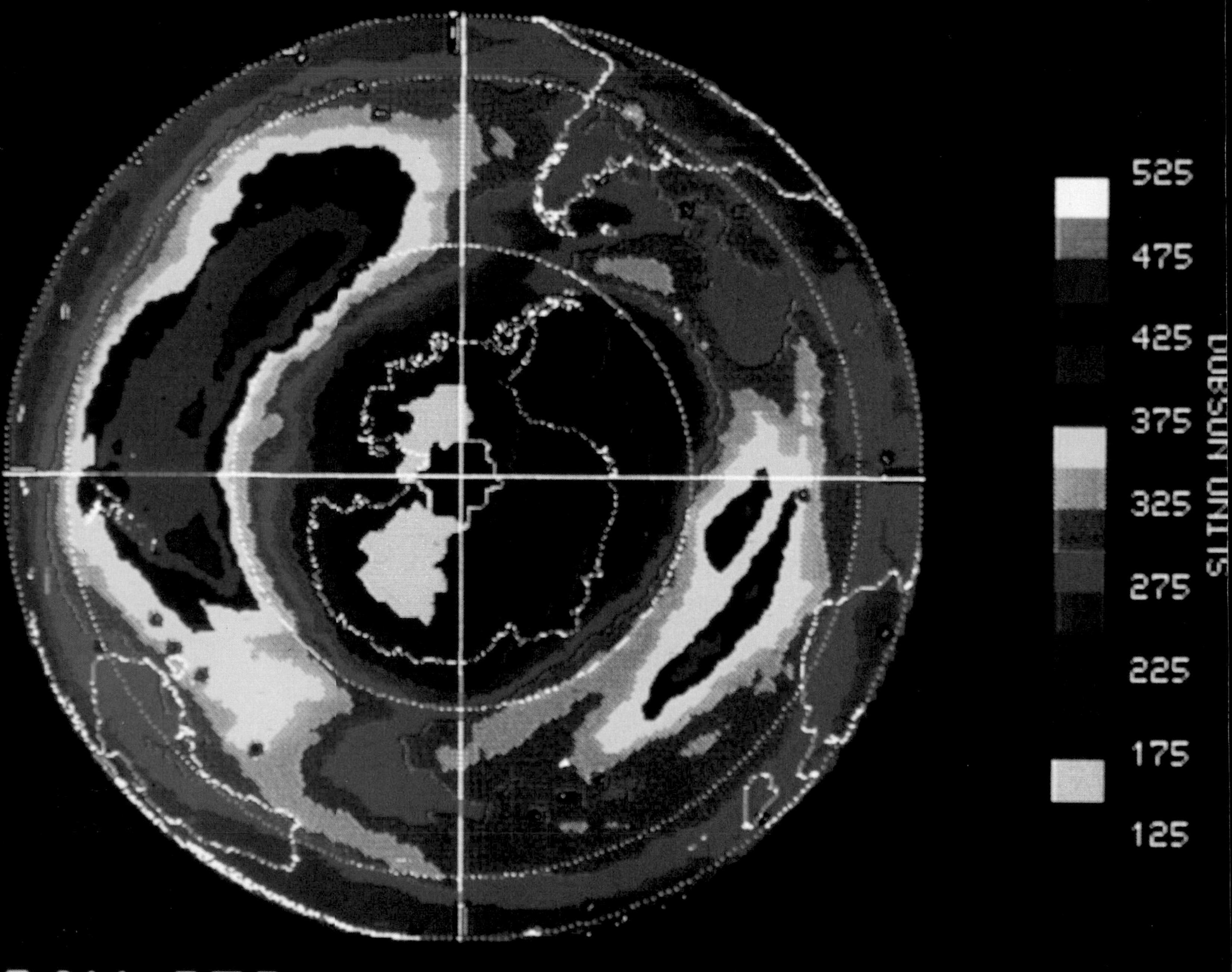

NIMBUS-7 : TOMS OZONE
525
475
425
375
325
275
225
175
125
DOBSON UNITS
DAY:258
SEP 15, 1987

The changing ATMOSPHERE

In 1984, the discovery of a hole in the ozone layer over Antarctica provided dramatic evidence that human activities are causing irreparable damage to the atmospheric systems upon which life depends. Although the size of the hole fluctuates from year to year, in 1987 it was as deep as Mount Everest and covered an area as large as the entire US.

Since the early 1970s, scientists have warned that the stratospheric ozone layer could gradually be eroded by chlorofluorocarbons (CFCs) and other gases, but they had no idea that the damage could occur on such a drastic scale. Indeed, for several years prior to 1984, American satellites had been measuring the appearance of the hole each Antarctic spring, but their computers were programmed not to accept this 'aberrant' data.

THE CAUSES OF OZONE DEPLETION

The ozone layer lies in the stratosphere, between 15 and 35 km (10 and 21 miles) above the surface of the Earth. Here, ultraviolet light from the Sun falls on oxygen molecules and changes them from being made up of two oxygen atoms (represented by O_2) to having three oxygen atoms (O_3) - a form of oxygen known as ozone.

Ozone is formed and broken down again all the time in the stratosphere, with the two processes cancelling each other out, so that there is usually a constant amount of ozone. But highly reactive substances such as chlorine (Cl_2) and nitric oxide (NO) cause ozone to break down much more quickly. These molecules set off a chain reaction, with a single atom of chlorine capable of breaking down thousands of ozone molecules.

◀ A computer-generated map showing the 'ozone hole' over than Antarctic, compiled by the Nimbus-7 weather satellite. The hole is visible as an area of deep blue, purple pink and black, positioned over the whole of the Antarctic continent.

Our industrial activities have released numerous compounds containing chlorine into the atmosphere. The most damaging ones are those such as CFCs which do not break down easily. It is that very stability that has made CFCs so useful in a wide range of industrial processes and consumer goods - from refrigerators and air conditioners to expanded foam packaging, insulation foam, aerosol spray cans and the production of microelectronic components.

The stability of CFCs in the troposphere, the layer of the atmosphere up to 15 km (9 miles) from the surface of the Earth, means that over a period of years, CFCs drift up to the stratosphere. Here they are broken down by ultraviolet light into free chlorine atoms and other compounds containing chlorine. Under normal conditions, most of the chlorine in the stratosphere is bound to other molecules, and thus does not damage ozone. However, during the Antarctic winter, the winds of the polar vortex, swirling around the South Pole, create unusually cold conditions. Ice crystals form in the clouds of the stratosphere, providing a surface on which the chlorine-containing compounds begin to break down. With the coming of spring and warmer weather, the damaging chlorine atoms are released as the ice crystals melt. The chlorine is then free to attack ozone.

It is not only CFCs that are implicated in the destruction of the ozone layer. Other gases must share the blame, including halons (used in fire-fighting equipment), methyl chloroform (used in cleaning solvents, adhesives and aerosols) and carbon tetrachloride (used for many purposes, including dry cleaning). All these release chlorine or the related gases bromine and fluorine (known collectively as halogens).

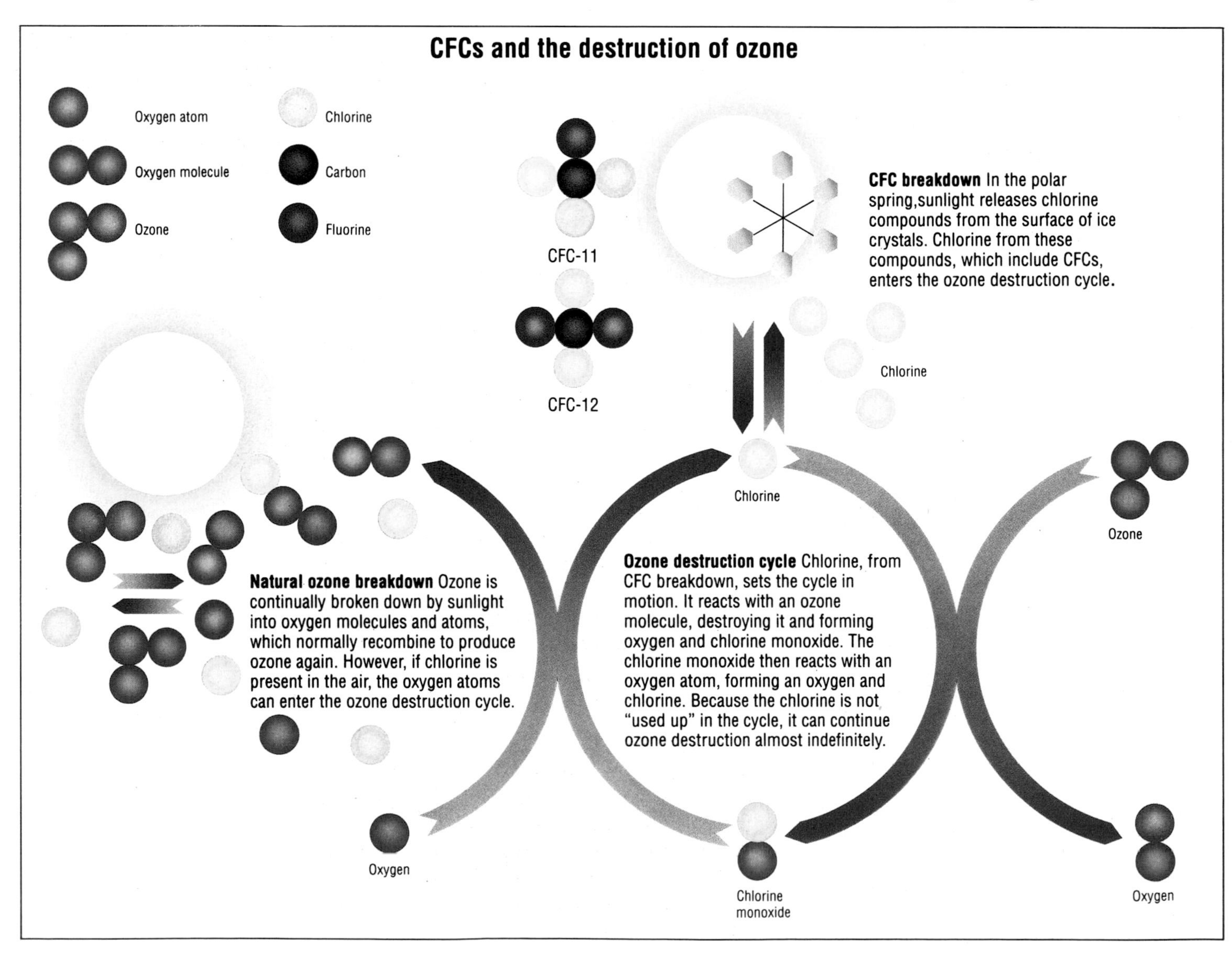

Nitrous oxide (N_2O), another culprit, is a naturally occurring gas, but one that is increasingly present in the atmosphere due to human activities. Its concentration is currently rising by 0.4 per cent a year. Like CFCs, nitrous oxide is stable until it reaches the stratosphere, where it breaks down to form nitric oxide (NO) which attacks ozone in a similar fashion to chlorine.

OZONE TRENDS

Predicting future trends in ozone depletion is fraught with difficulty. As research into stratospheric ozone continues, it becomes apparent how little is understood of this vital layer of our atmosphere. It was thought, for example, that the Antarctic ozone hole in 1989 would not be as large as in previous years, partly due to increased solar activity (which creates extra ozone) and partly to changes in the direction of the stratospheric winds, bringing warmer air to the polar vortex. In fact, the 1989 hole appeared to be at least as large as it was in 1987.

▲ Industrial pollution in the Tees Estuary, England. Many of the chemicals released by plants like this act as 'catalysts' for ozone destruction, allowing it to occur at many times the normal rate.

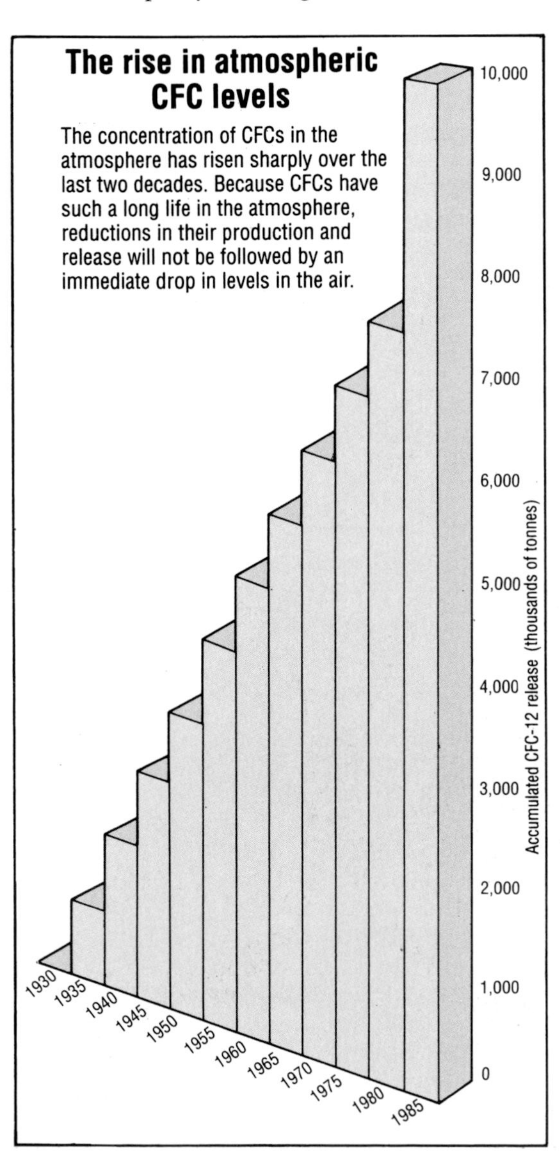

Also disturbing was the discovery that up to 30 per cent of the ozone had been lost in the latitudes between 50°S and 60°S - areas well outside the southern polar vortex. It is not yet understood why this unexpected depletion took place. There are few people living below 50°S, but if such depletion starts to occur on the same scale in the northern hemisphere, the areas affected could include all of the British Isles, a large part of mainland northern Europe and most of Canada and the USSR.

The outlook for stratospheric ozone is bleak. Although many governments now agree the need to phase out CFC use, stratospheric chlorine levels will continue to rise dramatically in the next few decades. Even if CFC emissions were to cease completely by 2000, the long atmospheric lifetime of these gases means that the concentration of chlorine in the stratosphere will not fall, even to the high levels of the late 1980s, until the latter part of the 21st century.

THE THREAT FROM OZONE DEPLETION

The main importance of the ozone layer is in protecting life from short-wave ultraviolet-B (UV-B) solar radiation. Most UV-B is absorbed by ozone, but the proportion which does reach the ground is responsible for causing sunburn and skin cancer in fair-skinned people. Excessive UV-B also causes eye cataracts and weakens the immune system.

The most harmful consequence of increased UV-B, however, may well be its impact upon food production and ecosystems. Of 200 plant species which have been tested, two-thirds showed sensitivity to UV-B. Plant sensitivity to UV-B appears to be greater if the phosphorus level in the soil is high. Heavily fertilized agricultural areas could therefore be badly affected.

The effect on marine life may be equally severe. Phytoplankton are microscopic plants which float in huge numbers near the surface of the ocean and supply all the food on which marine animals depend. It has been found that UV-B affects the phytoplankton and the larvae of some species of fish. Some scientists believe the effect of ozone depletion in the Antarctic's seas will be particularly dramatic, since krill, the tiny shrimps which are essential to the Antarctic food chain *(see p.221)* depend on phytoplankton. If the krill decline, the whole ecosystem of the Southern Ocean could collapse.

ATMOSPHERIC DISRUPTION AND GLOBAL WARMING

For over a century, scientists and environmentalists have warned that rising levels of carbon dioxide, due to the burning of fossil fuels, could bring about a catastrophic warming of the Earth's atmosphere. But it took a succession of unusual weather events around the globe, especially during the late 1980s, to alert public and politicians to the threat of global warming.

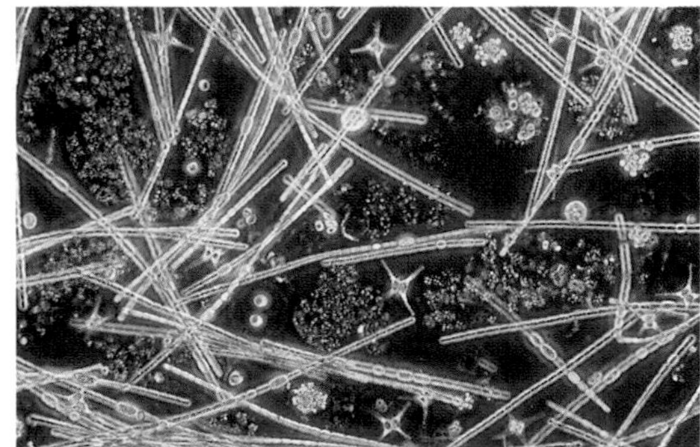

▲ A tangled mass of phytoplankton in a drop of water. Simple plants like these could suffer through ozone depletion.

◀ The azure sky above these French hills is not simply a blanket of life-giving oxygen. It is also a shield that protects us from damaging radiation.

These included a succession of terrible droughts in the Sahel and Ethiopia, a number of devastating tropical hurricanes, disastrous floods and drought in Bangladesh, the US drought in 1988 and a series of unusually mild winters and hot summers in Europe.

The Earth's atmosphere has warmed by 0.5°C (almost 1°F) over the last century; six of the ten warmest years recorded have occurred in the 1980s, and significant changes have occurred in rainfall patterns. Such data have convinced many scientists that global warming has already begun. When NASA climatologist James Hansen claimed in 1988 that "the evidence is pretty strong that the greenhouse effect is here", most of his fellow scientists thought this view extreme. Now it is close to becoming the accepted opinion in the scientific world.

THE CAUSES OF GLOBAL WARMING

As we have already seen *(p.17)*, conditions would not be suitable for life on Earth if there were no greenhouse effect. Without naturally occurring greenhouse gases, such as carbon dioxide (CO_2), water vapour, methane (CH_4) and nitrous oxide (N_2O), the planet would be some 30°C (55°F) cooler. These important gases allow short-wave energy from the Sun to pass through them, but trap the longer-wave infra-red or heat radiation which is reflected back from the Earth's surface.

Unfortunately, the levels of these greenhouse gases in the atmosphere are now rapidly increasing, mainly as a result of the burning of fossil fuels and destruction of forests. The increased use of nitrogen fertilizers, the expansion of irrigated rice production and rising numbers of cattle are also playing a role. The production of synthetic greenhouse gases has greatly worsened the problem. Around 30-40 trace greenhouse gases are known, and there may be many more - every year hundreds of new gases are produced with little idea of their greenhouse potential. By far the most potent of the man-made greenhouse

▲ Fuelwood on the move in Dire Dawa, Ethiopia. Ethiopia is one country where the effects of drought have been felt most severely in the last decade. Several factors have contributed to the famines that have ravaged this country. The extensive cutting down of trees, together with changes in climate, are foremost among them.

► A storm gathers over the baked mud of Etosha Pan in Namibia. The wildlife of arid areas such as this depends on seasonal rains, which replenish waterholes and promote a flush of plant growth. In this kind of environment, any change in the pattern of rainfall can have unpredictable results.

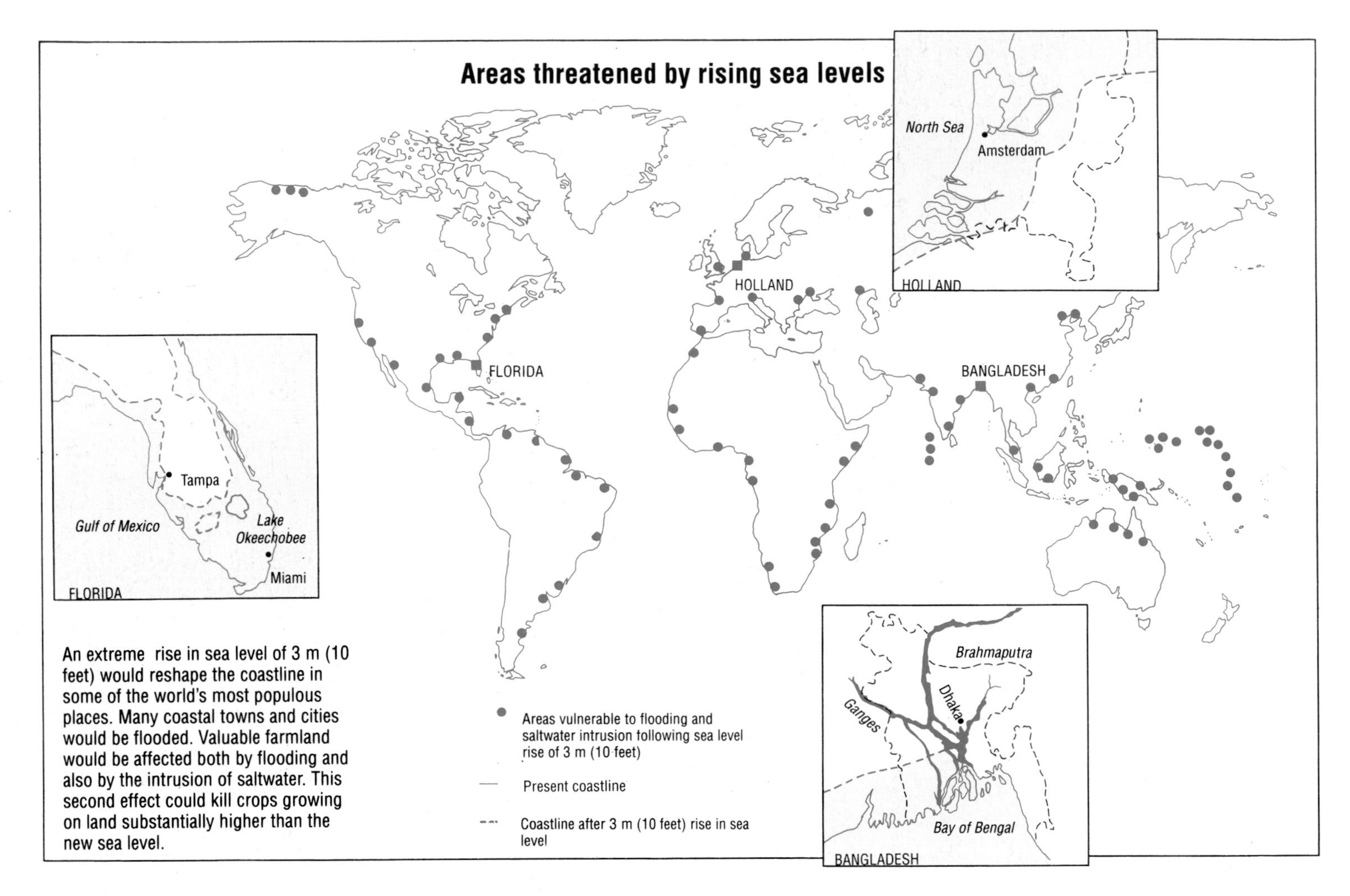

An extreme rise in sea level of 3 m (10 feet) would reshape the coastline in some of the world's most populous places. Many coastal towns and cities would be flooded. Valuable farmland would be affected both by flooding and also by the intrusion of saltwater. This second effect could kill crops growing on land substantially higher than the new sea level.

gases happen to be the CFCs, which are also a prime cause of ozone depletion. Each molecule of the most commonly used CFC is 10,000 times as effective at trapping infra-red radiation as a molecule of carbon dioxide. Again, the persistence of CFCs exacerbates the problem as they may survive for a century or more.

Because it is produced in such quantities the world over, carbon dioxide is expected to be responsible for over half of the total predicted global warming. Levels of carbon dioxide in the atmosphere have already increased by 25 per cent from pre-industrial levels and are now at their highest levels for 160,000 years. Without significant action to curb emissions, the atmospheric concentration of carbon dioxide will double before the middle of the 21st century.

Methane, a hydrocarbon gas, is responsible for around a fifth of global warming and its atmospheric concentration has already doubled from pre-industrial times. It is produced by organic material decomposing in the absence of oxygen, and large amounts are created in nature in swamps and marshes. Other sources of methane include rice paddies, landfill rubbish dumps, the burning of plant fuel and fermentation in the stomachs of cattle and the intestines of termites.

Ozone in the air near ground-level is another greenhouse gas. It is created by the action of sunlight on hydrocarbons and nitrogen oxides emitted by industry and road vehicles. Its global concentrations are difficult to estimate but they are usually assumed to be increasing at around 0.25 per cent annually.

Water vapour in the stratosphere, mainly due to aircraft emissions, is estimated to contribute around 2 per cent to global warming.

This cocktail of greenhouse gases is expected to cause a rise in global temperature of 1.5-4.5°C (2.5-8°F) by 2050. Factors such as changing cloud distributions, increasing water vapour in the atmosphere, and shifting snow and ice margins may slow down global warming or speed it up.

Other uncertainties in the precise details of climate change are due to the wide range of political, economic and social factors involved in the control of greenhouse gas emissions. Even if these forecasts are accurate, they represent only *average* changes in temperature for the whole planet. Wide regional differences are likely to occur, with a much larger warming expected at the poles than in the tropics.

RISING SEA LEVELS

Current estimates suggest that global warming could cause a sea level rise of anything between 7 cm and 165 cm (2½ ins and 5 feet 5 ins) by 2050, due to the expansion of the oceans as they warm up and the melting of glaciers and the Greenland and Antarctic ice caps.

A sea level rise of 1 m (3 feet 3 ins) could affect up to 300 million people. Especially vulnerable are the Nile and Ganges Deltas and island nations such as the Maldives. Many major towns and cities - including London, Bangkok and New York - could be be at risk. According to Gjerrit Hekstra of the Dutch Ministry of the Environment, if maximum storm surges and the effects of saltwater intrusion along river mouths are taken into account, a 1-m (3 foot 3 ins) rise in sea level could potentially affect all land up to the 5-m (16-foot) contour line. Worldwide, this would amount to an area of some 5 million km^2 (1.9 million square miles). This is only 3 per cent of the land area of the globe, but it includes a third of the cropland.

Rising sea levels and rising temperatures will take a heavy toll on wetland ecosystems. A 1-m rise in sea level would inundate 80-90 per cent of the coastal wetlands in the US. Coastal wetlands have in the past been pushed inland in response to natural sea level rise. But urban development along the coast, and the sea defences which will be needed to protect both agricultural land and human settlements, will restrict the areas available for new wetlands.

CHANGING THE FACE OF THE EARTH

Temperatures in polar regions could rise by up to 12°C (22°F) over the next 60 years - a totally unprecedented rate of change which could spell extinction for many of the best-known polar species, including polar bears and walruses. The melting of the Arctic ice sheet would destroy the algae which grow on its underside and which

► A natural gas processing plant in New Zealand. The use of natural gas, a fossil fuel, contributes to the greenhouse effect by producing carbon dioxide.

form the basis of the local food chain. It is on this food chain that vast numbers of fish, seabirds and seals depend. The seals also need the ice as a platform from which to hunt and on which to breed. Recent data show that the Arctic ice cap has thinned by some 30 per cent in the last decade.

A rise in the temperature of the seas may alter ocean currents which could deplete fish stocks. By bringing nutrients to the surface from the seabed, currents encourage the growth of plankton, producing planktonic 'blooms' which supply other animals with plentiful food. Many of the world's fisheries depend on such blooms.

Although carbon dioxide encourages the growth of vegetation (by providing more raw material for photosynthesis), many tree species which are adapted to cold conditions will prove unable to survive the warmer temperatures of a greenhouse world. As the northern landmasses warm, the zones of tundra, boreal forests and temperate forests *(see pp.56-57)* will all move northwards. The speed at which forests can 'migrate' in response to these changing conditions will be critical in determining the rate of species extinctions. For example, at the southern limit of their current distribution in the US, few beech trees would be able to reproduce if the climate shifted - seedlings would wither and the ageing trees which remain would be weakened by heat and drought, making them more susceptible to storm damage and acid rain. Over the next century, beech may thus have to move as much as 900 km (560 miles) further north if they are to survive.

Animals are more mobile and many could move quickly in response to changing conditions. Nonetheless, they will be restricted by the rate at which the vegetation of their natural habitat can move. The movement of both animals and plants will also be severely hampered by roads, cities and the 'deserts' created by modern intensive agriculture, as well as by natural barriers such as mountain ranges. Species which are currently restricted to national parks and other areas surrounded by human-dominated landscapes may well find themselves trapped.

Some migratory species may die due to changes to the times at which the food supplies for their long journeys are available. For example, in Delaware Bay on the east coast of the US, 1.5 million wading birds arrive in early May, en route from their wintering grounds in South and Central America to their breeding grounds in the Arctic. Their arrival coincides with thousands of horseshoe crabs emerging from the sea to lay their eggs. It is estimated that 340 tonnes of eggs are laid, providing a rich food source for the birds. But the crabs respond very precisely to the sea temperature. If the sea were to warm up earlier, well before the birds arrive, enormous number of birds could starve.

Changing rainfall patterns may also cause problems. In the tropics, the lack of a dry season, or a dry season that is longer than usual, can disrupt the plant-insect relationships necessary for pollination to occur. Evidence from Costa Rica suggests that seedlings are much less likely to succeed in seasonal tropical forests if the frequency of dry spells increases.

CHAIN REACTIONS

Perhaps the most frightening aspect of global warming is the number of possible 'feedback mechanisms' *(see p.21)* which could trigger a runaway greenhouse effect. The Arctic tundra, for example, contains billions of tonnes of methane, a greenhouse gas, much of which will be released to the atmosphere as the permafrost melts. Similarly, the boreal forest vegetation and soils contain a quarter of all the Earth's organic carbon. If the boreal forests are unable to adapt quickly enough to the warmer climate, widespread forest die-back will occur. As the trees rot and the soils dry out, yet more carbon dioxide will be added to the atmosphere, warming the climate still further. A shrinking Arctic ice cap will also reduce the amount of heat reflected back into the atmosphere, snow and ice reflecting some four times more than the ocean.

Plankton play a key role in absorbing carbon dioxide from the atmosphere, by incorporating it in their cells. A reduction in plankton productivity could therefore dramatically accelerate global warming. Plankton, which thrive in cold seas, are under threat both from rising ocean temperatures and from increased exposure to ultraviolet due to ozone depletion.

Recent research suggests that there could be a very strong 'positive feedback' from increased water vapour in the atmosphere. If the sea temperature rises above 27°C (80°F), which at present happens only in the tropics, the ability of water vapour to trap heat appears to increase dramatically. Water vapour is known to be a greenhouse gas, but it is not understood why its heat-trapping properties suddenly become more efficient at about 27°C (80°F). Should this happen, then global temperatures could rise three or four times higher than expected.

▲ The Taimyr Reserve in the Soviet Arctic. Permafrost underlies this flat landscape, preventing water from draining away, and thereby creating a network of lakes and pools. The frozen ground also acts as a carbon store, containing vast amounts of methane locked up by the ice. A global warming of just a few degrees would be enough to melt much of this ice, releasing methane into the atmosphere.

There is also the frightening possibility of a 'climatic flip'. The atmosphere may not react to increasing pollutants in a linear and (at least partly) predictable manner, but may reach a threshold point above which there would be a global change to an entirely new and unpredicable climate. According to James Lovelock *(see p.16)* "if stressed beyond the limits of whatever happens to be the current regulatory apparatus, Gaia will jump to a new stable environment where many of the current range of species will be eliminated."

THE NEED FOR ACTION

There is now overwhelming evidence to suggest that climate is going to change, but scientists will never be exactly sure by how much, or at what rate. Likewise with the ozone layer, we know that it is set for greater disruption but any predictions as to how much will be lost, or by what date, are pure speculation.

These uncertainties, however, must not be taken as an excuse for inaction in dealing with the problem. On the contrary, they should be seen as a reason to begin reducing our impact on the atmosphere *now* as an insurance policy against the massive disruption of the Earth's natural systems. It is these natural systems on which all our social and economic structures ultimately depend. The risks are too great not to start taking action immediately.

◀ Crabeater seals under the Antarctic ice. Species like this have evolved over millions of years to survive and flourish in an environment where temperatures are low. Rapid global warming may disrupt the food chains on which they depend.

▶ Forest being burned in Mato Grosso state, Brazil. As millions of tonnes of carbon are poured into the air through fires following deforestation, the world's atmosphere is undergoing changes at an unprecedented speed, and with unknown consequences.

FORESTS

When Christopher Colombus first sighted the islands of the Caribbean, he was "so overwhelmed at the sight of such beauty" as to be "unable to describe it". Landing in Cuba, he found "a multitude of palm trees of various forms, the highest and most beautiful trees that I have ever met with, and an infinity of other great and green trees". So dense were the flocks of parrots on some islands that, according to Colombus, they "obscured the sun". But the jewel of the islands discovered on that first voyage to the New World was Haiti. It jutted out of the sea, in all the splendour of its tropical vegetation, "its mountains higher and more rocky than those of the other islands, but the rocks rising from among rich forests". Truly, declared Colombus, it was "one of the most beautiful islands in the world".

Today Haiti stands stripped of its trees. Forty years ago, forests still covered 80 per cent of the country; now, the figure is down to less than 4 per cent. Nor is Haiti alone. Many tropical countries which were heavily forested only a few decades ago are now virtually denuded of trees. Africa has lost almost half of its tropical forests, while the Americas have lost a third of theirs. In Madagascar, whose forests harbour thousands of unique animals, 93 per cent of the island's original primary forest has been destroyed in the last 60 years.

"We have to feel the heartbeats of the trees, because trees are living beings like us."
Sunderlal Bahuguna, Spokesperson for the Chipko Movement

◀ Eroded hills in the interior of Madagascar, once covered with forests of a kind found nowhere else on Earth. In many parts of the island, all that is left today after extensive deforestation is bare ground. The grey subsoil can be seen where the red topsoil has been washed away.

BROADLEAVED AND MIXED FORESTS

Tropical and subtropical rainforest, including montane forest, cloud forest, heath forest and flooded forest. Very dense, evergreen. Mostly broadleaved but with conifers in some regions, especially Australia and New Zealand.

Dry tropical and subtropical forest. Fairly dense, semi-deciduous. Many trees shed their leaves before the dry season. Entirely broadleaved.

Monsoon forest. Fairly open, entirely deciduous. Leaves shed before prolonged dry season. Entirely broadleaved.

Savanna woodland, wooded steppe and eucalyptus woodland. Grassland where tree canopy covers less than 50 per cent of the ground. Few shrubs. Most trees deciduous.

Laurel forest, wet eucalyptus forest, subtropical river-bottom forest and similar. Very dense, evergreen. Trees have tough, oil-rich or resinous leaves. Mostly broadleaved but mixed with conifers in some regions, for example New Zealand. Also called 'wet hardleaf forest'.

Dry hardleaf forest and dry eucalyptus forest. Fairly dense to open. Mostly evergreen. Trees have tough, oil-rich or resinous leaves. Mostly broadleaved.

Dry tropical scrub and thorn forest. Often dense and impenetrable. Plants may be evergreen or deciduous. Stunted trees and shrubs, especially acacia (wattle) and succulents.

Temperate broadleaved forest. Usually dense, entirely broadleaved in most areas. Leaves thin and delicate. Deciduous: leaves shed before winter.

Temperate mixed forest. Dense. Broadleaved trees deciduous, losing leaves before winter. Conifers evergreen.

Mediterranean hardleaf forest and woodland. Fairly dense or open. Trees have very small, hard, waxy, oil-rich leaves. Most trees evergreen. Usually mixed, but broadleaved trees only in some areas. Oaks and pines are typical.

Chaparral, dwarf forest and similar. Fairly open and shrubby. Usually broadleaved, but may include some conifers. Evergreen; small, very tough, oil-rich leaves.

North American oak-pine forest. Fairly open. Broadleaved trees deciduous.

CONIFEROUS FORESTS

Boreal spruce-fir-pine forest. Very dense. Conifers entirely evergreen except where larches occur. Some deciduous trees may be present, especially birches.

Boreal larch forest. Fairly dense. Predominant conifers (larches) are deciduous. Found in areas with exceptionally cold winters.

Coniferous forest on mountains. Dense to open. Usually evergreen, but deciduous larches predominate in some areas.

Pacific coast coniferous rainforest. Very tall and dense. Evergreen.

North American redwood forest. Dense, evergreen.

North American subtropical pine forest. Evergreen. Adapted to poor sandy soils.

New Zealand kauri pine forest. Tall and dense. Evergreen.

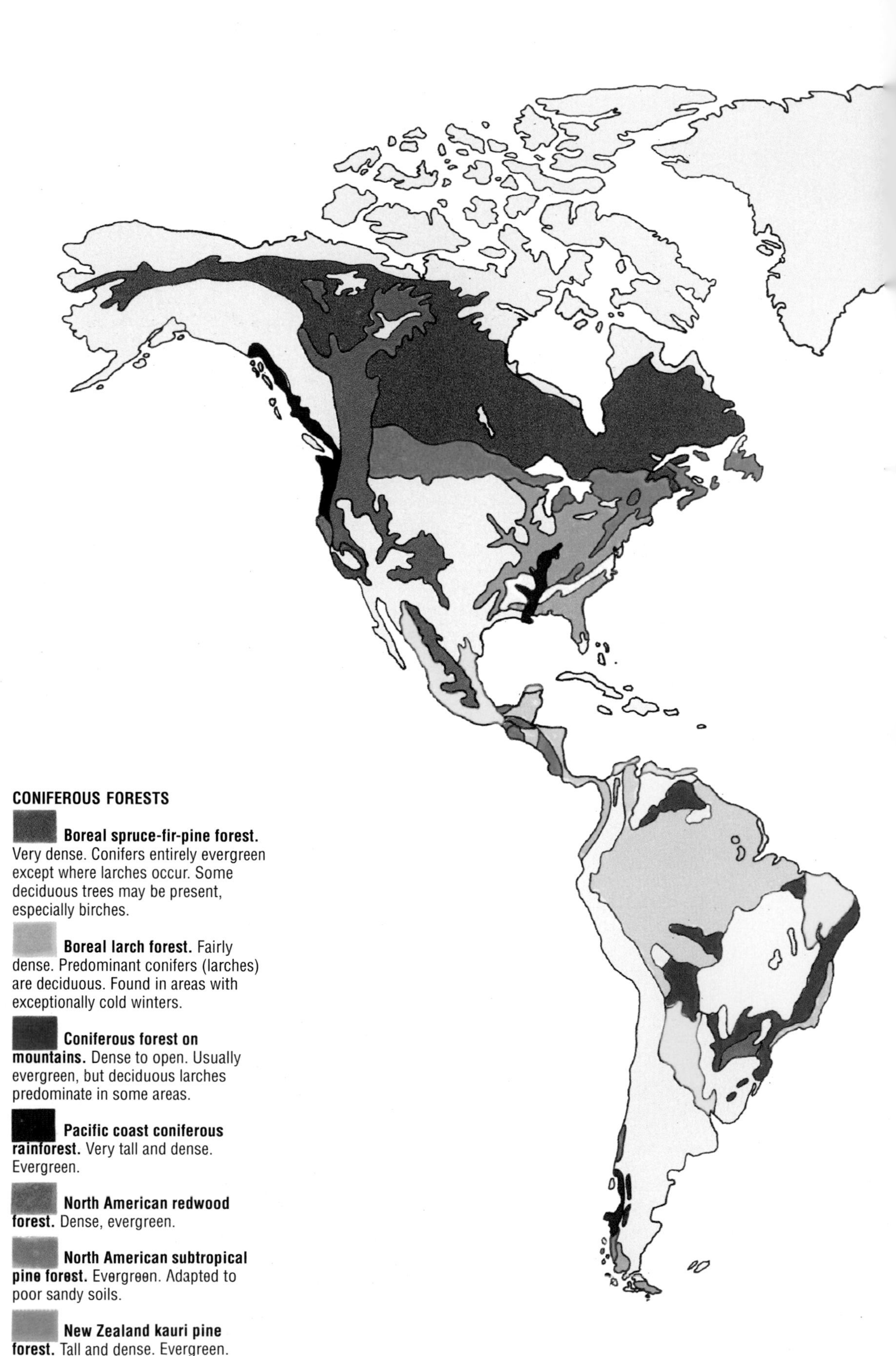

Naturally forested regions of the world

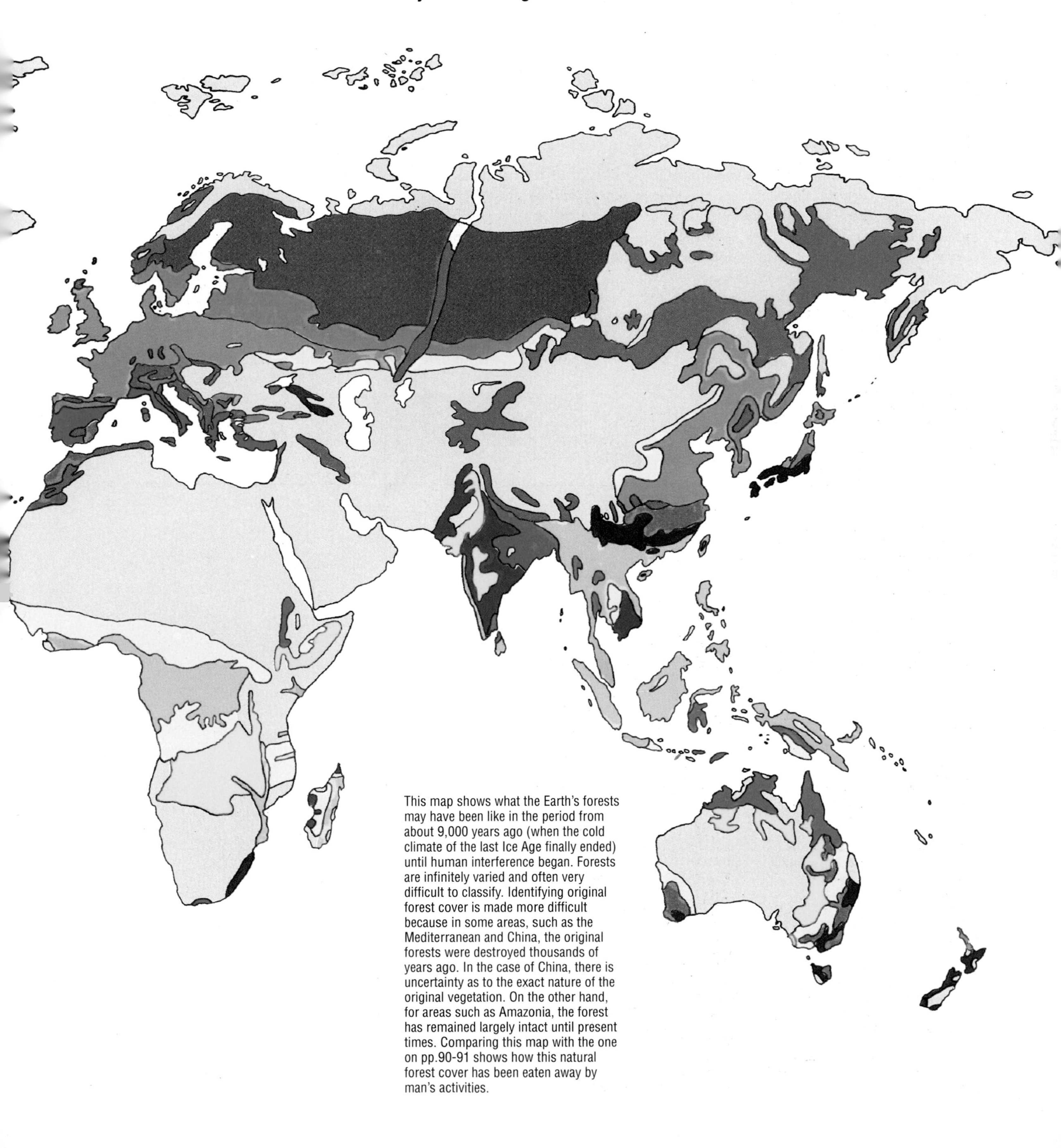

This map shows what the Earth's forests may have been like in the period from about 9,000 years ago (when the cold climate of the last Ice Age finally ended) until human interference began. Forests are infinitely varied and often very difficult to classify. Identifying original forest cover is made more difficult because in some areas, such as the Mediterranean and China, the original forests were destroyed thousands of years ago. In the case of China, there is uncertainty as to the exact nature of the original vegetation. On the other hand, for areas such as Amazonia, the forest has remained largely intact until present times. Comparing this map with the one on pp.90-91 shows how this natural forest cover has been eaten away by man's activities.

In clearing their forests, the countries of the Third World are following in the all too depressing footsteps of the northern industrialized countries. When the first European settlers arrived in what is now the United States, the continent was covered by an estimated 3.2 million km^2 (1.25 million square miles) of forest. In just 500 years, all but 220,000 km^2 (85,000 square miles) have been cleared. European colonization of New Zealand and Australia brought a similar pattern of destruction. In New Zealand, huge areas of forest were burned, without even the timber being salvaged, and the land they stood on is now given over to sheep. As for Europe itself, West Germany is now stripped of all but a third of its forests, while Britain has lost about 90 per cent.

At no time in history, however, has the assault on forests been as global or as systematic as it is today. It is an assault which is not only condemning thousands of species to extinction every year, but also destroying the livelihoods of millions of people. Moreover, the destruction of the world's forests has now reached the point where deforestation threatens the very climatic stability of the planet.

TREES: THE EARTH'S INHERITORS

Forests are the natural vegetation of a vast area of the world. In a newly cleared patch of land, provided that nature is allowed to take its course and that the soil structure is not seriously damaged, the ground will be colonized first by small 'pioneer' plants, then by larger plants, including grasses. Bushes will take over and shade out the grasses, and among them will be tree saplings. These will eventually grow tall enough to crowd out the bushes and form a dense stand of slender

► Orchids flowering above a rainforest stream.

▼ Mixed forest of conifers and birch in northern Finland, interspersed with swampy areas. This mosaic of habitats is particularly valuable for wildlife: many insects breed in the swampy regions, for example, providing food for birds. Finnish forestry companies frequently drain wooded wetlands to increase the timber yield, thus diminishing their value to wild animals.

trees which obscure most of the light. In time - and this may take several hundred years - some of these trees will lose out to others, creating a mature forest where large, ancient trees are interspersed with much younger ones, and in which there are occasional glades where an old tree has fallen, allowing sunlight to reach the forest floor.

This process is called succession and the final stage is described as climax vegetation. Except in the coldest latitudes, on mountain tops and in very dry regions, the climax vegetation is forest. Wherever it is warm enough and wet enough, trees are the ultimate inheritors of the land. In marginal areas, with relatively low rainfall, a mixture of trees and grasses may prove to be the climax vegetation, known variously as savanna forest, thorn forest, or (in Australia) woodlands.

But the process of succession can be fatally interrupted. Fires, grazing by animals, and repeated clearance can halt succession in its early stages. This is what happened to the forests around the Mediterranean. Where magnificent groves of holm oak and Aleppo pine once stood, there is now scrubland, known as maquis, and a more degraded heath-like vegetation, known as garigue. On the chalk downlands and upland heaths of Britain - clothed with woodland before Neolithic man set to work with axes - the tree cover cannot regenerate due to constant grazing by farm animals and rabbits.

FORESTS OF THE WORLD

Forests vary greatly from one region to another. In the rainforests, trees can stand over 80 m (265 feet) tall, while the elfin forests of some tropical highlands are scarcely 2 m (6 feet) high. Tangled undergrowth makes it difficult to penetrate the gallery forests that line tropical rivers, while some forests have a clear, uncluttered floor with little growing beneath the trees.

Above all the types of trees vary, influenced by temperature, rainfall and soil. The major division which botanists recognize is between two great plant groups: the conifers, or cone-bearing trees, such as pine, fir, spruce, cypress and redwood, and the broadleaved trees, such as oak, birch and maple, which belong to the flowering plants. A second important distinction is between trees that are clothed with leaves all the year round, the evergreens, and those which shed their leaves at the onset of a cold or dry season, described as deciduous.

Two distinctive bands of forest circle the globe *(see pp.56-57)*. One is a belt of tropical rainforest which runs like a girdle around the equator, wherever the rainfall is plentiful all the year round. The trees in such forests are almost entirely broadleaved, although there are a few conifers, such as the klinki pines and hoop pines (*Araucaria* species) of New Guinea and Australia. Owing to the constant rainfall, these forests are evergreen. The variety of trees is staggering, with over 200 different species being found in a single hectare in some Southeast Asian forests. Climbers, or lianas, and plants such as orchids, bromeliads, mosses and ferns growing on the tree branches add to the richness of the plant community.

The second great band of trees is the boreal forest of the far north, which is largely coniferous, consisting of pines, spruces, firs and larches. (If there were land at the appropriate latitude, the boreal forest would have a counterpart in the southern hemisphere, but the zone where this should occur lies in the vast Southern Ocean.) In marked contrast to the tropical rainforest, boreal forest has far fewer species of tree - often just two or three species in a hectare - and undergrowth is usually sparse. Almost all conifers are evergreen, but the larches drop their leaves with the dying of the summer, and they predominate in the bleak climate of eastern Siberia, where even tough conifer needles would be damaged by the extreme winter cold. Along its southern margin, birch and other hardy broadleaved species grow among the conifers, producing a mixed forest which is much richer in wildlife, owing to the variety of its trees.

Outside these two zones, forests are enormously varied. In the far eastern region of the

USSR, leopards hunt in the mixed forests of Ussuriland where Korean pines and Khinghian firs grow alongside purple-flowered maples, Manchurian walnuts and Mongolian oaks. On some slopes of the Himalayas are well-watered forests of tree-like bamboos, while the native vegetation of the warm, moist area between the Black Sea and Caspian Sea is a luxuriant mixture of trees and shrubs, including rhododendrons, laurels, hollies and oaks.

The Australian continent alone supports dozens of different types of forest, with wattles (*Acacia* species) and the native *Eucalyptus* species playing a prominent role in most types. These forests have many unique features, notably the important part played by natural fires in maintaining them, and the unusual adaptations to burning shown by eucalyptus trees. Owing to the dryness of the climate, many Australian trees are of the sclerophyllous or 'hard-leaved' type, with leathery, oil-rich leaves having relatively few stomata (breathing pores). Sclerophyllous forests are found in other dry regions of the world, including the Mediterranean, and parts of South Africa and California. The drought-resistant leaves allow the trees to be evergreen despite the aridity of the climate.

In most parts of the tropics, trees have a different approach to coping with dryness, losing all their leaves during the rainless months. Such forests are called seasonal tropical forests or dry tropical forests, and some show a spectacular flush of colour before the leaves fall, or when new leaves appear after the onset of the rainy season. Those with the longest dry season are the monsoon forests, where the rains, when they return, are torrential. Teak and sal are often the dominant trees in this type of forest.

◀ Tropical rainforest on Fraser Island, off the coast of Queensland, Australia. Australia's rainforests are limited in area but rich in wildlife, with many unique plants and animals.

▼ Dry tropical forest along the Limpopo River in Mozambique, showing a flush of spring leaves on its mahogany trees at the beginning of the rainy season. The red colour is associated with toxins in the leaves that deter insects from eating them while still young and soft.

▲▲ Beech woods in southern England. These trees have artificially short trunks because they were managed for timber production in earlier times. They have been repeatedly lopped off or 'pollarded', about 4 m (13 feet) above the ground, with the aim of providing a steady supply of long, straight branches.

▲ Deciduous forest of southern beech on the windswept island of Tierra del Fuego, at the southernmost tip of South America. These forests are the southern-hemisphere equivalent of the northern deciduous forests. If there were more land at this latitude in the south, such forests would be far more extensive.

The northern hemisphere has a unique type of forest, made up entirely of broadleaved, deciduous trees, such as oak, maple, sycamore, beech and poplar. The mild, moist summer and short, sharp winter apparently favour trees that lose their leaves. Because the leaves are replaced each year, and do not have to survive cold or drought, they can afford to be delicate. By contrast, the few evergreen shrubs growing in these woodlands have leathery leaves, be they conifers, such as juniper, or broadleaves, such as holly.

The success of conifers in both the warm moist conditions of the temperate rainforests,

and the hot aridity of the Mediterranean, shows that their northern stereotype of 'cold-weather trees' is mistaken. But conifers also do well on mountains all over the world, suggesting that they do have the competitive edge over most broadleaved trees in low temperatures. By keeping their hard waxy needles throughout the winter, they can resume growth rapidly in the spring, and not waste a moment of the short growing season. The shape of the trees, with pointed crowns, downwardly sloping branches and smooth, slender needles, helps them to shed snow, reducing the risk of broken branches.

◀ Canadian mixed forest at the close of summer, with the birches and aspens about to lose their leaves for the winter, while the spruces and firs hold on to their hard, waxy needles. The mix of species is especially valuable for wildlife, and far more species live here than in the pure coniferous forests.

▲ Pine trees growing in the Black Forest in Germany. Like most conifers, pines cast a dense shade which excludes other plants, so undergrowth beneath them is often sparse.

▶ A grove of maples and paper birch in Michigan, US, with brilliantly coloured leaves that herald the approach of winter.

Of all the temperate forests, the most impressive are the rainforests, which occur wherever rainfall is high all year round. These forests are widely scattered over the globe, and very varied in composition. Those of America's west coast consist mainly of conifers such as western hemlock and western red cedar, while those of New Zealand include magnificent stands of southern beeches, kauri pines and the yew-like podocarps or yellow-woods. In Tasmania, the rainforests are dominated by myrtle, sassafras and those delicate survivors from a distant era, the tree-ferns. In all these forests, the abundant moisture encourages mosses, liverworts and ferns to grow, turning every rock and tree-trunk green.

THE ARCHITECTURE OF THE FORESTS

All forests have a particular 'architecture' depending on what kind of forest they are. The tallest trees usually reach a similar height, and thus form a canopy of foliage. This is known as a closed canopy if the trees meet, or an open canopy where the trees are more widely spaced. The more open the canopy, the more light can reach the ground, and this encourages other plants to grow beneath the trees. In tropical rainforests this can mean two, three or more layers of vegetation beneath the canopy, with shorter trees, such as palms and tree-ferns, forming one or more understoreys. In broadleaved deciduous forest, shrubs such as hazel and holly, and small trees such as limes, sometimes form an understorey. Australian eucalyptus forests may have the native proteas, banksias and heatherlike shrubs of the family Epacridaceae filling the same niche.

In general, the richer the forest in terms of rainfall, warmth and soil nutrients, the more complex its structure. This is particularly noticeable in the tropical rainforests. The lowland forests are massive, complex structures, but shorter trees, fewer species and a simpler structure are found in the cooler cloud forests of the mountain tops and in the heath forests which grow on poor soils.

▲ Temperate rainforest in the Olympic National Park, Washington, US. These Pacific-coast forests of North America are the prime example of 'temperate rainforest', with their moist atmosphere and towering trees, draped with moss. The North American Pacific-coast rainforests are entirely coniferous, but other temperate rainforests are broadleaved or mixed.

◀ An ancient, moss-covered tree in the moist, high-altitude forest of Mt. Anne, Tasmania. The island's forests have recently been the subject of controversy as conservationists try to protect them from exploitation.

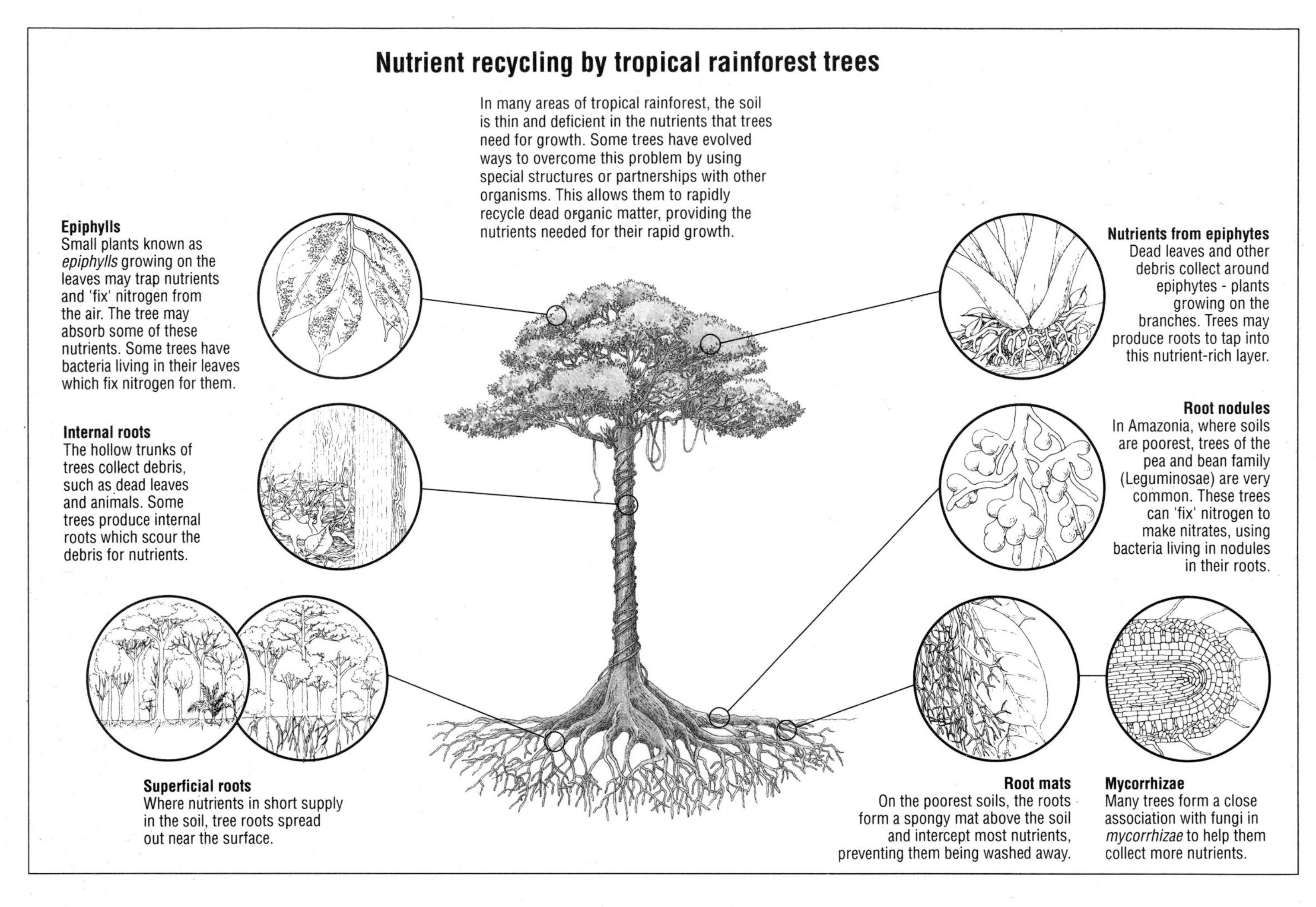

On a local scale, the composition of a forest is not static. Individual trees die and are replaced by others, although not necessarily by those of the same species. On a wider scale, the forest's structure remains unchanged.

THE NEED FOR NUTRIENTS

Trees, like all living things, need nutrients. Because of their great size, many nutrients are 'locked up' in their trunks and branches, and if the underlying soil is deficient in any minerals, this exacerbates the problem. Not surprisingly, decomposers such as fungi, earthworms and termites are of vital importance in all forests, breaking down fallen leaves and other debris and returning the nutrients to the soil for further use. Many forest trees have specialized fungi growing in close association with their roots, forming an interwoven structure known as a mycorrhiza. In return for the sugars provided by the plant, the fungus can make available essential materials, including nitrogen compounds for building proteins. Wherever decomposition is inhibited in a forest, as on the cold, waterlogged floors of tropical cloud forest, the shortage of nutrients stunts the growth of trees.

Some tropical rainforests, particularly those of central Amazonia, happen to grow on extremely poor, infertile soil. This has led to the evolution of some exceptional strategies for retrieving nutrients. In most nutrient-starved forests, trees are able to intercept the nutrients from decaying leaf litter even before they reach the soil, through a mat of surface roots. The tiny roots may grow up to enmesh fallen leaves and fruit, thus monopolizing their nutrients.

Mycorrhizae are very important in these forests, and many of the associations are highly specific, with a given species of tree dependent on a particular fungus. The clear-felling of forest can leave the soil bereft of these specialized fungi, preventing some species of tree from regenerating even though viable seeds are present and germinate successfully. This slows down the regrowth of the forest, so that it takes centuries to return to its mature state.

► Tropical rainforest in Zaire seen from the air, with some trees in flower. Zaire has the largest remaining area of rainforest in Africa.

"Here I first saw a tropical forest in all its sublime grandeur - nothing but the reality can give any idea how wonderful, how magnificent the scene is... I never experienced such intense delight."
Charles Darwin, writing from Brazil

▲ Tree kangaroos survey the rainforest floor. Monkeys and apes never reached the forests of New Guinea and Australia, but their place is taken by marsupial mammals such as these.

◀ A southern leaf-tailed gecko, New South Wales, Australia. The enlarged tail is for defence. By looking like the lizard's head it diverts the attacks of predators to a less vulnerable part of the body. If gripped, the tail can be shed, allowing the gecko to escape.

THE BENEFITS OF THE FOREST

Tropical forests are particularly rich in wildlife. Indeed, according to some scientists, they may contain well over half the species on Earth. Of the 250,000 species of plants described by botanists, at least 30,000 are to be found in Amazonia alone. Deforestation in the tropics, and in the temperate rainforests, is causing a loss of biological diversity on an unprecedented scale, with almost one species being condemned to extinction every hour. Moreover the rate of extinctions is accelerating: between 1990 and 2020, as deforestation eats into the heart of the remaining forests, it is predicted that 100 species could be lost a day.

Insects thrive in the constant warmth of the tropical atmosphere, and, over the millennia, rainforest plants and trees have developed an impressive array of chemical defences against pests. Many of these chemical compounds act as drugs, or have some other useful application. Among the natural chemicals yielded by the Amazon are an antihypertensive and tranquillizing drug from the root of the *Rauvolfia* plant, an anti-tumour drug from *Tabebuia*, the boxwood tree, and a drug from the plant *Calea pinnatifida* that attacks amoebic parasites. Recently, a drug extracted from the now-famous rosy periwinkle, a plant which originated in the forests of Madagascar, has been found to offer a 90 per cent chance of remission in cases of lymphatic leukemia. Yet fewer than one per cent of rainforest plants have been assessed for use in medicine. Unless deforestation is halted, many will disappear before their potential can be realized.

PROTECTING SOIL AND WATER

Forests play a vital role in protecting soils and regulating water supplies. Although soil erosion in deforested areas of the temperate countries is a problem, it is as nothing compared with the degradation caused by forest clearance in the tropics. To understand why, we need to look beneath the tropical forests, at the soils on which they grow.

Because tropical forests generally lie on infertile soils, deforestation quickly leads to essential nutrients being leached out of the soil, and can transform the land into a wasteland, denuded of all but the most unpalatable grasses. Typically, as much as half the soil's organic content is lost just three years after clearing the land for agriculture.

An additional problem in many tropical areas is that of 'laterization'. Many tropical soils are

▲ A keel-billed toucan from Belize. Toucans are playful birds, chasing through the treetops and sometimes 'fencing' with each other using their huge beaks. The beak is useful for reaching fruit at the tips of branches, and for scaring other birds, so that the toucan can approach their nests to steal eggs and nestlings. The colour and pattern of the bill probably allow birds to recognize their own species, but so rarely has courtship been observed that this is uncertain.

▲ A cornet moth from Madagascar. The huge feathery antennae show that this is a male. He must find a female for mating by detecting a special scent that she emits, and following the odour until he reaches her.

◀ The flower of *Rafflesia pricei*, Borneo. The rafflesias are parasitic on lianas, and only the flower is normally seen, the rest of the plant being within the liana root. Rafflesias produce the largest flowers in the world.

“The issue is so simple. Rainforest conservation and deforestation is one of our major environmental concerns - the erosion of the Earth's genetic diversity.”
Dr Aila Keto, President of the Rainforest Conservation Society, Australia

▶ Surveying the devastated remnants of Madagascar's rainforests from a newly cleared hilltop. The red soil that underlies the forest has already been washed into the Indian Ocean in vast quantities. As in other rainforests, the loss of tree cover leaves the bare soil vulnerable to the torrential rain. The more hilly the terrain, the more easily soil is lost. Tragically, much of the land cleared for agriculture is now barren due to erosion.

rich in iron and aluminium oxides. If completely exposed for a prolonged period, these turn to a hard, brick-like substance called laterite, on which it is impossible to grow anything. In no time at all, previously rich forest is converted into unworkable rock.

Stripped of their forest cover, the soils of the tropics are also increasingly vulnerable to being either washed or blown away. Scientists working in the Ivory Coast have recorded massive differences between the rates of soil erosion in forested and deforested areas. They report that, even in mountainous areas, soil erosion in forest is as low as 0.03 tonne/ha (27 lb/acre) per year. Once cleared of vegetation, the rate rises to 90 tonne/ha (37 tons/acre). In Amazonia, Harald Sioli of the Max Planck Institute has warned that if the destruction continues, "There is a danger that the region may develop into a new dust-bowl".

In those tropical regions that experience torrential rains, adequate forest cover is vital for human welfare. Clearing the forest dramatically increases surface run-off from rainfall, because far more rain reaches the ground rather than being caught in the canopy and thus returned to the atmosphere through evaporation. The soil itself is also unable to absorb as much water after clearance, largely as a result of compaction by heavy machinery. Forested soils are now known to absorb 10 times more water than pasture - water which, once the forest has gone, simply cascades over the denuded soil straight into the local streams and rivers.

All too often, the consequence is massive flooding. As a result of deforestation in India's river catchments, the amount of land now classified as flood-prone doubled between 1971 and 1980. By 1984, the figure had trebled. Paradoxically, the destruction of forests can produce drought as well as flooding, especially in monsoon regions, with their seasonal rainfall. Forest soils hold water well, releasing it slowly into local streams and rivers. They thus smooth out the extremes of climate, spreading the supply of water evenly throughout the year.

Without the buffering action of the forests, there is often a 'drought-flood cycle', with massive floods during the monsoon periods, alternating with devastating droughts during the dry season. When this happens, wells dry up. In the Indian state of Maharashtra, deforestation is largely responsible for the drying up of traditional water supplies in 23,000 villages, an increase of 6,000 villages during the last five years.

▲ Excavations for a road in Venezuela reveal how extremely thin the rainforest soil is, and how shallow are the roots of rainforest trees.

"...people vied with each other to build houses, and wood from the southern mountains was cut without a year's rest. The natives took advantage of the barren mountain surface and converted it into farms... If heaven sends down a torrent there is nothing to obstruct the flow of water. In the morning it falls on the southern mountains; in the evening, when it reaches the plains, its angry waves swell in volume and break embankments causing frequent changes in the course of the river... Hence Ch'i district was deprived of seven-tenths of its wealth."

Chinese manuscript, sixteenth century

▲ Honey is prized by rainforest tribes throughout the world. Gathering the sweet honeycombs is far from easy - here smoke is being used to sedate bees nesting inside a hollow tree-trunk.

"The white man is walking in the dark, blinded by the glitter of gold. That is why he doesn't see us."

Davi Yanomami, Yanomami Indian leader, Amazonia

▲ ▶ South American Indians gathering tubers in the rainforest. The forest can supply all that is needed for a complete diet. This includes protein from animals including fish, carbohydrates from tubers such as the ones shown here, and fats and oils from animals, seeds and fruit.

THE PEOPLES OF THE FOREST

As a direct result of deforestation, hundreds of millions of people in the Third World are being condemned to living a degraded and impoverished existence, with very little prospect of ever improving their lot. But, for the 50 million tribal peoples who live in the forest itself, the impact of deforestation is far worse. It extends beyond the ecological devastation caused by the loss of their forests to the loss of their culture, their identity and their whole way of life.

Such peoples rely on the forest for their food, the building materials for their houses, the wood for their agricultural implements, the herbs for their traditional medicines, the fibres and dyes for their clothes, and the materials for their religious and cultural artefacts. Just as important, they have deep-rooted cultural ties with the forest - ties which extend far beyond the economic and which give meaning to their lives and cohesion to their cultures. Not surprisingly, for most tribal peoples, the destruction of their forest world spells physical and social doom. Many succumb to disease while others drift into the slums, where they fall prey to alcoholism, prostitution and mental illness.

FORESTS AND CLIMATE

Remote as the destruction of tropical forests and their peoples might seem to the inhabitants of faraway cities such as New York, London or Tokyo, they too may be gravely affected by the

wider climatic implications of deforestation. Forests contain massive amounts of carbon, and deforestation, especially when the forest is burned as happens on such a large scale in Latin America, adds considerable quantities of carbon dioxide to the atmosphere. Deforestation is thus adding appreciably to the greenhouse effect *(see pp.40-53)* with devastating consequences for both North and South. So extensive is the forest clearance in Amazonia that Brazil alone could be releasing 500 million tonnes of carbon a year, far more than previously estimated.

Deforestation may also disrupt another vitally important climatic mechanism. Between half and three-quarters of the rain falling over Amazonia is returned promptly to the atmosphere, both through evaporation from wet foliage and through transpiration - the process whereby plants take up water from the soil and lose it through the stomata (breathing pores) in their leaves.

In the tropics, this vapour, like perspiration, serves to cool the plants down so that they can continue to photosynthesize. It also provides at least 50 per cent, and sometimes as much as 80 per cent, of the rain that precipitates downwind, as the clouds are carried westwards towards the Andes. The same moisture falls as rain, evaporates and falls again, many times on its journey across the forest. Deforestation in eastern Amazonia could fatally disrupt this recycling process, causing the remaining forests of western Amazonia gradually to dry out.

The clouds formed by water vapour above the tropical forests have a high reflectivity, or albedo, so they reflect sunlight back into outer space, thus cooling the tropics. Some of the water vapour is transported to higher latitudes, where its warmth is imparted to the atmosphere. The Amazonian rainforest, as a consequence of its size, is therefore an integral part of a giant solar heat-pump that moderates the climate worldwide, keeping the tropics cool while transporting heat to colder climes. Some climatologists believe that widespread deforestation over the Amazon Basin may disrupt the transfer of heat from the tropics to the northern hemisphere, which, in turn, could become cooler.

▲ Lianas and epiphytes in the cloud forest of Venezuela. Cloud forest usually has fewer animal species than lowland rainforest, but its inaccessibility can make it a valuable haven for some species.

“When they have finished, there will be no more rubber trees, no Brazil nut trees, not even their cattle - just dust.”
Rubber-tapper, Amazonia

The death of the RAINFORESTS

▼ Millions of years of evolution go up in smoke as an area of Brazilian rainforest is set on fire.

Flying over the forests of the tropics today, it is hard to accept that the green mantle, stretching as far as the eye can see, could conceivably be destroyed within a few decades. But travel along the roads or down the major rivers leading into the forest and the destruction quickly becomes evident. Charred stumps, scrub and bare soil scar the landscape. Even where trees and bushes have begun to recolonize the land, the dense foliage is deceptive. This secondary forest lacks the rich variety of species of the primary forest which was there before. Where the land has been converted to pasture - frequently the case in Latin America, though not in the rest of the tropics - the chances of its recovery diminish with each year that it is grazed. Annual burning of the land to encourage grasses kills the seedlings that would permit regrowth, transforming the land into savanna where few plant species can survive. It is useless for crops, and, after a time, useless even for cattle.

With an area the size of West Germany being deforested or severely damaged every year, the prospects are grim. Experts now warn that, unless action is taken to halt the destruction, the world's original lowland rainforests will have been obliterated by the end of the century, except in inaccessible sites and a few reserves.

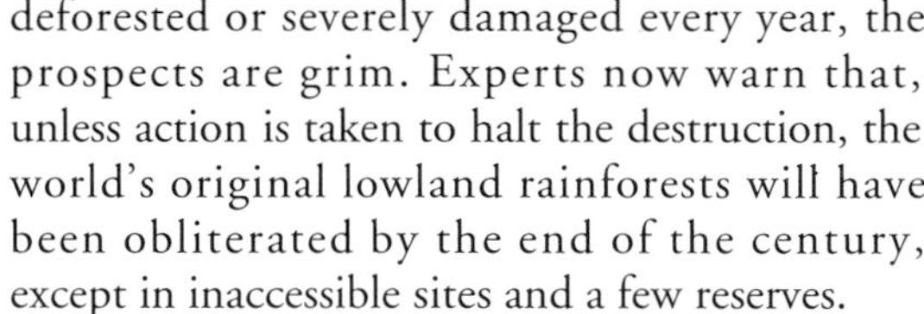

BLAMING THE POOR

The UN Food and Agriculture Organization (FAO) blames shifting cultivation by landless peasants for causing 50 per cent of tropical forest clearance worldwide, and names fuelwood collectors as the second major cause of deforestation. The impression given is that the forests are being cleared by those too poor and hungry to care about protecting the environment.

On the face of it, FAO has a case. From Amazonia to Indonesia, it is indeed landless peasants who light the fires and operate the chainsaws that annually destroy vast areas of forest. Around the major cities in Africa or India, trees stripped of all but their highest branches bear terrible testimony to the fuelwood crisis. But blaming landless peasants for deforestation is tantamount to blaming conscripted soldiers for causing wars. As the ecologist James Nations puts it, "peasants carry out much of the work of deforestation, but they are mere pawns in a general's game."

Firewood shortages, for example, are the consequence, rather than the cause, of deforestation. The main destruction is in those areas which have already been grossly deforested, or otherwise impoverished, generally through commercial pressures. In many parts of the Himalayas, commercial logging, which started under the British colonial administration, has resulted in rich forests of sal being replaced by monocultures of chir pine. This allows little undergrowth and deprives villagers of a source of fuel.

In many cases, the peasants who move into the forests have been forcibly dispossessed of their own land to make way for development projects and have no option but to encroach upon the forest if they are to survive. Unequal access to land is a major cause of the problem. In Brazil, where 42 per cent of the country's cultivated land is owned by just one per cent of the population, rocketing rents and the encroachment of the big estates are causing daily dispossession among the rural poor. Yet there is no shortage of fertile soil: if the large estates in the south were to be broken up, there would be enough land for everyone without having to cultivate a single hectare of Amazonia.

Rather than institute land reform, many Third World governments have looked to the forests as convenient dumping grounds for the landless. Since the late 1960s, the Brazilian government has been actively colonizing Amazonia under the slogan "land without men for men without land". The most notorious colonization programme - the Polonoroeste Project - saw 70,000 to 80,000 settlers invading the Amazonian state of Rondonia in 1987 alone. The area of forest being cut or burned in Rondonia is doubling every other year. Despite the promise of "good land, appropriate land", the settlers have generally been reduced to penury through the infertility of the soil. As new roads spread through the state, the abandoned land is often bought up by entrepreneurs, who use it to graze cattle, preventing any chance of its recovery.

"...it is social relations, not simply the pressure of numbers, which are destroying the tropical forests. These same processes are bringing about the genocide of indigenous forest dwellers."
Jack Westoby, former Senior Director of Forestry, FAO

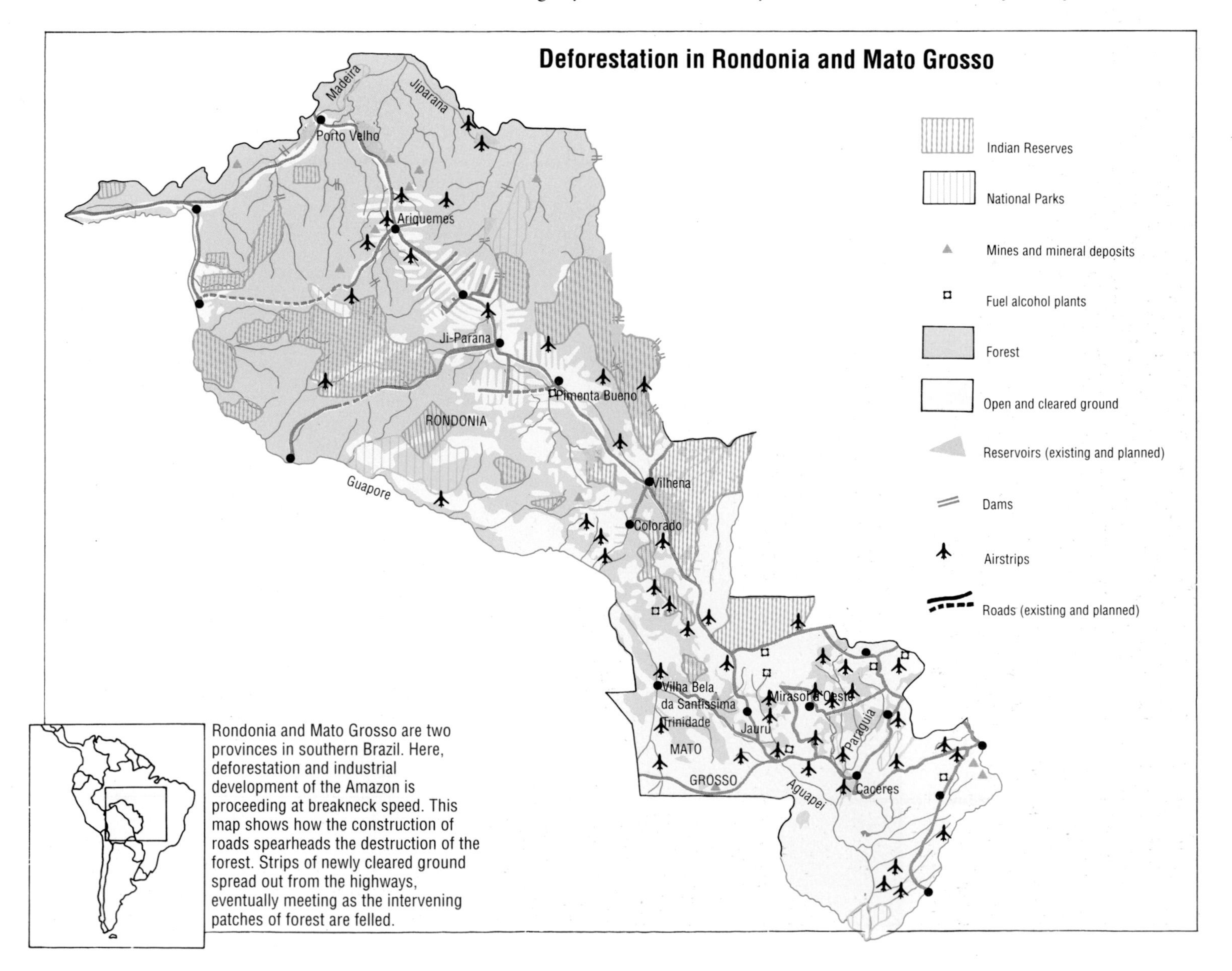

Deforestation in Rondonia and Mato Grosso

Rondonia and Mato Grosso are two provinces in southern Brazil. Here, deforestation and industrial development of the Amazon is proceeding at breakneck speed. This map shows how the construction of roads spearheads the destruction of the forest. Strips of newly cleared ground spread out from the highways, eventually meeting as the intervening patches of forest are felled.

On the other side of the globe, Indonesia has been carrying out the largest colonization programme in the world - the "Transmigration Programme" - intended to ease the pressure on land in the densely populated island of Java by moving Javanese peasants to the outer islands of the Indonesian archipelago. Over 100,000 km^2 (39,000 square miles) of tropical forest could be at risk as a result of the influx of settlers.

Between the early 1970s and the mid-1980s, some 3,600,000 people were moved. If the Indonesian government has its way, a further 65 million will be moved in the next 20 years. Predictably, where the forest has been cleared, the soil has frequently proved too barren to support agriculture, with the result that colonists are forced to clear yet more forest. Some colonists have starved. Others have cut their losses and returned home to Java, leaving behind them a degraded landscape.

▲ An Indonesian transmigrant ploughs his land near Sekayu in southern Sumatra. The success of these subsistence farmers depends very much on the soil. In some parts, such as central Kalimantan, the soil is so poor that transmigrants have been reduced to starvation.

► Planting maize on a slope cleared by burning, Madagascar. When the land is abandoned it may, if not eroded, be colonized by spiny thicket, a type of vegetation known as 'savoka' in Madagascar. After about a hundred years, something resembling primary forest returns, but the regular reburning usually prevents this from happening. The animal life of the native forest does not always return to regenerated patches like these. A slight drying of the climate, and the repeated use of fire to clear the native vegetation, has reduced the original forest cover drastically, and the central part of the island is now completely deforested. The pace of forest destruction is currently increasing.

The programme also threatens numerous tribal groups, forcing them to participate in development projects which can only destroy their cultures. Many have had their land taken from them without any compensation. By the end of 1984, West Papua alone had 24 major transmigration sites, appropriating a total of 7,000 km^2 (2,750 square miles) of land belonging to traditional tribal peoples. In an attempt to assimilate the West Papuans, whole communities have been broken up, in some cases even being bombed by the Indonesian airforce, and individual families have been dispersed to separate transmigration colonies.

Concerns over national security have also played their part in colonization programmes. Eager to secure the forests against territorial claims by her neighbours, Brazil has encouraged colonization into border areas, particularly in the northwest.

But more important, colonization is part of a general strategy designed to open up the forests to commercial exploitation and to integrate forest peoples, often against their wishes, into the wider nation state. In the pursuit of that strategy, the forests have been plundered.

Thousands of hectares have been cleared for plantations, transformed into pasture, logged or torn apart by mining. Rivers have been dammed, flooding some of the most remote areas of forest in the world, and the products of millions of years of evolution have been bulldozed aside to create the necessary 'infrastructure' to support the development process.

FORESTS VERSUS AGRIBUSINESS

In the Philippines, the island of Negros - once a carpet of forest - is now little more than a vast sugar estate. Meanwhile, those who previously farmed the land have been forced to clear more and more marginal lands in the forested uplands of Negros, which are currently losing their forests at the rate of more than 20,000 ha (50,000 acres) per year.

In Thailand, many of the most fertile soils have been used to grow cassava, mainly for export to feed Europe's livestock. In the 10 years between 1973 and 1982, exports of cassava from Thailand to the EC rose from 1.5 million tonnes to 8 million tonnes. Almost all the increased production took place in the east and northeast of the country, the bulk of it at the expense of forest. Again, the peasants displaced by cassava plantations have little option but to encroach on the forests.

▲▲ Flowers being grown for export in Doi Inthanon National Park in Thailand. Forest once covered these hillsides, but gradually it has been cut down for growing cash crops. One of the traditional crops of this area is opium. Flowers and other cash crops are now being encouraged as an alternative to opium production.

▲ In Brazil, an area once covered by dry tropical forest is ploughed for agriculture. As tractors move across the field, the dry soil churned up by their wheels is blown away on the wind.

In Latin America, cattle ranching is responsible for the destruction of at least 20,000 km^2 (7,700 square miles) of forest a year. Since 1950, two-thirds of the lowland tropical forests in Central America have been cleared, mostly for pasture, and in several countries cattle now outnumber people.

The expansion of cattle rearing has been promoted by the major international development banks and leading agencies - the Inter-American Development Bank, the World Bank and the UN Development Fund. Governments have also promoted ranching through local fiscal incentives. In Brazil, grants of up to 75 per cent were paid on ranch development costs, along with 100 per cent tax holidays if the ranches were in the Amazon region or the dry northeast. One company to take advantage of the fiscal incentives offered was Volkswagen, which cleared a vast tract of Amazonia, although it has now withdrawn from the area. Large corporate ranches now cover some 8,750 km^2 (3,400 square miles) of Amazonia. In the states of Para and Mato Grosso they have been responsible for 30 per cent of clearings.

The ecological destruction caused by ranching programmes is long-term and often irreversible. The land is rapidly depleted of nutrients and invaded by toxic weeds. In a few years, it is so degraded that it must be abandoned - a fate that has befallen one in three of the large cattle ranches established in Amazonia. The cost to the government has been some $2.5 billion.

In Central America, most of the beef raised is exported, 80 to 90 per cent of the exports going to North America. Cheap beef imports have reduced the price of hamburgers by a few cents. However, the cattle trade has done little to help the poor in the cattle-exporting countries, few of whom can afford to eat beef.

In Brazil, the situation is different. Because foot-and-mouth disease is endemic in Amazonia, little beef can be exported. Even with government subsidies, cattle-rearing is rarely economic, only 3 out of 100 large ranches in a recent government survey making any profit from their livestock. Rather, it is land speculation that has encouraged the explosive growth in cattle ranching. Under Brazilian law, anyone who clears an area of forest can lay claim to the land. Cattle permit large amounts of land - and the mineral rights below it - to be claimed with minimal labour. It is thus no coincidence that those areas where clearance is most vigorous are frequently closest to gold strikes. It is this quest for quick (and easy) profits - rather than a hunger for beef - that provides the key to much of the deforestation in Brazil. Indeed, at least half of the largest ranches in Brazilian Amazonia have never even sent a cow to market.

THE TIMBER TRADE

Worldwide, the tropical timber industry is held responsible for degrading some 50,000 km^2 (19,300 square miles) of primary rainforest annually. Much of the wood is used to make cheap, throwaway goods. Eight out of every ten logs imported into Japan, for example, are made into cheap plywood, much of which is used by the building industry to make frames for moulding concrete and scaffolding. These are discarded after being used once or twice. Tropical hardwoods are also used to make throwaway chopsticks and paper. In Europe and the US, tropical hardwoods are primarily used to make doors and window frames, furniture, plywood, blockboard and veneer sheets - uses for which native hardwoods and softwoods, harvested on a sustainable basis, could well act as substitutes.

“In the area they destroyed there, the last harvest produced 1,400 cans of Brazil nuts, a good crop. We challenged the owner of the land and the Governor himself to work out the annual income per hectare produced by forests products such as Brazil nuts and rubber and then compare it with that produced by grazing cattle there. They refused because they knew we could prove the income... is 20 times greater.”
Chico Mendes

▼ Forest destroyed by burning in Amazonia, apparently to create pastureland for cattle. The real reasons for this wanton destruction are more complex.

The tropical timber industry is particularly sensitive to accusations of causing deforestation, arguing that it is only responsible for some 10 per cent of forest destruction worldwide. However, that global average obscures the fact that logging is the major cause of primary rain-forest destruction in much of Southeast Asia, Africa, Australia and the Pacific islands.

The industry's figure also ignores the widespread forest loss caused indirectly by logging, mainly through opening up previously inaccessible areas to landless peasants. If that subsequent forest loss is taken into account, the timber industry is probably responsible for 40 per cent of forest destruction.

The damage caused by logging is extensive. Current logging practices in Sarawak, Malaysia, leave 33 trees damaged for every 26 trees removed as timber. In some areas, up to 70 per cent of the remaining trees die from their injuries. The extensive network of roads and skid trails needed to get the logged timber out of the forest adds to the damage. Erosion is a particular problem, increasing the silt load of rivers

“We need to tell the world that Brazil is eating itself up. We feel we are losing our health, which is obviously dear to us. But we're also losing nature.”
Timber worker, Amazonia

◀ Timber being sawn by hand in the forests of Madagascar. Small-scale exploitation of this sort, without the use of bulldozers or other heavy machinery, could be sustainable, but it cannot give the high short-term returns demanded by logging companies.

▼ A logger with a chainsaw can quickly cut through even the largest of trees.

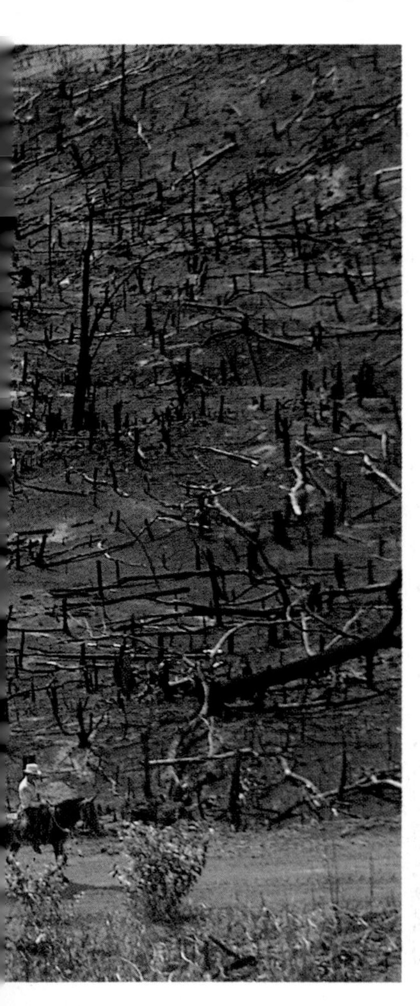

with consequences that extend far beyond the forests. Two out of every three rivers in Sarawak are now officially classified as 'polluted' due to soil erosion, leading to a dramatic reduction in fish catches. Fish provide a major part of the native people's diet. In the Philippine island of Palawan, erosion due to deforestation has all but destroyed coastal fisheries.

Faced with mounting criticism over the destruction caused by commercial logging, the timber industry has argued that, with proper management, tropical rainforests could supply a sustainable timber yield. Unquestionably, the tropical timber industry could be made more efficient and less destructive. However, 'sustained yield management', as promoted by the timber industry, is above all a commercial concept. It is not, in its present form, concerned with preserving the rich ecosystem of primary rainforest, conserving biological diversity, stabilizing water supplies and climate, or ensuring the livelihoods of forest peoples.

Significantly, there is hardly a single example of sustained industrial timber extraction from tropical rainforests. Indeed, the International Tropical Timber Organization (ITTO), the industry's trade organization, has itself concluded that successful sustainable tropical timber operations cover less than 0.8 per cent of rainforest lands and are "on a world scale negligible".

That is not to say that timber cannot be extracted on a sustainable basis - forest peoples have been doing precisely that for millennia. Several new schemes show promise, but they are generally small-scale, rely on minimal machinery (some use draught animals to extract the logs) and are rigorously managed. Moreover, few could provide the rates of return demanded by commercial companies. Under present political and economic circumstances, sustainable commercial logging in the rainforests would appear to be a pipe-dream. Commercial pressures, corruption, the ravages of heavy machinery and the roads that service timber operations would all militate against it.

MINING, INDUSTRY AND DAMS

The world's tropical forests are known to contain large quantities of minerals and oil. Schemes to exploit these resources are increasingly important as causes of deforestation. In Brazil, the Grande Carajas programme is opening up a vast area of eastern Amazonia to industry. One-sixth of Brazilian Amazonia will be affected - an area the size of France and Britain.

▲◄ A newly created road slips down a hillside in Malaysia, as the slope below becomes eroded following deforestation. Ironically, many of the vehicles that use the road are logging trucks. The more timber they take away, the worse the problem becomes.

The centrepiece of the project is the Serra dos Carajas open-cast iron ore mine. To fuel the iron-ore smelters, the forest is being cleared to produce charcoal, and by the time the mine is in full production, some 6,000 km^2 (2,300 square miles) of forest will be lost a year - a figure that will accelerate even further the rate of deforestation in Amazonia. Other projects planned under the programme include a massive bauxite mine and an aluminium smelter. In addition, some 54,400 km^2 (21,000 square miles) will be cleared to make way for export-oriented plantations and plantations growing sugar cane for the production of synthetic fuels. A further 30,000 km^2 (11,500 square miles) will be given over to cattle ranches.

In the northwest of the Brazilian Amazon, where gold, diamonds, uranium, titanium and tin have been discovered, tens of thousands of 'wildcat' miners from all over the country have now invaded the area. Particularly threatened by the project are nearly 10,000 km^2 (3,860 square miles) of tropical rainforest which form the homeland of the Yanomami. These Indians have been systematically deprived of their territorial rights to ensure unrestricted access to the minerals beneath their lands. To date, 27 concessions for mining have been granted in the area and 363 further applications have been made, covering in total over a third of Yanomami territory.

◀ Construction of the massive iron-ore extraction plant in Serra dos Carajas, in the Brazilian Amazon. The production of charcoal for iron smelting will destroy an area of forest the size of the US state of Wisconsin.

▼ Charcoal burners in Haiti. Charcoal is a fuel created by burning wood with insufficient air present for complete combustion. Here the wood is burned under a pile of earth. Over the past 40 years, charcoal-burners and farmers have been forced to turn to Haiti's remaining forest for fuel, destroying over 80 per cent.

"I am extremely worried about Balbina. I don't talk about it. There are going to be serious problems there."
Official of Eletronorte, northern Brazil's state electricity company

The quest for cheap electricity, presumed essential to economic development, has seen river after river being dammed throughout the tropics *(see pp. 130-141)*. Massive areas of forest have disappeared beneath the floodwaters of the dams' reservoirs. In Brazil, the Tucurui Dam flooded some 216,000 ha (535,000 acres) of primary rainforest. The Balbina Dam, near Manaus, is now filling and will eventually flood 234,600 ha (580,000 acres) of rainforest. The dams planned for Amazonia are expected to flood an area the size of Montana; 68 dams are planned for Brazilian Amazonia alone.

The state electricity authorities of many tropical countries insist that consumers will suffer an energy crisis unless new dams are built. Yet studies of the situation in Brazil by the World Bank suggest that sufficient generating capacity already exists to satisfy the expected rise in demand over the medium term, provided that energy is used more efficiently.

THE TROPICAL FORESTRY ACTION PLAN

Several international programmes have now been put forward to 'solve' the deforestation crisis. Of these the most advanced in terms of implementation is the Tropical Forestry Action Plan (TFAP), now being promoted by - among others - the World Bank, the UN Food and Agriculture Organization (FAO) and the UN Development Programme.

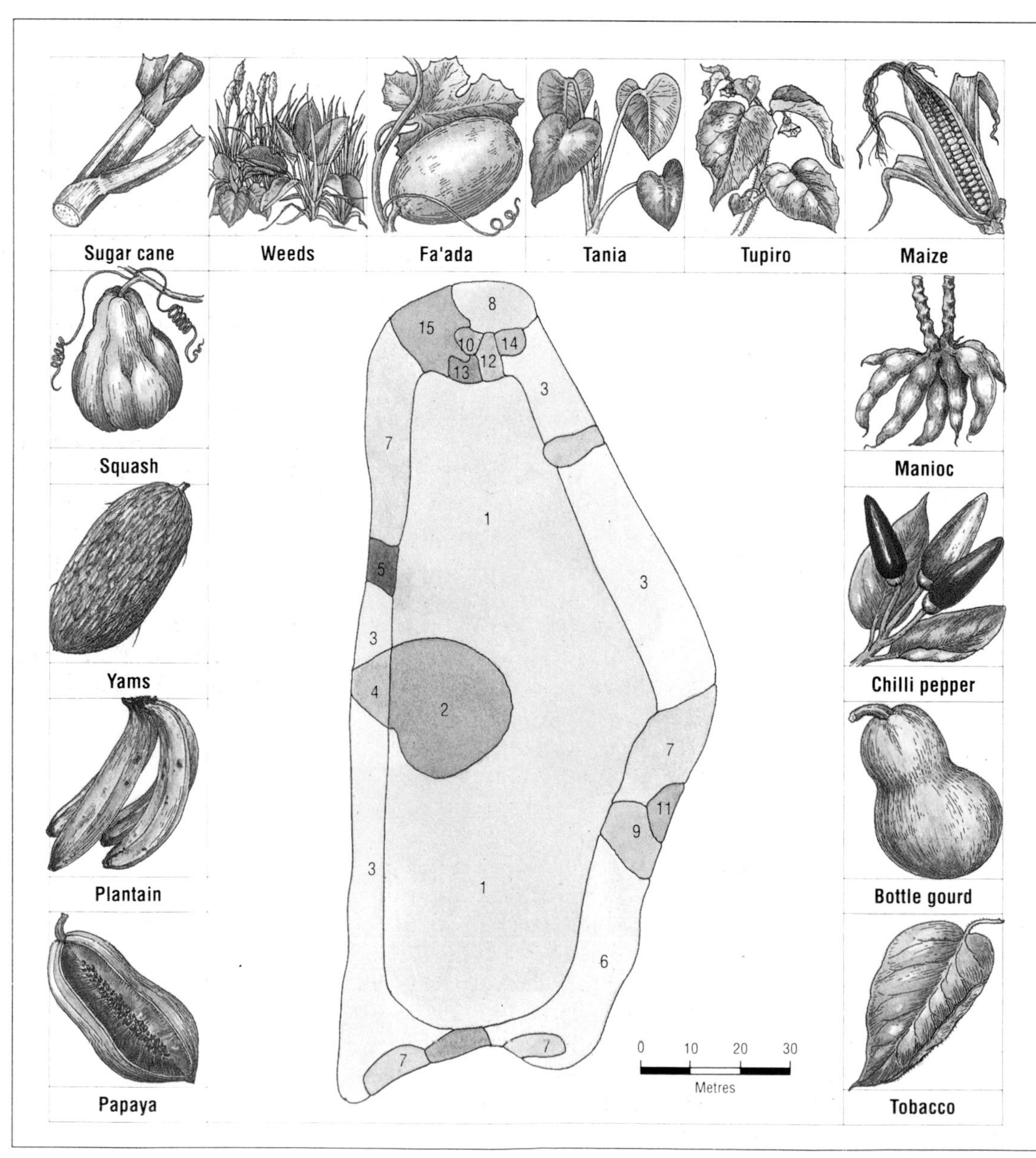

Yekuana Indian forest garden

The Yekuana Indians practice shifting cultivation in the rainforest of southern Venezuela. They rely on expert knowledge of local conditions to provide them with enough food on poor rainforest soils. After a suitable site for a 'garden' has been selected, it is then cleared and burned, releasing nutrients to fertilize the soil. The crops are then planted, some on their own, others mixed together. The Yekuana use over 70 different food plants, choosing which ones to plant according to soil. A garden remains productive for between two to four years, with crops maturing at different times, after which it is abandoned.

1. Manioc (cassava) with maize, yams
2. Manioc with maize, yams, squash
3. Plantain with maize
4. Plantain with maize, squash
5. Plantain with maize, bottle gourd
6. Plantain with maize, tobacco
7. Sugar cane
8. Fa'ada
9. Bottle gourd, tobacco
10. Papaya, maize
11. Chilli pepper
12. Tobacco
13. Tania
14. Papaya
15. Tupiro
16. Weeds

The plan lays out five objectives: to increase firewood supplies, to promote agroforestry, to reforest upland watersheds, to conserve tropical forest ecosystems, and to upgrade forestry practices in the Third World.

Despite these aims, the plan is deeply flawed. Indeed, many environmental groups fear that, if implemented, the TFAP will exacerbate the problems of deforestation. There has been official criticism too. A study for the West German Federal Chancellery recently concluded that the aims of the plan "are not compatible with the need to protect intact tropical forests, their ecology or species diversity."

The plan appears less concerned with the preservation of forests than with ensuring the growth of the timber industry and the setting up of commercial plantations of fast-growing species, such as eucalyptus. When alien trees such as this are planted they offer little food to native animals which are not adapted to feed on them, and they may even disrupt local water cycles or harm the soil. In Latin America, the TFAP recommends the investment of between $2 billion and $2.8 billion a year for the next decade in the industrial development of the region's forests. In Africa, the Cameroons intends to use TFAP money to double the production of logs from the country's rainforests, at present some of the best preserved in Africa. Meanwhile in Thailand, it is planned to increase the area under eucalyptus twenty-fold.

But perhaps most damaging of all is the failure of the TFAP to address the prime causes of deforestation. Although it acknowledges that colonization schemes, dams, plantations, logging and industrial development programmes have played a major role in the destruction of forests, it puts forward no measures to curtail such projects. If the plan is put into effect, the destruction can only continue.

USING THE FOREST WISELY

The havoc now being wrought upon the forests in the name of development might give the impression that it is impossible to exploit the forests on a sustainable basis. This would be wrong. Over millennia, the indigenous inhabitants of the forests have developed a wide range of sophisticated strategies for living off the forest without destroying it.

Apart from hunting and gathering, shifting agriculture - also called 'swidden' or 'bush fallow' - is the most common means of food production amongst forest peoples. Long decried as 'unproductive' and 'destructive', it is now increasingly recognized as one of the few successful means of farming the forests sustainably. Small areas of bush are cleared, the smaller trees being cut down and the rest burned, and gardens are then planted. The burning unlocks the nutrients stored in the vegetation and provides the soil with fertilizer in a form which can be readily taken up by plants.

In areas where there is little competition for land, the gardens tend to be abandoned as soon as weeds begin to invade the plots - the job of keeping them at bay being more onerous than moving to a new clearing. The abandoned clearings are then left fallow for 20 years or more in order for the forest to regenerate and the fertility of the soil to be restored.

Forest peoples are well aware that the regeneration of forest is crucial for the recovery of the soil and future use of the area. They are also aware that a large buffer area, perhaps a hundred or even thousand times greater than their gardens, needs to be retained to provide them with game and wild fruit, as well as other essential materials, for construction, for tools and weapons, for medicines and for ritual purposes.

In areas which are densely populated, however, a number of intensive techniques have been developed so that the forest can be cultivated for a longer period without degrading it. One strategy is to intermingle a wide variety of crops so that they mature at different times and grow to different heights. In the Philippines, for example, the Hanunoo people grow as many as 40 different crops in a single 1-ha (2-acre) garden, while over 70 different crop species are used by the Lacondan Mayas of Mexico.

The effect is to create a multi-layered garden which, in many respects, mimics the natural forest. Soil erosion and the loss of nutrients are thus kept to a minimum, and pest problems are avoided. The Kayapo Indians of Amazonia maintain 'corridors' of natural forest in between their cultivated plots, thus providing islands from which the forest can readily regenerate.

Observing strict fallow periods is vital to the sustainability of swidden agriculture, and shifting cultivators have devised a number of built-in controls to ensure that the land is not overtaxed. A classic study of the Tsembaga of New Guinea reveals that 90 per cent of cleared areas are eventually indistinguishable from primary forest. Indeed, the very existence of the forests today is a testament to the success of swidden systems over the millennia.

"The forest is one big thing - it has people, animals and plants. There is no point in saving the animals if the forest is burned down. There is no point in saving the forest if the animals and people are driven away. Those trying to save the animals cannot win if the people trying to save the forest lose."
Bepkororoti, Kayapo Indian, Brazil

"Our parents give us life but the forest sustains it. From it we get the four necessities of life - food, shelter, clothing, medicine. It balances the air we breathe, cleanses the water we drink, produces the soil we grow our crops in. It nourishes the spirit in the same way as it nourishes the body. We should be endlessly grateful to it - every grove, every tree, every leaf."
Ajaan Pongsak, Dhammanaat Foundation Reforestation Project, Thailand

"My dream is to see the entire forest conserved because we know it can guarantee the future of all the people who live in it. Not only that, I believe that in a few years the Amazon can become an economically viable region not only for us, but for the nation, for all of humanity, and for the whole planet."
Chico Mendes

The methods used by swidden cultivators are age-old. But, more recent settlers - such as the rubber-tappers of Brazil - have also evolved a way of life that enables them to live off the forest without destroying it. Their success lies not in cultivating the forest but in harvesting its 'minor products'. Apart from rubber, these minor products include fruits, nuts, herbs, spices, oils, fibre and medicines.

Recent research has revealed the enormous economic potential of such products. Studies in Peruvian Amazonia, for example, have shown that they produce an income two to three times higher than that from either timber plantations or ranching - and, of course, such harvesting it leaves the forest intact. Research from Southeast Asia has reached similar conclusions. Several countries in the region already earn substantial sums from exporting minor forest products - Indonesia earns some $238 million a year.

SAVING THE FORESTS

International trade, and the consumer society that feeds it, ensures that we are all parties to the destruction of the rainforests. The beef mountains of Europe have been fed on soyabeans grown in southern Brazil, on land from which peasants have been dispossessed and sent as colonists into the forest. Aluminium cans which end up on our rubbish dumps come from smelters fed with bauxite from Amazonia and powered by dams that have flooded forests. The hardwood timber in most homes and public buildings comes from the forests of Southeast Asia and West Africa, forests that have been destroyed by logging.

▲▲ Wild fruits from Amazonia on sale in Manaus, the 'rubber-boom city' in the heart of the rainforest. Such fruits and nuts are harvested and eaten by thousands of people and could be sold more widely.

▲ Gathering Brazil nuts in Amazonia. At present, these trees cannot be grown in plantations - the bees they need for pollination are only found in unspoiled forest. Brazilian law forbids the felling of Brazil nut trees, and they are often left, surrounded by pasture, but unable to bear nuts.

◀ Smoking latex to dry it, before it is taken to market. Natural rubber is collected from rubber trees growing wild in Amazonia by rubber-tappers or *seringueiros.* These tappers are descended from Indian tribes, and from nineteenth-century immigrants from Brazil's northeastern regions, where sugar-growing had led to destitution and famine. They have worked under the notorious system of debt-bondage until relatively recently, and lived in extreme poverty as a result. Now their livelihood is being taken away altogether as the forest is destroyed by ranchers.

In that respect, saving the tropical forests relies, not just on measures taken in the rainforest countries themselves, but also on the international community adopting policies which reduce its ecological impact.

Given the enormity of the changes required - changes that will affect everything we do - it would be understandable if the determination to save the forests gave way to despair. But, despite the gloom, there is light at the end of the tunnel. The case for exploiting fruit and other minor products is now gaining ground, a large area of Amazonia recently being put over to this use. And in Colombia, the government has set a remarkable precedent by granting inalienable land rights to its indigenous Amerindian peoples. Some 180,000 km^2 (70,000 square miles) - two-thirds of Colombian Amazonia and an area equivalent to the size of Great Britain - have been handed over.

In doing so, the Colombian government has recognized that forest peoples have a key role to play in saving the rainforests. This role can be fulfilled if those peoples have rights to their land and a decisive say in what happens to their environment. At least in the Amazon, one country appears to have got its priorities right. Who will now follow?

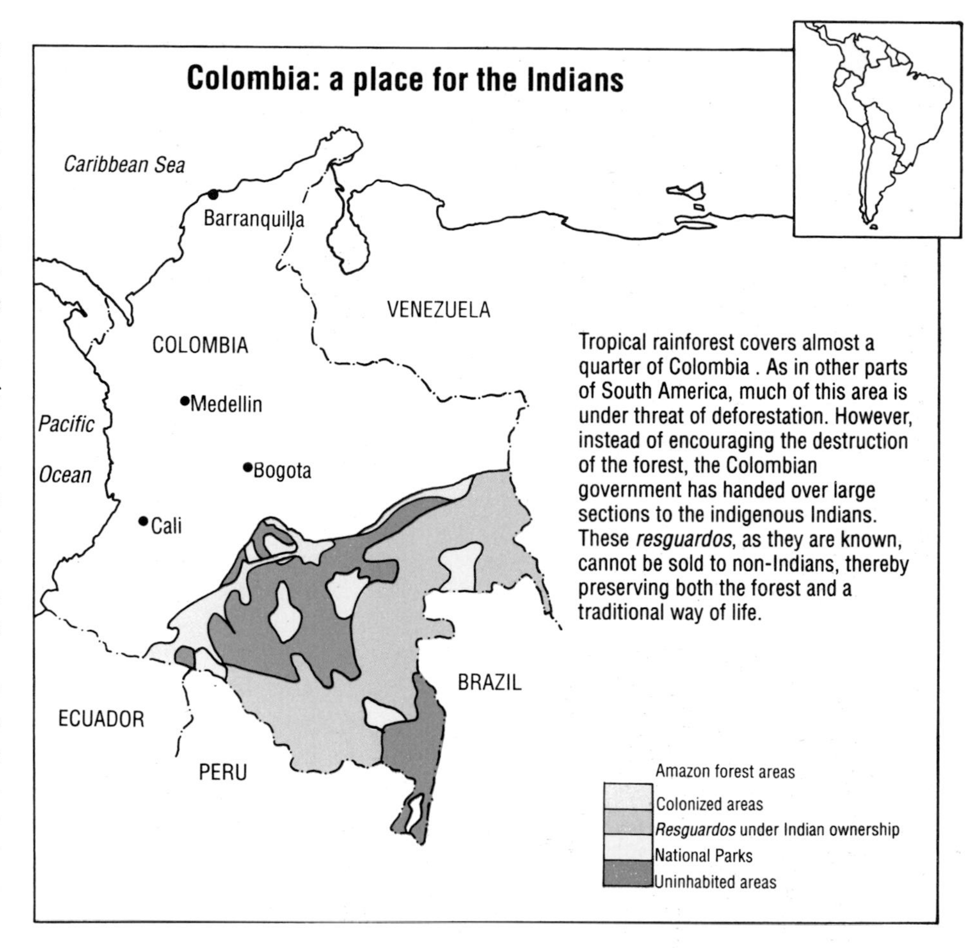

Tropical rainforest covers almost a quarter of Colombia . As in other parts of South America, much of this area is under threat of deforestation. However, instead of encouraging the destruction of the forest, the Colombian government has handed over large sections to the indigenous Indians. These *resguardos*, as they are known, cannot be sold to non-Indians, thereby preserving both the forest and a traditional way of life.

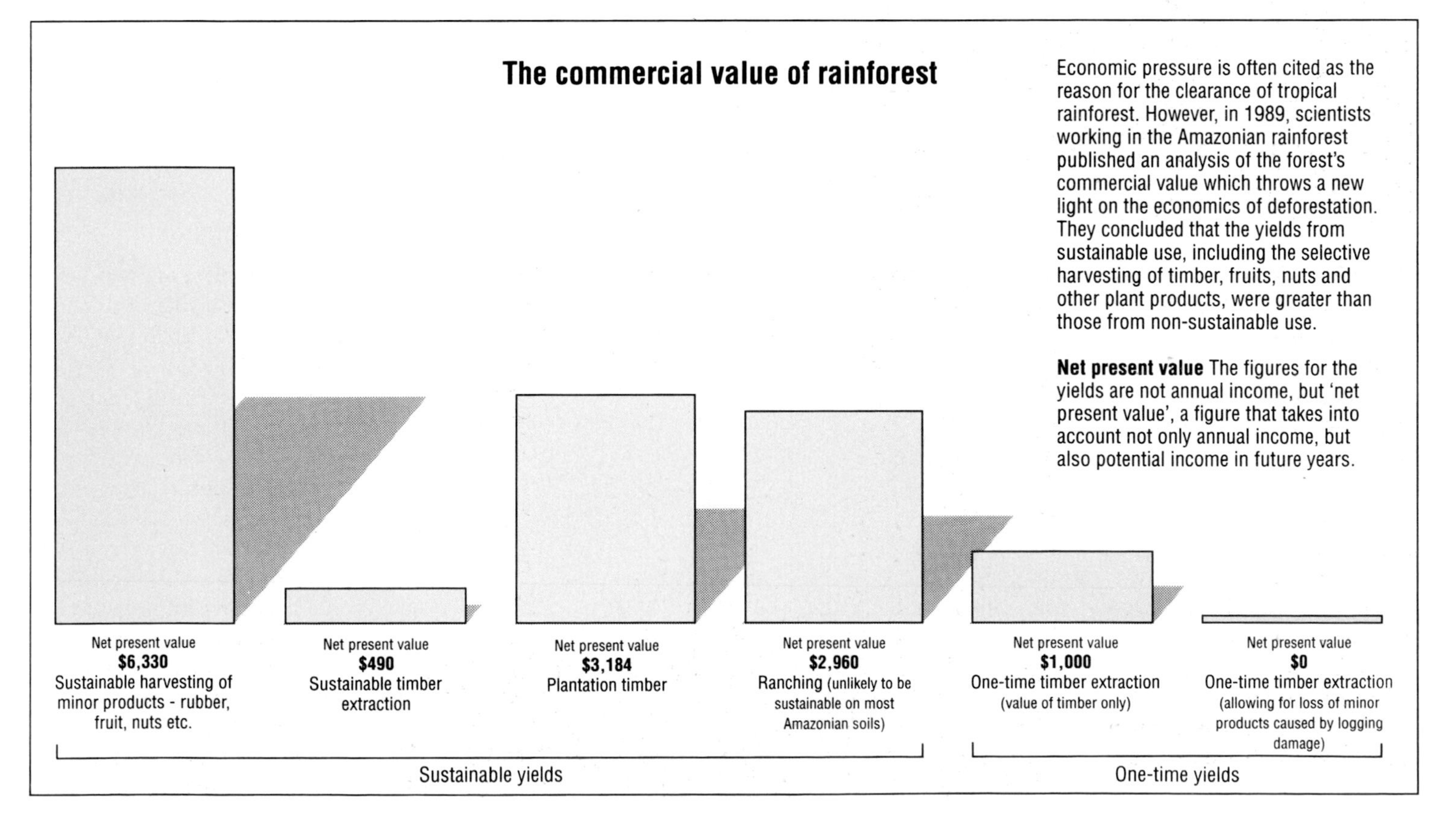

Economic pressure is often cited as the reason for the clearance of tropical rainforest. However, in 1989, scientists working in the Amazonian rainforest published an analysis of the forest's commercial value which throws a new light on the economics of deforestation. They concluded that the yields from sustainable use, including the selective harvesting of timber, fruits, nuts and other plant products, were greater than those from non-sustainable use.

Net present value The figures for the yields are not annual income, but 'net present value', a figure that takes into account not only annual income, but also potential income in future years.

The destruction of TEMPERATE FORESTS

Towering over 90 m (300 feet) tall - higher than a 30-storey skyscraper - the Carmanah Giant stands like a sentinel at the mouth of Vancouver Island's Carmanah Valley, on the Pacific coast of Canada. At 400 to 500 years old, this massive western red cedar is a comparative youngster in the forest. Many of the others, though smaller than the Giant, are far older and some were saplings when William the Conqueror invaded England in 1066. Many of the trees have trunks 3-4 m (10-13 feet) in diameter, and few are less than 70 m (230 feet) tall. Nurtured by the warm air of the Pacific and by heavy rainfall (up to 760 cm - 300 inches - in a year), Carmanah is one of the finest remaining tracts of the mighty conifer forests that stretch down the Pacific coast of North America, forests so lush that they have earned the description temperate rainforests.

To environmentalists, the valley is "a monument to way Canada used to be - lush and green." To MacMillan Bloedel, the company that owns the timber rights to the valley, Carmanah is an asset to be cashed in. With each spruce worth up to $40,000, the entire valley, if logged, would be bring in some $13 million. And, if the company has its way, all but a 538-ha (1,345-acre) reserve, less than 10 per cent of the valley, will indeed be logged.

The Carmanah Valley is the most recent wilderness area in western Canada to come under threat from the logging industry. Every year, over 200,000 ha (500,000 acres) of 'old-growth' forest in British Columbia are clear-cut for timber - an area the size of Carmanah being lost every ten days. On Vancouver Island alone, all but 11 of the 67 rainforest watersheds on the island have been clear cut. Even those forests set aside as parks have not escaped the onslaught. The Strathcona Park, British Columbia's first provincial park, has already seen two of its most splendid valleys clear-cut and then flooded to make way for the Strathcona Dam. Other areas are earmarked for mining.

With current government proposals committing 97 per cent of old-growth forests to logging, environmentalists warn that all forests with useful timber will have been destroyed within just two decades. Areas under threat, or in the process of being destroyed, include the temperate rainforests of the Stein Valley, the largest major unlogged watershed in British Columbia, Lower Mainland, and South Moresby, home to both the endangered Peale's peregrine falcon and the world's second largest population of bald eagles.

A PATTERN OF DESTRUCTION

It is a familiar tale of destruction, one that is being repeated, daily, throughout the forests of the temperate and subtropical world. Most of the old-growth forests of Ontario and eastern Canada have long since disappeared. The native hardwood forests of the eastern US are but remnants of their former glory, 80 per cent of the valley forests having been cleared for agriculture.

Moreover, the onslaught looks set to continue: plans that were announced by the US Forestry Service in 1989 would, if implemented, see logging extended to all of the country's 76 million hectares (191 million acres) of national forest - this despite clear evidence that much of the logging in national forests in the US is uneconomic. In the USSR, where pine, spruce, larch, cedars and birches still cover about half of the country, recent reports suggest that forests are being lost at a rate equal to that in Brazil.

► The massive trunks of giant sitka spruce in Carmanah's temperate rainforest. Moss grows in profusion on their trunks in the damp atmosphere.

▼ Clear-felling in British Columbia. Despite the fact that 'old-growth' forests like this contain trees of all ages, none are spared as the lumberjacks move in. Cutting swathes up valleys and across mountainsides, they leave behind them a mass of sawdust and shattered roots.

Several regions that were heavily forested until recently are now denuded, including Buryatia to the east of Lake Baikal. In Finland, all but a few patches of old-growth spruce forest have been destroyed. In Australia, whole areas of eucalyptus forest are being converted to wood chips for export, approximately 1 km^2 (0.4 square mile) being chipped every day.

In the industrialized world, few areas of pristine forest remain. Even forests that look untouched by man are usually the product of successive generations of human use, be it for timber or firewood or to provide cover for game. For centuries human intervention, rather than the natural process of succession, has determined which species of trees should grow, where and in what numbers. Two-thirds of the forests in the US are managed forests and 58 per cent of those in Europe. In some countries, such as Great Britain, virtually all remaining ancient woodlands are managed to some degree.

Where the aim is to provide timber, the forests are either selectively logged (the most valuable timber being extracted tree by tree, care being taken to minimize the damage to surrounding trees) or clear-cut, in which case whole areas are felled and then replanted or left to regenerate. Of these two methods, clear-cutting is by far the most common. In Virginia's George Washington National Forest, clear-cutting will be used on 85 per cent of the acreage destined for logging over the next 50 years.

Foresters justify the logging of old-growth on the grounds that logging 'improves' the forest. They argue that clear-cutting simply mimics the natural clearings created by lightning, windfall and the like - the chief difference being that the fallen timber does not 'go to waste'. However, in natural clearings, the essential structure of the forest remains unaltered, trees of differing ages and types surviving. By contrast, clear-cutting removes all the vegetation, regardless of age or species, reducing the forest to a moonscape. Regrowth, particularly when achieved through replanting, results in secondary forest containing even-aged blocks of trees. Although less damaging, selective logging also alters the structure of the forest.

There are other critical differences between natural and man-made clearings. In natural clearings, the damaged trees remain where they have fallen, the dead wood providing a habitat for a wide range of species and supplying the forest with nutrients as it rots. In clear-cut areas, however, the 'debris' is removed, leading to a loss of nutrients and to reduced soil fertility. Moreover, man-made cuts are connected by roads which lead to the gradual fragmentation of the forest and open it up to hunters. Large mammals, such as bears, avoid human activity and rarely cross clear-cut areas, so clear-felling restricts them to smaller and smaller territories. In addition, roads increase erosion rates and exacerbate the problem of windthrow by effectively creating wind-tunnels through the forest.

The world's forests today

After centuries of felling, the world's forest cover has shrunk dramatically from its former extent *(see pp.56-57)*. The continuous forest that once covered much of the temperate zone in the northern hemisphere has been fragmented, and in some regions has disappeared altogether. In the tropics, the process of deforestation has started much more recently, but is taking place at a much faster rate.

▲ Felling plantation hardwoods in France. In Europe, most temperate hardwoods come from plantations or managed woodland and are therefore sustainably produced.

► An area of South Island, New Zealand, once clothed with forests of southern beech, now felled for wood chips. Gorse - introduced from England - is invading the land and crowding out the native vegetation.

“We shake down the acorns and pinenuts. We don't chop down the trees. We only use dead wood. But the white people plough up the ground, pull down the trees, kill everything.”
Holy woman of the Wintu Indians, California

The impact on wildlife is severe. In North America, a third of all the species listed by the US government as threatened or endangered live in the country's national forests, in addition to 800 other species that are 'candidates' for listing. Although logging favours some species, such as deer and rabbits, they tend to be common species anyway, and the increase in their numbers is frequently at the expense of other species. In eastern Canada, for example, logging has encouraged the rampant growth of buckthorn and honeysuckle to the detriment of numerous other plants.

The direct physical impact of logging is considerable and cannot compare to that in natural clearings. Erosion is a particular problem, leading to landslides, to the silting up of rivers and to the destruction of fish spawning grounds. In Idaho, logging has increased erosion rates in some areas 220 times. The impact on local salmon spawning grounds has been devastating. Salmon production in the Salmon River, which once supported more than half of the entire summer Chinook salmon population in the Columbia River Basin, has been halved. Many fear the salmon runs will never recover.

By simplifying the forests, modern forestry practices have also increased the problem of insect pests. In undisturbed forest, insect populations are kept under control both by the predators that feed on them and by the mix of tree species, as most pests feed on one type of tree. Transforming the forests into homogenized blocks of single-aged trees not only expands the niche available to certain insect species but also interferes with the natural controls on their population. The result is that minor pest infestations are replaced by major epidemics. In New Brunswick, for example, the area infected by spruce budworm grew from 89,000 ha (220,000 acres) in 1952 to 3.8 million ha (9.5 million acres) in 1976.

Attempts to control the infestations by chemical spraying have generally proved ineffective - and in most cases have exacerbated the problem, in addition to killing wildlife such as squirrels and damaging the health of local people. One problem is that insects breed and evolve quickly, and many have developed resistance to the pesticides used. Another is that the pesticides have frequently proved equally effective against the predators of the pests as against the pests themselves. By killing the pests' natural enemies, they have allowed the surviving pests to proliferate in even greater numbers.

ACID RAIN

The loss of forests to logging, agriculture and urban development has a long history in temperate areas. Today, however, there is a new threat - one that affects not only old-growth forests but plantations also. In many areas of the temperate world, trees of all types are rapidly being afflicted by pollution and by pollution-related diseases - a syndrome known generically as waldsterben or 'forest death'.

Many of the symptoms of waldsterben had never been seen before the early 1970s. Affected trees lose some leaves as well as suffering a reduction in the photosynthetic efficiency of those remaining. The leaves often show discoloration and, through becoming twisted, expose the lower surface upwards, thus reducing substantially the light-receiving surface. The production of new leaves and branches falls substantially. Underground, the fine roots shrink and the mycorrhizae *(see p.68)* dwindle and disappear. Finally the trees become prematurely senile, losing their crowns and showing distortion in the growth of branches.

Waldsterben has claimed more than 70,000 km² (27,000 square miles) of forests in 15 European countries. In West Germany, 52 per cent of the country's forests - some 38,000 km² (1,470 square miles) - are affected. In Britain, nearly half the oak and three-quarters of the yew in southern England have lost over a quarter of their natural foliage.

On the other side of the Atlantic, the picture is not much different. Acid rain - just one component of the waldsterben phenomenon - has been devastating, particularly for the conifer forests stretching up the Appalachian Mountains from Georgia to New England. In Vermont, New Hampshire and New York, 60 per cent of high-elevation red spruce are reported dead, while, in southern California, 87 per cent of the Ponderosa and Jeffrey's pines in the San Bernardino Forest are damaged. In Canada, sugar maples are dying over wide areas, due in large part to the increasing acidification of soils in eastern Canada.

Waldsterben is undeniably a new disease of the post-war era. The enormous increase in the use of motor cars and aircraft, the emission of thousands of man-made chemicals, the massive increase in power generation and the industrialization of agriculture have all added dramatically to the number of pollutants in the environment. Among those which have caused most damage

"...it is not Christ who is crucified now; it is the tree itself, and on the bitter gallows of human greed and stupidity. Only suicidal morons, in a world already choking to death, would destroy the best natural air-conditioner creation affords..."
John Fowles

▲ A spruce showing signs of 'waldsterben' in its yellowing needles.

to trees are sulphur dioxide, nitrogen oxides and volatile hydrocarbons, which react with water and sunlight to form sulphuric and nitric acids, ammonium salts and other mineral acids. These then fall to earth, often thousands of kilometres from where they were emitted, as dry particles or as acidified rain, snow or fog.

The strategy of building higher and higher smokestacks in order to reduce local pollution has led to pollutants being dispersed over wide areas. Emissions of nitrogen oxides and other industrial pollutants are also accelerating the formation of ozone in the air around us. Ozone is extremely toxic to plants, and in the US, ozone damage to forests has been documented in the San Bernardino Mountains of California, the Appalachians and the Blue Ridge Mountains.

So widespread is the destruction that Professor Peter Schutt, editor of the *European Journal of Forest Pathology*, has come to the depressing conclusion that the entire forest ecosystem is breaking down.

FROM FORESTS TO TIMBER FACTORIES

As the old-growth forests are logged out or cleared for urban and agricultural expansion, so monocultures of pines and other commercial species have taken their place. In the southeast of the US, vast expanses of natural pine have been replaced by even-aged industrial plantations. In Britain, more than a third of the country's ancient woodlands have been turned into conifer plantations during the past 40 years.

Plantation forestry now covers 13 million ha (32 million acres) in northern Europe, 11 million ha (27 million acres) in North America and 17 million ha (42 million acres) in the USSR and eastern Europe. Moreover, in several countries, the area under plantations is being expanded. In Britain, for example, 400,000 ha (1 million acres) of flow country - an area of peatland famed for its rare plants and wading birds - are to be planted with pine.

Such plantations cannot be considered forests in any meaningful sense of the word. In reality, they are industrial timber stands. Aesthetically displeasing - they have been dubbed "forestry's equivalent to the urban tower block" - they are also ruinous to wildlife, detrimental to the soil and destructive of water supplies. In Sweden, for example, the replacement of old-growth forests, especially mixed forests, by uniform stands of Canadian lodgepole pine has led to a severe reduction in many forest species - despite 57 per cent of the land area being covered with trees. According to Ingmar Ahlen, Professor of Ecology at the Swedish University of Agricultural Sciences, 40 vertebrates that either breed or feed in the forest are now seriously endangered. Of the fungi, lichens and flowering plants, about 50 species are on the verge of extinction, and another 220 are in some danger.

Growing conifers on unsuitable lowland soils leads to the acidification of soils in planted areas. Many pines are mountain species with the ability to secrete acid from their hair roots, thus breaking down rocky soils and enabling the roots to get a hold. Where grown close together, as in plantations, the acid is used to destroy the roots of competing trees. As a result, a hard, impermeable layer, known as an acid pan, builds up. In addition, conifers scavenge some polluting chemicals from the air and moisture around them and concentrate them in the soil. The experience in Czechoslovakia suggests that, for these and other reasons, conifers that are grown on unsuitable soils so degrade the land that after seven generations of trees the soil is no longer capable of supporting commercial forestry.

▲ Native forest in New Zealand - already reduced to scattered remnants - is attacked with napalm. Once the forest has been burned, it will be replaced with plantations of non-native trees, which offer little food or shelter to New Zealand's rare endemic species. The destruction of such forests and their inhabitants, by a relatively wealthy nation for marginal financial gain, shows how little the environment is valued by politicians and commercial interests.

▲ ▶ Conifer plantations on moorland in Scotland. These dense blocks of trees are like a dark green desert, virtually devoid of wildlife, and often damaging to the soil. They have been planted on some environmentally important areas, such as the flow country of the far north.

The adverse impact of conifer plantations on water quality is well known. In Wales, for example, watercourses running through or near conifer plantations have been found to be more acidic than those on open moorland, with disastrous consequences for wildlife. In some areas of upland Wales, populations of salmon have been depleted and attempts to reintroduce them have failed: even the normally hardy American brook char has succumbed in some waters. Aquatic plants and birds have also been affected.

Ironically, in those areas where deforestation has been most extensive, reforestation is now a potential cause of further forest loss. In Thailand, for example, the government plans to plant at least 30,000 km^2 (11,580 square miles) with eucalyptus, a programme that could, according to official estimates, lead to six million peasants being displaced from their land. With nowhere else to go, many will be forced into the country's remaining natural forests.

Likewise, in India, where a 'social forestry' programme is being actively promoted by the central government, plantations have been sited on village common lands, thus denying peasants land which was previously used to grow food. The choice of eucalyptus as the major tree crop is also causing serious social and environmental problems. Unlike traditional tree crops, eucalyptus provides no fodder for grazing animals, gives little protection against soil erosion or run-off during the rainy season, and consumes large quantities of water. Moreover, what ecological services the plantations do provide are short-lived, the trees being cut every few years for pulping or to make wood chips. In many areas of both India and Thailand, local villagers have resorted to uprooting the eucalyptus seedlings.

The peasants know that their future security does not lie in planting trees for pulping, but rather in regenerating the genuine wealth of forests. For them, trees are not merely a source of profit, but protectors of the soil, providers of fodder for their animals, and sources of nuts, fruit and other products. As the forest activists of the Chipko movement in India put it, "What do the forests bear? Soil, water and pure air." With the need for trees to combat global warming now more urgent than ever, it is a message we would all do well to learn.

▲ The ghost orchid, a plant that parasitizes trees. In Sweden it is found only in forests where broadleaved trees occur. These are being systematically replaced by pure conifer stands as part of government policy, with the loss of many plants and animals.

AGRICULTURAL LANDS

In the late 1960s, two engineers at the Soviet Ministry of Water drew up a master plan to transform the arid plains of Soviet Central Asia into productive cotton fields. It was to be the project of the century. Thousands of hectares of canals would criss-cross Uzbekistan and Turkmenia, irrigating over 10 million ha (25 million acres) of land previously fit only for pasture. Every available drop of water from the Amu Dar'ya and Syr Dar'ya rivers, the largest in Soviet Central Asia, would be abstracted, and this water would be replaced by reversing the flow of the great Siberian rivers to the north.

The plan was accepted and the two engineers, Konstantin Kakitan and Lev Litvak, were rewarded with rapid promotion. But for the people of Uzbekistan and Turkmenia, the project of the century has proved anything but a success. Irrigation, year in, year out, has ruined their land, causing natural salts in the soil to rise to the surface, where they dry out, turning the fields into a salt-encrusted desert. Two-thirds of the land under irrigation is affected by such salinization, and much of it can no longer support crops. To add to the devastation, dust storms whip up the dried-out bed of what remains of the Aral Sea *(see p.178)*, carrying millions of tonnes of salt onto the land.

In those areas which have been spared salinization, the land is so overexploited that crops will only grow if massive amounts of chemical fertilizers are applied. In Uzbekistan, the amount of fertilizer which must be applied on every hectare is over 10 times the national average. Pest epidemics are kept at bay only by spraying tonne after tonne of pesticides. To make it easier to harvest the cotton by machine, the leaves are first killed by herbicides. These include butifos, a herbicide known to cause genetic damage, affect the nervous system and lower immunity.

◀ A wheatfield in Montana, heavily treated with fertilizers and pesticides, being harvested by a flotilla of combine harvesters.

Local drinking water is now heavily contaminated with pesticides and herbicides. The impact on the health of those who still live and work in the region has been devastating. One in 10 children die in infancy and there has been a sharp rise in the incidence of stomach and respiratory problems, typhoid and throat cancer. Malnutrition is also rife.

By their utter disregard for nature, the Soviet planners have beggared the environment and peoples of Central Asia. Much of their land is effectively dead. And, tragically, it is a story that is being repeated - to varying degrees - throughout the world.

PUTTING THE LAND TO WORK

No one knows for certain why and when humans first began to practise agriculture. It seems unlikely that the move from hunting and gathering to farming came about as a result of population pressure or food shortages. The available evidence suggests that our hunter-gatherer ancestors were well fed. Furthermore, one has only to reflect on the desperation and hopelessness on the emaciated faces of the starving of Ethiopia to realize that if Stone-Age man was short of food, he would have had neither the means nor the time to experiment with growing crops. Initially at least, cultivation probably provided no more than a supplement to hunting and gathering.

Whatever its origins, the adoption of agriculture has been the single most important force in shaping our environment. By harnessing natural biological productivity and directing it to one end - that of feeding humans - even the simplest garden radically alters the natural ecosystem.

In clearing land for agriculture, the farmer interrupts the natural process of ecological succession *(see p.59)*, arresting the development of

▼ A rare oasis of uncultivated prairie land in Texas. The variety of plants, with clumps of trees scattered among the grass and flowers, is typical of the natural prairie. Its diversity contrasts sharply with the stark, monotonous expanses of wheat raised by modern farmers.

the ecosystem and preventing it from developing its natural 'climax' vegetation. Instead, through human management, the land is kept artificially at the 'pioneer' stage of development, ready for plants to exploit.

There is a good reason for this. In a climax ecosystem - a mature oak forest, for example - the bulk of the energy received from the Sun is used to maintain the system. Very little energy goes into new growth. In a pioneer ecosystem, by contrast, all the available energy and nutrients are used to promote growth, as the plants and animals in the ecosystem take advantage of the open ground and seek to establish themselves. The ecosystem is thus at its most productive stage - and it is this productivity that farming systems are designed to exploit.

But the increased productivity of pioneer ecosystems comes at a price. As an ecosystem develops and matures, its plants and animals

▲ An aerial view of farmland in Kansas. The sheer scale of modern intensive agriculture defies the checks and balances of the natural world. Monocultures - fields of single crops - are a bonanza for insect pests, turning them from minor problems into major ones.

▲▲ Part of the cotton crop grown in Soviet Central Asia. The extravagant use of water for irrigating the cotton fields has reduced the Aral Sea *(see p.178)* to a fraction of its former size. Heavy applications of pesticides have combined with the parched soil to form toxic dust storms.

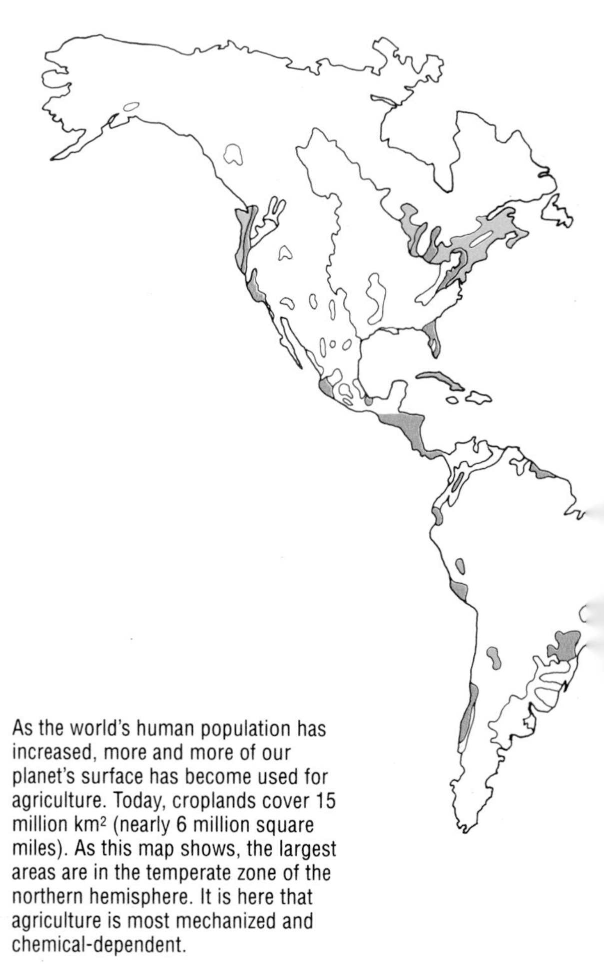

As the world's human population has increased, more and more of our planet's surface has become used for agriculture. Today, croplands cover 15 million km² (nearly 6 million square miles). As this map shows, the largest areas are in the temperate zone of the northern hemisphere. It is here that agriculture is most mechanized and chemical-dependent.

▲◀ Fertile farmland along the Indus valley in Ladakh, Kashmir. Throughout the world, flood plains like this have been cultivated for millennia. New land is now being brought under the plough, much of it unsuitable for agriculture, to satisfy the needs of the growing population.

◀ Traditional farmland in North Wales. The mix of fields, woods and uncultivated hilltops creates a rich, varied habitat in which many wild plants and animals can survive. Such diversity also minimizes problems with pests and diseases.

The world's croplands

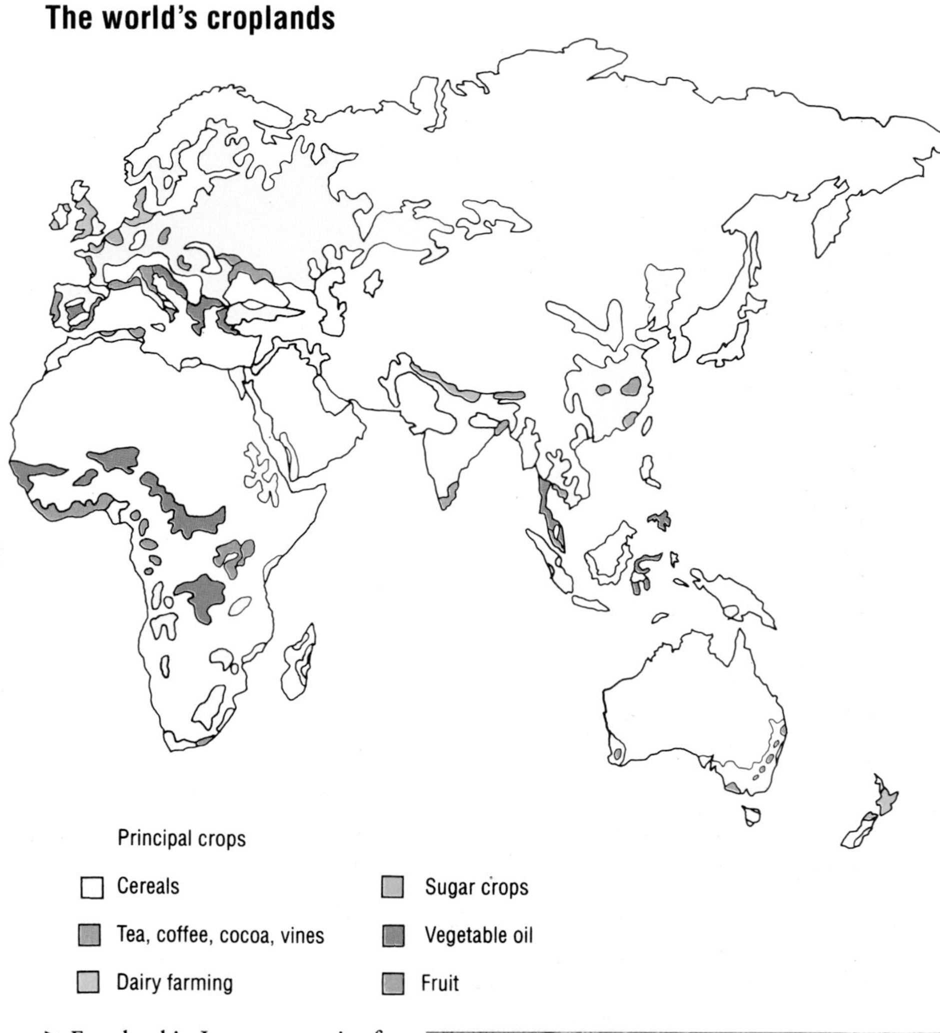

form relationships that contribute to its overall stability. The resulting checks and balances minimize the likelihood of pest epidemics, population explosions and the like. They also influence the natural water and nutrient cycles, making the ecosystem more resilient to droughts, floods and other disturbances.

By contrast, the checks and balances in a pioneer ecosystem are extremely crude. As a result, it is more vulnerable to sudden population explosions of animals and plants, followed by equally sudden population crashes. Pests and diseases are more common, and floods and droughts more likely and more damaging. Pioneer ecosystems are simply less stable.

THE EFFECTS OF AGRICULTURE

The instability of open ground is accentuated by agriculture. In order to allow crops to flourish, farmers remove anything that might compete with them. The rich mixture of plants and animals that would normally live on the land is swept away, and is often replaced by a monoculture - a vast, uninterrupted expanse of a single crop. This presents pests with an ideal environment in which to proliferate. A field of cotton, for example, provides a banquet for insects such as the boll weevil, because none of their natural predators are left to keep them under control.

Land that is managed for farming has other features that tend to increase its vulnerability to

▶ Farmland in Japan, a mosaic of small fields with varied crops, including bananas in sheltered groves, and trees for timber. The mountainous terrain requires an adaptable approach to farming, and tiny fields are made wherever there is enough level ground.

ecological disruption. The plants in a natural ecosystem differ widely in their ages. Their root systems are at different stages of development, with the result that they exploit different nutrients at different depths in the soil. In agricultural land, however, the annual cycle of planting and harvesting dictates that the crops grown tend to be of the same age. Consequently, their roots compete for the same nutrients and moisture at the same depth and, often, in the same limited patch of ground. The risk of exhausting the soil, and the vulnerability of the whole crop to drought, is greatly increased.

In a natural ecosystem, nutrients are constantly recycled, becoming first part of one organism and then another during the processes of life, death and decay. Little of the organic material and nutrients are lost, ensuring the land's fertility for future generations. When crops are harvested, however, the plant matter is removed. Unless steps are taken to put it back on the land, the soil will become slowly degraded. Even then, growing the same crops on the same plot of land year after year exhausts the soil of specific nutrients, making it increasingly infertile.

When forested land is cleared for agriculture, the cycles that regulate water supplies in natural ecosystems can be fatally disrupted. In tropical areas, in particular, deforestation robs the land of its capacity to absorb water, leading to floods and droughts. Unless vegetation is quickly allowed to recover, erosion is also increased, stripping the land of its fertility as its topsoil is washed or blow away.

In arid areas, where rainfall is insufficient to grow crops, farmers must irrigate their land. Here the potential for ecological damage is enormous. All soils contain a variety of mineral salts which, if allowed to accumulate, are toxic to plant life. In areas of high rainfall, the salts are regularly flushed out of the soil, eventually reaching the groundwater and ultimately the sea. In arid areas, however, the lack of rain and the high evaporation rates lead to high natural levels of salt in soils. The groundwater, too, tends to be more saline than in temperate areas.

Irrigation can drastically increase the salinity of soils. As the land soaks up the irrigation water, so the salts in the soil are dissolved. Drawn to the surface by the heat of the Sun, the irrigation water evaporates, leaving behind its burden of salts. If the land is allowed to rest and the salts are flushed out of the soil, no harm is done. But where poorly drained land is irrigated year after year, the land becomes increasingly waterlogged, causing groundwater levels to rise. The groundwater adds to the salt load of the soil and eventually the land becomes so salinized that crops can no longer grow.

Pests, declining soil fertility, erosion, waterlogging and salinization are just some of the many challenges that have confronted farmers since the dawn of agriculture. But farming would never have survived - let alone taken on the dominant role it plays today - if our ancestors had not evolved numerous strategies for tackling these problems. These strategies all have one feature in common - they attempt to mimic nature.

▲ Farmland alongside a lake in China. The demands of China's huge population have meant that little of the countryside remains untouched by agriculture. Almost nothing remains of the natural vegetation in some areas.

▶▲ Farmland in France, with vines, mustard, wheat and pasture. Crop diversity reflects the natural diversity of the living world and is more sustainable without high inputs of chemicals.

▶ Tulips being grown for the cut-flower trade. The use of fertile agricultural land for crops other than food, while food for livestock is imported from the Third World, reflects the imbalance created by unfair distribution of wealth.

▲▲ White storks nesting on a church roof in Spain. These birds have long been regarded as bringers of good luck in Europe, and this has helped them to survive. Foraging through the fields after harvest-time, the storks eat many insect pests and improve the farmer's yields in following years. In northern Europe, stork numbers are now declining with the intensification of agriculture.

▲ Hares, once common in European farmland, are now rare, due to changing methods on most farms.

MAINTAINING DIVERSITY

The hallmark of traditional agriculture is diversity - not only a wide range of crop species, but also in the enormous number of varieties of the same crop that are grown. In India, until the advent of the 'Green Revolution' *(see p.109)*, some 30,000 varieties of rice were cultivated. Each had its special use: some were rich in minerals and thus well suited to nursing mothers, other more suitable for the elderly and so on. Similarly, the Indians of the Amazon, whose elaborate forest gardens *(see p.84)* are acknowledged to be among the most sophisticated agro-ecosystems in the world, have a knowledge of 70 or more varieties of manioc. They would consider the person who relied on just one variety to be an extremely poor gardener and one likely to fail. And with good reason. Different crops and different varieties of the same crop have different needs and can survive in different conditions. A drought that kills one species, or even one variety, may not affect another. Likewise, while one species may succumb to pests, others will be left unaffected. The more crops and varieties that a farmer grows, the less the risk that his entire harvest will fail.

Just as farmers knew the value of growing a wide range of crops, so they knew the value of retaining uncultivated areas within their farms. These fallow areas acted as refuges for the birds and other animals that would feed on pests in the fields. In Sri Lanka, large stretches of jungle were preserved in the high ground above paddy fields, where they not only prevented erosion but also provided a habitat for wildlife.

INTERPLANTING

Still other strategies enhance the farmer's security. Some crops are toxic to the pests of others and by interplanting them, such 'companion' species can be made to protect each other. Leafhoppers, for example, are a very destructive pest of rice. Research in Colombia has shown that their numbers are dramatically reduced when rice is grown together with beans. In Central America, farmers have interplanted maize, beans and squash for as long as recorded history. The squash reduces erosion and weed growth by providing ground cover, the beans enrich the soil with nitrogen, and the maize provides a climbing frame for the beans to grow up.

What goes on below the ground is just as important as what goes on above it. Thus farmers frequently interplant shallow-rooted cereals, such as maize or sorghum, with longer-rooted plants, such as cow peas and rye grass. Because the plants use nutrients and water at different levels in the soils, the risk of exhausting the land is minimized.

Interplanting reaches its most sophisticated form when trees are introduced into the cycle. The trees not only provide food in the form of fruit and nuts, but also provide fodder for animals, organic matter to enrich the soil and shade for the plants growing below. In the fertile multi-storied gardens of the Indonesian island of Java, trees - often jack fruits, coconut palms, mangoes or bread fruits - protect the cinnamon, cardamom, tea and coffee shrubs growing below. These, in turn, shelter the crops at ground level, such as cassava and mountain rice. The trees reduce the impact of rain on the soil, control erosion, provide humus and control weed

growth. Pest outbreaks rarely occur in the gardens, nor do the gardens require large amounts of water or labour.

In Senegal in West Africa, farmers have traditionally interplanted food crops with acacia trees, the land also being used to raise livestock. Such systems can support 50 to 60 people per ha (20 to 25 people per acre) on a sustainable basis - several times the number of people that can be supported when only a single crop is grown. When the acacia sheds its leaves in the rainy season, the nitrogen-rich leaf litter adds to the organic content of the soil. The acacia 'fixes' nitrogen in nodules in its roots. In the dry season, the acacia leaves and seed pods make up fodder for the animals, which then provide dung that further improves the soil. The trees also create the shade which the animals badly need during the dry season.

▲ A Dogon village in West Africa, with straw-roofed granaries where grain is stored. Traditional forms of storage resulted in little wastage, but in modern granaries, especially in the tropics, much of the crop is lost to moulds, rats and weevils.

◀ Clove trees and bananas being grown together in Indonesia. Such intercropping mimics natural diversity and the layering of vegetation that is found in forest or scrub. It can reduce pest problems, suppress weeds and benefit the soil.

MAINTAINING FERTILITY

In all systems of farming and gardening, maintaining soil fertility is critical. Unless something is returned to the soil each year, the farmer finds his crops failing. The most common way of replacing lost fertility is by manuring the ground with animal dung or with composted vegetable remains. But there is more to fertility than simply adding lumps of manure, and very early on farmers discovered the benefits of leaving land fallow, that is, allowing it to rest between crops.

From Roman times until the eighth century, European farmers operated a 'two-field system', in which one half of the tilled area was left fallow each year, the other half growing the crop. In the eighth century, the 'three-field system' was introduced. In both instances, animals were grazed and kept on the fallow land to help manure it and restore fertility.

The three-field system permitted a greater diversity of crops to be grown - spring corn as well as winter corn, for example - thus insuring against total crop failure. The farmers also tilled spring oats, supplying a source of feed for their horses. By the Middle Ages, farmers also planted more green crops and vegetables than their ancestors, and by all accounts, the nutritional standards of the time were remarkably high.

By the eighteenth century, farmers in Europe had all but abandoned the fallow system and instead maintained fertility through rotating crops with livestock. Typically, a field might be used to grow wheat, then turnips, then put down to grass interplanted with leguminous plants such as clovers and vetches, and then used as pasture. The planting of clovers and vetches was critical to the rotational cycle, since unlike other plants, legumes - or members of the pea and bean family - can 'fix' nitrogen from the air.

Crop rotation not only maintains soil fertility but also helps to combat pests and diseases. Moving crops from field to field prevents a 'reservoir' of disease-causing organisms building up, particularly those that can survive in the soil. Most pests and crop diseases are specific to individual plants and will not attack other species. A fungus that might devastate potatoes, for example, will not affect beans. If potatoes are grown on the same patch year after year, the fungus has a permanent source of food and may quickly become established as a major pest. But if the potatoes are moved to another field every other year and beans are grown instead, the fungus is not given an opportunity to establish itself.

◀ Hay-making on a mixed farm in France. In Europe and North America, such small-scale farms are largely disappearing as machinery takes the place of people.

▼ Vines and olives growing in Tuscany, a system of agriculture thousands of years old. Although vines are regularly dosed with pesticides, land like this escapes the much heavier chemical burden of arable land.

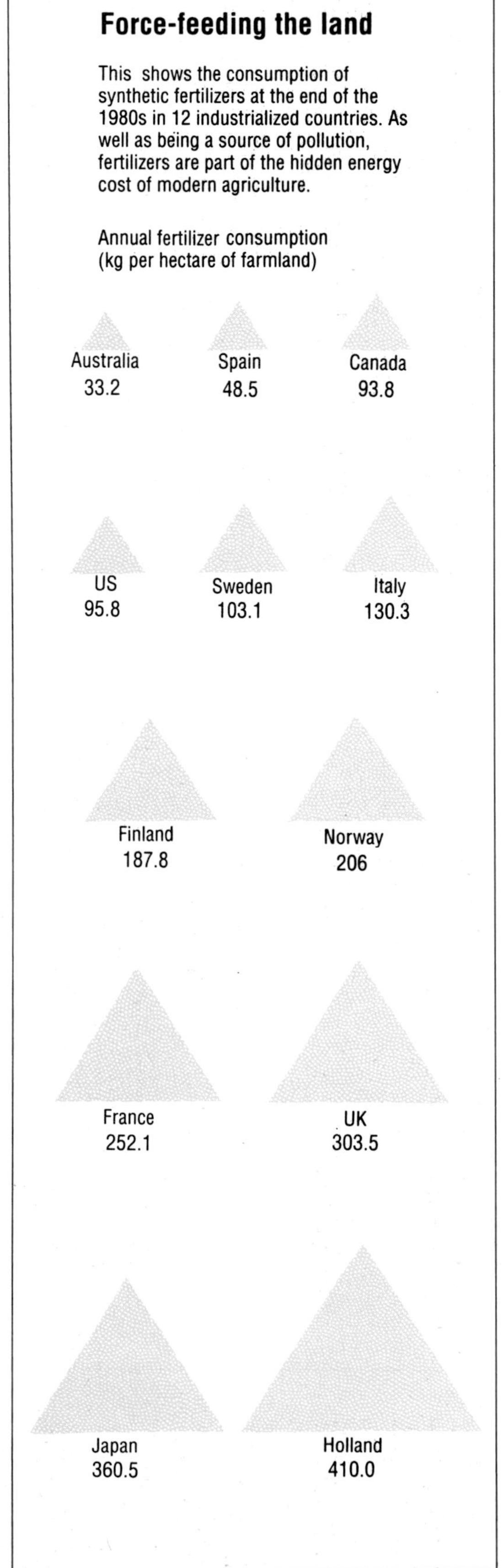

There are few short-cuts to good husbandry. Wherever man has attempted to take too much out of the land or has abandoned those strategies that ensure the soil's fertility, the consequences have been disastrous.

Travelling the route from Antioch to Damascus, one passes the ruins of houses whose doorsteps are a metre or more above the bare rock, the soil having eroded away through the abuse of the land. The oldest city in the world, Jericho, is believed to have been abandoned because its hinterlands could no longer support the demands of its inhabitants, and succumbed to overgrazing and devastating erosion. In Mesopotamia, the cradle of western civilization, the demands of the great city states that emerged in the fourth millennium BC led to such widespread salinization of land, through over-irrigation, that many archeologists now ascribe their eclipse to ecological degradation. Indeed, it is a great irony that while agriculture enabled cities to emerge, growing cities have repeatedly brought about the destruction of the agricultural land on which they rely.

MODERNIZING AGRICULTURE

The nineteenth century saw the beginnings of a revolution which has transformed agriculture beyond recognition. As chemists began to learn more about soils and the precise mineral requirements of plants, so farmers abandoned rotation and instead applied the minerals directly to the land. Just three main elements - phosphorus, potassium and nitrogen - were commonly applied, far fewer than the land in fact requires.

The phosphorus and potassium were obtained through mining, while, initially, the nitrogen was supplied from manure. Huge quantities of bird droppings, or 'guano', were imported from islands which supported large bird colonies. Subsequently, however, advances in industrial chemistry enabled artificial fertilizers to be manufactured directly from atmospheric nitrogen. The same crop could thus be grown year after year on the same patch of ground, the soil being kept fertile (at least in the short term) by the farmer adding the right mixture of chemicals.

The advent of powerful synthetic pesticides and herbicides brought with it a second revolution in agriculture. Instead of relying on planting a wide range of crops to minimize the risk from pests, farmers were now able to grow hectare upon hectare of the same crop, dowsing such monocultures with chemical sprays to prevent pests or to combat infestations.

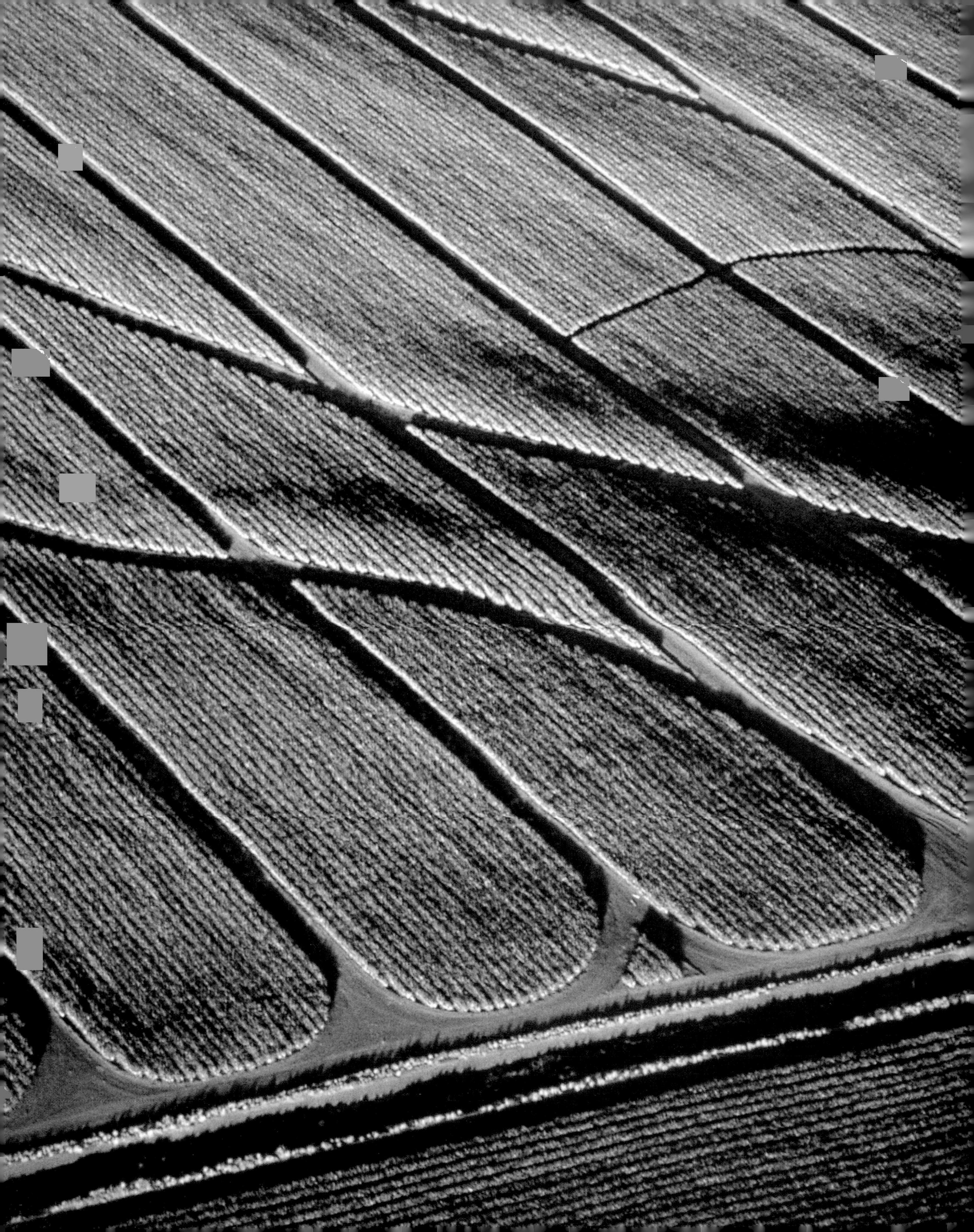

▲ **Ploughing rice fields in Indonesia. Traditional forms of agriculture, using animals for ploughing and transport, are sustainable provided the human population stays within reasonable limits. The animals are fed by fodder crops produced on the farm. Modern agriculture, which is reliant on fossil fuels to power tractors and other machinery, and to produce fertilizers and pesticides, cannot be sustained indefinitely.**

◀ **Pineapple fields in Hawaii. Like other monocultures, these pineapples are vulnerable to pests and diseases, particularly pineapple mealybug, pineapple wilt disease and nematode worms. Heavy doses of pesticides are required to sustain production on this scale.**

THE 'GREEN REVOLUTION'

Over the last 40 years, all but a small number of farms in the industrialized world have abandoned non-chemical or 'organic' agriculture in favour of farming with chemicals and modern machinery. The Third World, too, has seen its traditional agricultural patterns transformed. In the name of development, Third World countries have been persuaded to specialize in a few export crops in order to develop a competitive advantage in the world markets. Small farmers have been squeezed out of production, or resettled, to make way for vast mechanized monocultures of tea, coffee, tobacco and other crops. In West Africa, for example, 70 per cent of Gambia's arable land is used to grow peanuts, while in Africa as a whole, the area under coffee has quadrupled since the mid-1960s.

At the village level, the greatest changes to agriculture in the Third World have come as a result of the so-called Green Revolution, a massive campaign launched in the 1960s by the UN Food and Agriculture Organization to boost Third World food production. Farmers have been encouraged to abandon their traditional crop varieties and instead to grow new high-yielding varieties of cereals.

These new varieties require massive quantities of artificial fertilizers and so are better described as 'high-response' varieties. In India, the Green Revolution has increased wheat yields by 50 per cent and rice yields by 25 per cent, but this has only been possible because of a twenty-fold increase in fertilizer use. Because the new varieties are grown in monocultures, and are particularly susceptible to pests and disease, pesticide use has also rocketed.

The soil, instead of being treated as a living layer, has become a factory floor, with the modern farm worker increasingly isolated from the earth he tills. He sits encased in the cabin of his tractor, and what goes on behind the tractor has more to do with chemical technology than with the wisdom of countless generations of his predecessors. Traditional husbandry has been forsaken for the use of great quantities of artificial fertilizers, chemical sprays and animal feedstuffs. The result is a global loss of agricultural land on a stupendous scale.

"...with handkerchiefs tied over our faces, and Vaseline in our nostrils, we have been trying to rescue our home from the wind-blown dust which penetrates wherever air can go. It is almost a hopeless task, for there is rarely a day when at some time the dust clouds do not roll over. Visibility approaches zero and everything is covered again with a silt-like deposit which may vary in depth from a film to actual ripples on the kitchen floor."
Letter from Oklahoma resident, June 1935

▶ A drought-affected cornfield in North America. The US has been hit by a succession of droughts in recent years, cutting yields in the grain-producing states. Where farmers rely on a single crop, droughts can be particularly damaging.

▶ Salinized land in Australia. Rising salt levels in the ground have killed trees and grass, making it useless for grazing. Salinization is now a major problem that faces Australia's farmers.

DECLINING FERTILITY

Farmers are utterly dependent on topsoil, the thin skin of earth, often only a few centimetres deep, in which plants are able to grow. Should that topsoil be lost through erosion or degraded to the point where it loses its fertility, then agriculture ceases to be possible. But the topsoil is more than just a collection of minerals and organic particles. It is a living community, alive with insects, fungi and a host of bacteria. A healthy field contains up to 30 million bacteria in every gramme of soil.

All these organisms play their part in ensuring the soil's health and fertility. Burrowing creatures, such as moles and earthworms, aerate the soil, allowing it to breathe, and keeping it drained. Bacteria and fungi decompose the animal wastes and decaying soil organisms, transforming them into humus, the rich organic material that binds the soil together and provides still other bacteria with food. Those bacteria in turn break down the humus, releasing nitrogen in a form which can be taken up by plants. The organisms that live in the soil, the plants that live off it, and the soil itself are so dependent upon each other that some compare them to a single living organism.

Modern chemical agriculture, with its artificial fertilizers and chemical sprays, has undermined the natural fertility of soils. When a farmer stops applying manure and other organic material to the land, and instead spreads artificial fertilizer by the sackful, the soil's structure gradually begins to break down.

Without organic waste, the soil cannot support the micro-organisms that produce humus, and without a supply of humus, the population of nitrogen-fixing bacteria declines. The number of earthworms decreases (a problem made worse by pesticide spraying), and the soil becomes less well aerated, reducing the supply of oxygen available to soil organisms. The roots of plants become shallower as the nutrients in the soil become scarcer, thus weakening the plants. The soil's capacity to store water is also affected, making the land dry out in arid areas - which is why droughts increasingly occur even in areas where rainfall has not declined. Meanwhile, in areas where rainfall is heavy or which are under irrigation, the decline of organic matter makes soils more vulnerable to waterlogging, since drainage is greatly impaired.

To compensate for the soil's dwindling natural fertility, farmers must apply yet more fertilizers, perpetuating a downward spiral of increasing degradation that eventually ends in the ruination of the land.

In many areas, particularly in the Third World, the process is accelerated by the introduction of high-yielding varieties of crops. These are often voracious consumers of trace elements such as zinc, iron, copper, boron, manganese and molybdenum - none of which are replaced by artificial fertilizers. For many farmers, the end of the road has already been reached. Their lands too degraded to farm, they have little option but to sell up (if they can) and move elsewhere. They thus join the ranks of an increasing army of 'environmental refugees'.

SOIL EROSION

As the organic content of soil is reduced, so the land becomes more vulnerable to erosion. In the Third World, where many soils have a naturally low organic content, the problem is especially severe and has been greatly compounded by growing populations and the drive to boost export crops. With the best land increasingly taken over by plantations, peasant farmers have been pushed onto more and more marginal soils, often being forced to clear forest to create their fields. The result is the loss of soil on an unprecedented scale. Nor is it only the developing countries that are affected. In Britain, erosion has increased significantly over the past 30 years. A prime cause has been the three-fold expansion of land under winter cereals, which leaves the land unprotected by vegetation during the wettest parts of the year.

Nearly half the arable land in England and Wales is now classified as being at risk from erosion. Local people in the fens of East Anglia, the best wheat-growing area in the country, suffer regular dust storms known as 'fen blows', caused by the wind sweeping away the dust-like soil from the degraded fields. The soils of Illinois and the other states of the US cornbelt - once among the most fertile in the world - are also being severely eroded. At the current rate of erosion, much of this land will have lost nearly all its topsoil within 40 to 50 years.

SALINIZATION AND WATERLOGGING

Flying over the Indus Valley, the land below gives the appearance of being covered with snow. But appearances can be deceptive. For it is salt, not snow, that glistens on the surface. Year after year of irrigation on poorly drained land has turned vast stretches into a salt-encrusted desert, the land salinized beyond redemption.

Irrigation agriculture is one of the most productive forms of farming known to man.

“In China, erosion of the croplands is so severe that scientists at the Mauna Loa research station in Hawaii, over 5,000 km away, can detect the start of spring cultivation from the increase in atmospheric dust fallout.”
Geographical Magazine, *March 1990*

“The potential for erosion now is worse than in the 1930s. If it gets as dry as it was in the 1930s, we're in for some real trouble. You're in country now that man in his infinite wisdom did not improve upon.”
Bill Fryrear, head of the US Department of Agriculture Research Service Station, Big Spring, Texas

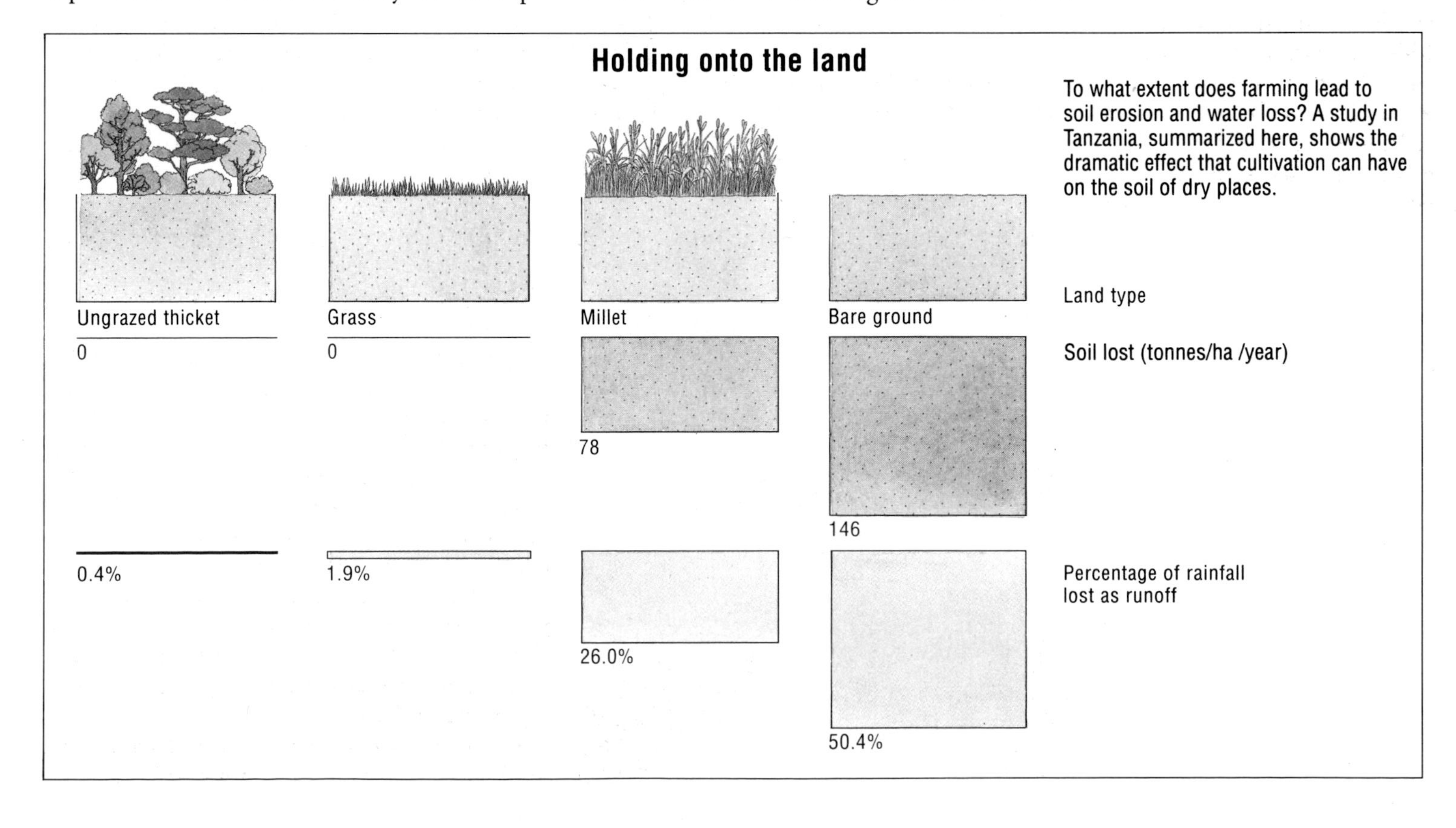

Without irrigation, a US corn farmer in Nebraska can produce 100 bushels per ha (40 bushels per acre) a year. By introducing irrigation, that yield can be trebled. Not surprisingly, UN agencies such as the Food and Agriculture Organization have the seen the extension of irrigation as a major means of increasing production, both of food crops and non-food crops.

But despite the clear evidence that most irrigated land requires periods of fallow, during which accumulated salts can be washed from the soil, modern irrigation schemes are designed to grow crops on the same land year after year. The consequence is an accelerating loss of land to salinization. In Egypt, a country where traditional irrigation systems stretch back to the time of the Pharaohs and beyond, the introduction of perennial irrigation has salinized a third of the country's cultivated land in little more than 30 years and condemned 90 per cent to waterlogging. The problem is also severe in the US, where a quarter to a third of the country's irrigated land suffers from salinity.

In many areas, irrigated land is now so severely degraded that it is unfit for agriculture. In China, more than 900,000 ha (2.2 million acres) of saline and waterlogged land has been abandoned since 1980. In the Soviet Union, 2.9 million ha (7.2 million acres) of degraded irrigated land were taken out of production between 1971 and 1985, and in India, as much irrigated land is abandoned every year as new irrigated land is brought into production.

Salinization and waterlogging are generally blamed on bad management. To an extent this is true. As traditional irrigation farmers have long been aware, irrigated land must be well drained if the land is not to become waterlogged and salts are to be prevented from accumulating. But modern irrigation systems are expensive. Because they cover huge areas, they require not only storage dams but also elaborate networks of canals, and so drainage is rarely installed. Even if it is, the problem of salinization is not solved, but rather postponed. However good the drainage systems, some salt remains in the soil and if the land is not allowed to rest, the accumulation of salt is unavoidable.

TOXIC POLLUTANTS

Agricultural land is being contaminated with other poisons besides salt. Many result from the use of pesticides and other agrochemicals; others from industrial plants, nuclear installations, mines, motor vehicles and toxic waste dumps.

◀ Icicles hang from an irrigation wheel on a frosty morning in Utah. Large-scale irrigation schemes can dramatically increase yields, but only at the expense of draining finite sources of groundwater.

▼ Small-scale irrigation in Mali. When the water is provided by human labour, there is a natural limit to the amount applied and the area irrigated. This tends to restrict the damage done and make such forms of irrigated agriculture sustainable.

For a long time it was denied that pesticides could contaminate the soil for more than a brief period, since it was assumed that they would either break down or become bound to soil particles and thus immobilized. It is true that pesticides bound to the humus in the soil cause little harm to plants or to those who eat them, but they can be released by micro-organisms. For example, proponil, a pesticide used for fumigating soil in rice paddies, breaks down very quickly into two decay products. One is relatively harmless, but the other, 3,4 - dichloroaniline, or DCA, is extremely toxic. Unlike proponil itself, DCA can be liberated from soil particles by various fungi and bacteria.

Many pesticides have now been found to persist in the soil far longer than previously thought possible. A case in point is the carcinogenic pesticide ethylene dibromide, or EDB, now banned in several countries. It can survive in the soil for more than 20 years. Its persistence results from traces of the pesticide becoming locked away in the soil's micropores, where they are out of reach of the micro-organisms that would normally break them down. Numerous toxic chemicals, including PCBs and many industrial solvents, behave similarly in the soil. They are like chemical booby-traps, waiting to be triggered. The moment the soil is disturbed, they can be released to re-enter the environment. The implications for the future contamination both of agricultural lands and underlying groundwaters are horrendous.

For the farmer, the problem is compounded by the so-called 'pesticide treadmill'. Because monocultures are vulnerable to heavy losses from pests, large quantities of pesticides need to be used. It is a battle, however, which the farmer cannot win. The more pesticides that are sprayed, the greater is the likelihood that the natural predators of the target pests will be killed. The result is that any target pests that do survive can proliferate uncontrolled, creating a new generation of pests that requires another application of spray. In addition, many pesticides upset the natural metabolism of plants,

"As crude a weapon as the cave man's club, the chemical barrage has been hurled against the fabric of life - a fabric on the one hand delicate and destructible, on the other miraculously tough and resilient and capable of striking back in unexpected ways."
Rachel Carson

▼ An agricultural worker in Thailand spraying pesticides onto crops without any protective clothing. Deaths and chronic illness from pesticide poisoning are alarmingly high in the developing world. Pesticides that have been banned in the West because they are too hazardous are often exported to Third World countries instead.

▶ Poppies growing in the uncultivated margins of arable fields. Where herbicides are not applied, wild flowers can flourish, providing food for butterflies and other wildlife.

“A few of my neighbors share my belief that ‘as you sow, so shall you reap’, and if you sow pesticides you reap poison. I gave up chemicals in 1970, and if I had to go back to them I'd quit farming.”
Organic farmer, Illinois

causing proteins to break down, and making them increasingly susceptible to attack by pests. The farmer thus becomes locked into a cycle in which spraying encourages pests which encourages yet more spraying. And the land becomes more and more contaminated.

Many pollutants, however, are beyond the farmer's control. In the southwest of Holland, for example, air pollution from factories along the Belgian-Dutch border has caused such extensive contamination of the soil that local residents have been advised not to grow vegetables. Dumping toxic wastes on land has also contaminated thousands of hectares throughout the world. Once polluted with toxic chemicals, agricultural land is difficult to decontaminate. In severe cases, the soil may need to be incinerated - a process that costs between $500 and $1,000 per cubic metre of soil. At Seveso in northern Italy, thousands of tonnes of soil had to be scooped up and incinerated after the land was contaminated with dioxins following an accident in 1976 at a chemical plant.

Incineration is only possible where the area contaminated is both limited and clearly defined. In some cases, however, the pollution is too widespread for remedial action. In the aftermath of the accident at the Chernobyl nuclear reactor in the USSR *(see p.36)*, radioactive fall-out contaminated 10,000 km^2 (4,000 square miles) of land to levels considered dangerous for human habitation. Some reports suggest that as much as 10 times that area in the European part of the USSR may be contaminated to levels unsuitable for agriculture. To clean up such a massive area is clearly impossible, even if the technical means to do so were available.

THE ORGANIC ALTERNATIVE

Although it may give the illusion of productivity, in the long run, modern farming is a failure - not only ecologically, but also economically and socially. Throughout the developed world, and especially in the US, small farmers are finding that they can no longer afford to live on their land. They are bought out by the large landowners and big corporations who can afford the expensive machinery and chemicals necessary to maximize short-term yields. Nearly a million maize farmers in the US have moved off their lands since 1969, and US farmers as a whole owe some $300 billion to the banks.

But, at long last, changes may be on their way. In 1989, the US Department of Agriculture - for many years the world's leading promoter of ‘modern’ farming - astounded the farming community by issuing a report which admitted that organic farms are as productive as those where pesticides and synthetic fertilizers are used. The report concluded that the wider adoption of

◀ Heavy horses drawing a plough. Once a common sight in Europe, these are now rarely seen, their place having been taken by tractors. As a result, farm output has rocketed, but only because input - in the form of fossil fuel - has also rocketed.

▼ A farmer weeding his crops in Niger. Much of farming in the Third World is carried out on a sustainable basis. However, it is increasingly being replaced with large-scale chemical-intensive agriculture, at the cost of community cohesion and environmental degradation.

organic farming would result in "ever increasing economic benefits to farmers and environmental gains to the nation."

One hundred thousand organic farmers in the US and elsewhere have long known this to be true. However, successive governments, encouraged by the manufacturers of agrochemicals, have given generous incentives to farmers to adopt chemical farming instead. In 1989, farm subsidies cost the US government $30 billion. with maize farmers being subsidized by roughly $1 for every bushel they produced.

The question now is whether or not governments will act to promote organic agriculture and wean agricultural land away from the chemicals on which it has become increasingly dependent. During this century, the combination of mechanization, the chemical revolution and agricultural policies aimed at maximizing the production of a few crops have degraded the world's agricultural land on a scale unprecedented in history. The growth in world food output, which had seen dramatic increases from 1950 to the mid-1980s, is now declining as a result of this prolonged and ruthless quest for higher yields. Modern agriculture can now be seen for what it is - the short-sighted plundering of a resource on which we all depend. Only organic farming can ensure that future generations have land that is fit to farm.

RANGELANDS

The vast expanses of the North American Great Plains were once home to 60 million bison, single herds covering hundreds of square kilometres. For the Plains Indians, the bison provided meat, oil for cooking and lighting, and hides for clothes and tents. The Indians - Apaches, Comanches, Araphos, Kiowas and Sioux - knew that if they killed more bison than the herds could support, they would be destroying their own livelihoods. They respected the bison and their environment and lived within its means.

In the 1860s, the Europeans arrived and a wholesale slaughter began. Killing bison for their hides and tongues, they reduced the seemingly unlimited herds to just 500 animals within 40 years. The Indians and their rich cultures were similarly devastated by the white men's quest for land and profit. By 1876, they had been pushed onto reservations, a defeated and broken people.

The Great Plains are just one example of the huge expanses of land across the world that are either too dry or too cold to support forests or agriculture. Such rangelands provide grazing for large numbers of wild animals and livestock. The driest rangelands include the semi-arid scrub of southern Argentina, Australia, the northern part of the Sahel and central Asia. Tropical savanna grasslands are found in East Africa, the southern Sahel, India and Brazil, while temperate grasslands form the Eurasian steppes, the North American prairies (which include the Great Plains), the pampas of Argentina and the South African veld. In the far north, another quite different kind of rangeland, the tundra *(see pp.226-235)*, covers vast areas of the sub-Arctic.

◀ The prairies of central Wyoming, once home to huge herds of bison. These grasslands tend to look very uniform from a distance, but in their natural state, they contain up to 80 different species of grass, together with many other small plants. Vast areas of land like this have now been ploughed up and are used for growing grain.

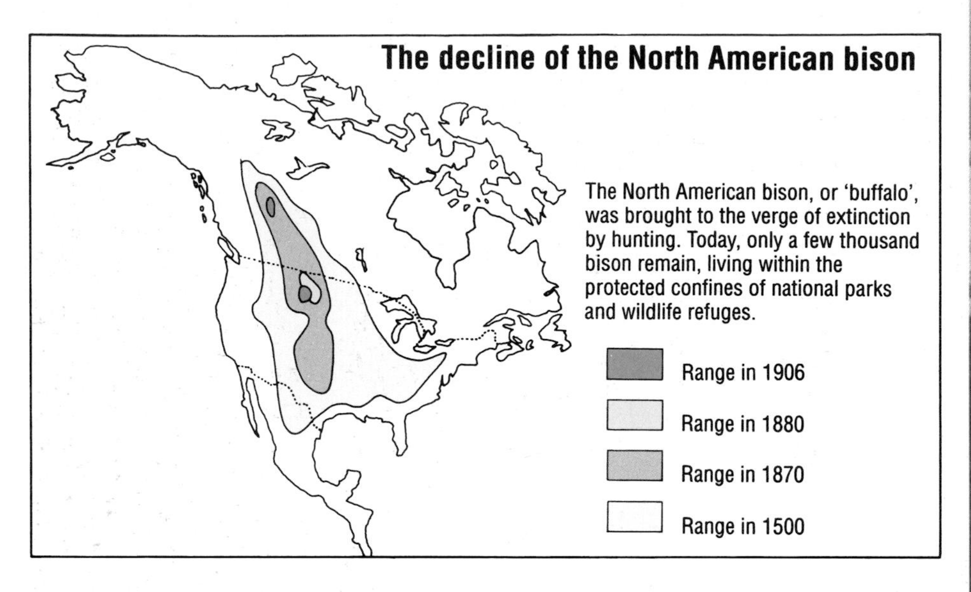

SEMI-ARID SCRUBLAND

Semi-arid scrubland is a demanding habitat. It endures long, hot summers and infrequent rainfall, while in some areas, such as western China and Iran, the winters can be very cold with heavy falls of snow.

The vegetation of these regions is dominated by low, thorny shrubs, such as the creosote bush and the tamarisk. These have small leathery leaves which minimize the amount of moisture lost through evaporation. The larger bushes are widely spaced, with root systems which extend over large areas to make full use of the little moisture available.

Semi-arid scrublands are, in biological terms, the poorest of the rangeland ecosystems. However, pastoral peoples such as the Tuaregs of the northern Sahel have managed to use them to support their herds for many centuries. They are also home to some of the world's last surviving nomadic hunter-gatherers - peoples such as the Australian Aborigines and the !Kung bushmen of Namibia and Botswana, who survive entirely by hunting and collecting wild food.

Although to an outsider the lives of hunter-gatherers appear harsh, they are healthier and live longer than most peoples in the Third World. In fact, anthropologists have termed hunter-gatherers 'the original affluent society', as they can so easily satisfy their needs from their environment. The wealth of their ecological knowledge enables them to live off the nuts, seeds, leaves, fruits and roots of the hundreds of edible plants to be found in these least inhospitable of lands, the remainder of their protein coming from hunting wild animals.

▼ A Tuareg family in Niger travels on with its livestock, hoping to escape the drought. The effects of the Sahel drought of the 1970s were made much worse by government-sponsored schemes to grow groundnuts on marginal land, which had previously been used by nomads for grazing.

▲ Guanacos on upland pastures in southern Chile. Alpacas and llamas are domesticated forms of the guanaco that can still interbreed with these wild creatures. The delicate appearance of the guanaco belies its toughness. It can withstand great heat and extreme cold, allowing it to live in semi-deserts as well as on high mountain peaks.

◀ An Australian aborigine at a waterhole, waiting for animals to be flushed out of scrubland by fire. As hunter-gatherers, Aborigines living in a traditional way were able to make the most of the limited resources of the arid Australian bush. Today, much of this land is now farmed with the help of artesian wells.

▼ **Hundreds of elephant tusks, confiscated from poachers, were publicly burned in 1989 in an attempt to raise awareness of Kenya's tough stand on the ivory trade. Some other African countries believe that controlled trade in ivory from legitimately culled elephants offers more protection in the long run, because it makes the elephant commercially valuable and therefore worth protecting. However, the legalized trade offers an easy outlet for the disposal of poached ivory. This accounts for over 90 per cent of current ivory exports from Africa.**

SAVANNAS

Savannas are dry tropical grasslands which are warm all year round, with alternating wet and dry seasons and occasional severe droughts. They support tall grasses and a scattered covering of trees such as flat-topped acacias, and baobabs, with their remarkable swollen, water-storing trunks and short, gnarled branches. During the dry season, the savanna is a parched, golden colour but with the first rains the vegetation bursts into life and the plains are transformed into a sea of green dotted with bright flowers.

The savannas of East Africa are one of the most famous wildlife habitats in the world. Over 90 species of mammals live in the Serengeti plains of Tanzania, providing prey for hunting dogs, hyenas and lions. The grazing and browsing animals of the plains make the most of the plant food available by being selective in what they eat. Thomson's gazelles and wildebeest, for example, prefer short grass, while zebras choose longer grass and stems. Elands and Grant's gazelles eat mostly leaves and shoots on bushes, while giraffes eat the leaves and new shoots much higher up on trees.

Such has been the pattern of life on savannas for hundreds of thousands of years. Today, however, things are very different. As a result of human interference, some savanna species are now threatened with total extinction. In the 1980s, the African elephant population declined by half, from 1.25 million to about 625,000, as a result of hunting for ivory. Most of the continent's elephants are theoretically protected within parks and game reserves, but the high price of ivory on the world market makes poaching a lucrative business.

In the game parks of East and southern Africa, an undeclared war is being fought between park wardens and the heavily armed and ruthless poachers. In October 1989, amid much controversy, the United Nations Convention on the International Trade in Endangered Species (CITES) approved a total ban on all trade in African ivory. Enforcing the ban will be difficult, especially as only three months after voting in favour of the ban, the British government announced that it would allow dealers in Hong Kong to sell their stockpile of ivory to buyers in China. The stockpile of 670 tonnes represents the tusks of 75,000 elephants. The decision is a

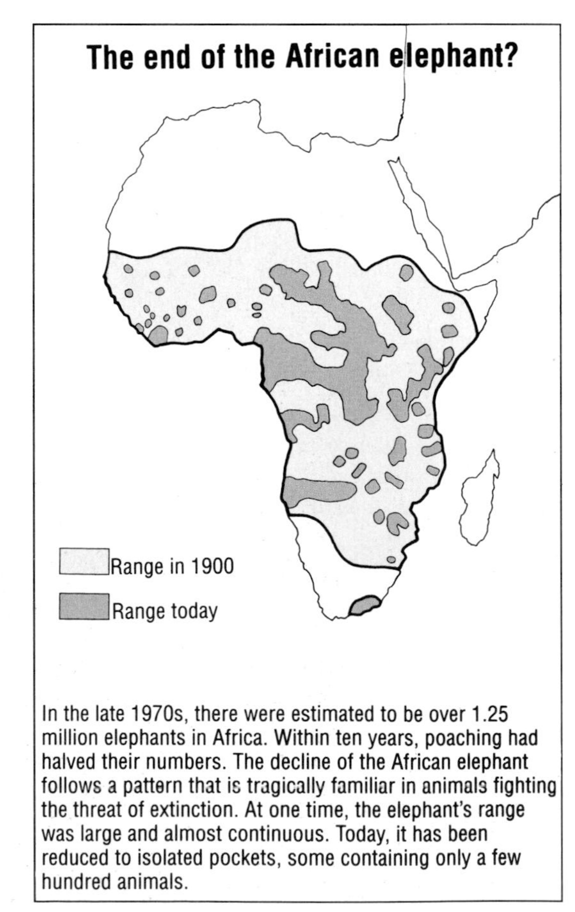

In the late 1970s, there were estimated to be over 1.25 million elephants in Africa. Within ten years, poaching had halved their numbers. The decline of the African elephant follows a pattern that is tragically familiar in animals fighting the threat of extinction. At one time, the elephant's range was large and almost continuous. Today, it has been reduced to isolated pockets, some containing only a few hundred animals.

great boost to black market ivory dealers who will be able to claim that their tusks are from the legal Hong Kong hoard.

Black rhinos are in an even more perilous situation, prized for their horns which are used to make medicines or handles for ornamental daggers. In 1987, only 3,800 black rhinos roamed the plains of Africa, compared to 60,000 in the mid-1970s. Recently conservationists started an unprecedented experiment when they removed the horns from some rhinos in Namibia so as to make the animals worthless to poachers. The fact that the trade in rhino horns has been illegal for several years does not auger well for the future of either the rhino or the elephant.

Despite their often parched appearance, savannas are biologically highly productive. Studies of savanna in Thailand and Mexico have shown that these grasslands produce yields over twice those from wheatfields intensively farmed with chemical fertilizers and pesticides. Although the biomass (the amount of living matter) of savannas is only a fraction of that of a tropical rainforest, savannas are much more efficient in converting carbon dioxide to carbohydrates through the process of photosynthesis. They consume as much, if not more, carbon from the atmosphere as tropical forests.

Much of the carbon that enters the savanna ecosystem becomes locked up in the soil, in roots and underground stems and also in dead plant matter. Burning the savanna and converting it into farmland releases much of this carbon as carbon dioxide, adding to the greenhouse effect *(see pp.40-53)*. Although it receives little publicity, burning the savanna may contribute even more to the greenhouse effect than burning rainforests.

TEMPERATE GRASSLANDS

Temperate grasslands are places of great extremes - hot and dry in summer and bitterly cold in winter. The flat or undulating landscapes of the steppes and prairies are covered in a dense growth of many different species of grass. Together with the effects of drought, fires and intense grazing, the grass helps to prevent bushes and trees from growing. The grazing animals of temperate grasslands range from tiny caterpillars and grasshoppers to deer, antelope and bison.

In temperate grasslands, grass not only provides food; it also protects the soil. Where the

▼ A fire sweeps through wooded grassland in East Africa as the vegetation is cleared for agriculture. With an expanding human population, there is increasing pressure to turn land over to farming.

soil is ploughed up, as has happened in large areas of North America, the consequences can be disastrous. The infamous Dust Bowl of the 1930s, created when farmers attempted to grow wheat with unsuitable technology on unsuitable land, was one of the worst ecological disasters ever caused by humanity. The thick mat of grass which held the thin plains topsoil together was ploughed up, and the soil laid bare to be dried out by the Sun and then blown away by the wind. The naive belief of the first settlers in the plains that "rain would follow the plough" was proved tragically wrong.

In 1935, when the 'black blizzards' of dust were at their worst, 13 million ha (33 million acres) of the southern plains area of Kansas, Colorado, New Mexico, Oklahoma and Texas, had been stripped of their grass. A single storm in March 1935 carried away from the plains twice as much earth as had been scooped out by men and machines to make the Panama Canal. For the western United States, the Dust Bowl proved to be a more significant economic disaster than the Great Depression.

PASTORALISM - LIFE ON THE MOVE

Around 50 million people worldwide are estimated to be true pastoralists, that is they are almost completely dependent on the rearing of livestock on rangelands. Around half of this number live in Africa, concentrated in the countries on the southern edge of the Sahara, such as Sudan, Somalia and Chad. Pastoralism is also the mainstay of the economy in parts of Asia, especially Mongolia, the Soviet Central Asian republics and Tibet. In many areas pastoralism is the most effective way - and indeed sometimes the only way - to use rangelands. Without the pastoralists, large expanses of land would be devoid of human habitation.

If there is one factor more than any other which has shaped the way of life of pastoral societies, it is the need to survive drought. The pastoralist's most important weapon against drought is mobility. When the rains fail, or

▶ Herding horses on the steppes of Inner Mongolia. Almost all traditional pastoralists survive by being nomadic, moving on from one grazing ground to another, according to the seasons. Mongolian nomads live in large tents known as *yurts*.

"...in certain areas of East Maasailand permanent water supplies were established for the Maasai and their cattle. The water was obtained by drilling and by piping spring water from the mountains. This 'improvement' has had two serious consequences: one is that the supplies of water for wild animals have been reduced, and the game has suffered as a result... The second is even more serious: the perennial water supplies have allowed the Maasai to reduce their annual migration so that grazing is concentrated throughout the year in certain limited areas. In these areas... the original grass has now deteriorated completely and it has been replaced by thorn scrub and trees: between the trees the soil is bare. An erosion pan of blowing sand and dust - a thorn bush wasteland - is the result."
Professor Bernard Campbell

when the local grazing resources are exhausted, they can take down their tents or huts and move in search of greener pastures.

Most pastoralists move in the same general direction each year with the coming of the different seasons. Their movements are similar to those of the wild grazing animals which live on the grasslands, except that different tribal groups have traditional rights to specific pastures and water sources. Pastoralists have a detailed knowledge of the ecology of their regions. They know where the best pasture and water is likely to be at any time, how long they can remain in an area before it becomes overgrazed, and how long the vegetation is likely to take to recover.

To make the best use of the land around them, nomads carefully manage their herds. The Tuaregs, for example, who spend most of their time in the scrubland on the southern fringes of the Sahara Desert, raise camels in addition to cattle, donkeys, goats and occasionally sheep. Further south on the Sahelian savanna, cattle are by far the most important animals for pastoralists such as the Fulani.

These animals all rely on different kinds of vegetation. Cattle and sheep require grass. Camels, which can survive for up to a week without food or water during the hottest time of the year, eat grass and the branches and leaves of trees and shrubs. Goats, the hardiest and fastest breeding domestic animals in Africa, can survive entirely on leaves and branches, requiring no grass at all. Should the food resources for one species become scarce, the other animals in the pastoralists' herds will often be able to survive.

THE THREATS FROM DEVELOPMENT

The pastoral way of life is the result of generations of experience in the art of survival. It is small wonder then that outside interference - even with the best of motives - can have harmful results, upsetting the delicate balance of people, their animals and their habitat.

The international development institutions, such as the World Bank and the United Nations Food and Agriculture Organization (FAO), have undertaken numerous projects in the last two decades with the aim of raising the living standards of pastoral peoples and the productivity of their herds. Between the late 1960s and the early 1980s, $625 million was invested in African livestock projects. Almost without exception, these projects have been unmitigated failures and have largely succeeded in achieving the opposite of their original aims.

Governments and aid agencies have actively promoted the digging of wells for the nomads. They reason that the rangelands are dry and so water must be provided. But wells also establish a physical reminder of the government's presence, and can be listed in aid agency reports, so there is further incentive for action.

For pastoralists, however, the new wells can be a very mixed blessing. They are generally sunk along the main routes to markets and so encourage overcrowding. They enable herds to be built up to levels which are unsustainable without the new water sources, thereby discouraging the herders from keeping on the move. The pastures around wells are therefore overgrazed and trampled by thousands of hooves. The wells also break down. In northeast Kenya, for example, out of 54 boreholes drilled since 1969, only 14 were still working in 1979. In Botswana, 40 per cent of wells never functioned for more than a short period.

In Botswana, the World Bank has promoted a number of notorious ranching schemes. As a result of a trade agreement with the European Community, Botswana has a ready market for as much beef as it can produce, and it has set out to maximize beef production with little regard for its huge social and environmental costs. The EC requires that its beef imports be free of foot-and-mouth disease, and as a result, Botswana has erected more than 1,300 km (800 miles) of fencing to separate its cattle from the wild animals which are thought to transmit the disease. The fences disrupt the migratory patterns of wildebeest, which now die in their thousands each year as the fences prevent them from getting to their dry season water sources.

The unrelenting pressure of two million cattle on Botswana's grasslands has caused rampant overgrazing and desertification on the ranches. The !Kung hunter-gatherers, who traditionally depend on wild animals and access to water for their survival, face a bleak future as their land is invaded by cattle. The livelihood of small animal herders has also been jeopardized. Their communal lands are taken over by huge private ranches and they are forced to graze their herds on ever smaller areas of pasture. Like the ranches, these inevitably become overgrazed.

Nonetheless, the World Bank continues to promote 'modern' livestock rearing methods in Africa. This is in spite of a mass of evidence showing that traditional pastoralism is the best method yet devised of using Africa's dry lands. Indeed, the 1977 United Nations Conference on

◀ Cattle herds belonging to Maasai tribesmen, gathering around watering holes in Kenya. The land here is dangerously overgrazed, making it vulnerable to soil erosion.

▼ Goats grazing near a nomad encampment in Mali. Goats can survive on extremely dry vegetation, and are immensely valuable in arid regions such as the Sahel. But they are also extremely destructive and can devastate the vegetation of an area, hastening desertification.

► Cattle in Botswana, destined to supply the tables of Europe, gather around a waterhole. Botswana's arid land is often overgrazed in the drive to produce meat for export.

"We see all round us a flat land, its horizon a perfect ring of misty blue colour where the crystal-blue dome of the sky rests on the level green world. Green in late autumn, winter, and spring... but not at all like a green lawn or field... In places the land as far as one could see was covered with a dense growth of cardoon thistles, or wild artichoke, of a bluish or grey-green colour, while in other places the giant thistle flourished, a plant with big variegated green and white leaves, and standing when in flower six to ten feet high. There were other breaks and roughnesses on that flat green expanse caused by the vizcachas, a big rodent the size of a hare, a mighty burrower in the earth. Vizchachas swarmed in all that district where they have now practically been exterminated..."

W. H. Hudson, describing the Argentinian pampas in 1850

Desertification came to the conclusion that nomadic pastoralism was the only sustainable method of husbandry in the fragile lands bordering the Sahara.

DESERTIFICATION IN THE DEVELOPED WORLD

Given the record of modern ranching system in the western world, it is hardly surprising that the attempts of the development experts to 'modernize' African animal husbandry have proved a dismal failure.

In South Africa, where commercial ranching has been practised by white farmers for generations, only 10 per cent of the veld rangelands remain in good condition. Some 3 million ha (7.5 million acres) of rangeland have been overrun by woody plants to such an extent that they are useless for normal stockbreeding. Over 2 million ha (5 million acres) of veld in the Northwestern Cape are so degraded that the government has declared that they are beyond restoration for many years.

In the United States, desertification, largely caused by the mismanagement of rangelands, is estimated to affect 90 million ha (225 million acres). Desertification is sometimes thought of as resulting from the encroachment of nearby desert, with sand dunes blowing onto the grasslands, but it usually takes the form of a gradual process of change in the composition of the grassland itself. Hardier, less nourishing plants, such as tamarisks and thistles, become more common at the expense of the grasses which protect the soil from erosion by wind and rain. Rather than advancing across a broad front, desertification tends to begin with small patches and spread outwards like a skin disease. Eventually the land completely loses its ability to support vegetation as the topsoil is blown or washed away and the ground baked rock hard by the Sun.

The principal causes of desertification in the US are ploughing and overgrazing by livestock. As early as 1878, John Wesley Powell, director of the US Geological Survey, prophetically warned that although the grasses of the arid West were nutritious they would be "easily destroyed by improvident pasturage and... replaced by noxious weeds." Powell's warnings, however, went unheeded and the effects of "improvident pasturage" can be seen throughout southern Nevada, Arizona, New Mexico, western Texas and Oklahoma, and southeastern Kansas. Over much of this land the soil's productive capacity has been virtually destroyed. Where pioneers found grass "as high as the cows' backs", now only sagebrush and pebbles are to be seen.

In the intensive system of livestock rearing now practised in the US, most cattle are only grazed on rangeland as calves, and on reaching maturity are taken to pens known as feedlots. Here they are fed large amounts of grain and protein-rich feeds, such as fish-meal, to fatten them for slaughter. Instead of being allowed to eat grass - something that we cannot make use of - they are fed with products such as cassava and fish, foods that we ourselves could eat. A third of the world's fish catch is fed to western livestock in this highly wasteful form of food production. Much of the plant-derived feed is imported from the developing world, and some of it is even grown on land on which pastoral peoples once grazed their own herds.

Not only does the overconsumption of beef in the developed world lead to desertification and the feeding of scarce food supplies to cattle rather than people, but it also creates pollution. Concentrating large numbers of cattle in pens produces vast quantities of manure. Where previously this manure would have fertilized the land, it now poses huge problems in disposal and water and air pollution. Methane, a potent greenhouse gas, is produced by fermentation in the stomachs of cattle. Cattle feedlots have been estimated to pump 80 million tonnes of methane a year into the atmosphere - around 20 per cent of the total methane annually emitted into the atmosphere - adding significantly to the greenhouse effect.

SAFEGUARDING THE FUTURE

If the degradation of the world's rangelands and the impoverishment of its pastoral peoples are to be halted, it is essential that cattle numbers are reduced worldwide. This is the opposite aim to that of many development agencies, and many big corporations which profit from selling animal pharmaceuticals and feedstuffs or from beef processing.

The pressure on pastoral peoples can only be reduced by stopping ranching schemes and the expansion of agriculture onto their lands. The pastoralists have shown that they are the best managers of arid lands and they should be allowed to manage them with their traditional methods. Indeed, stockmen in the developed world could do well to learn from their example, which, unlike modern cattle ranching, has stood the test of time.

"A lot of range has been put in cultivation that shouldn't have been. There's no body to the soil."
Texan farmer

"It's not an act of God. It's an act of greed. God doesn't have a plow."
Texan rancher on dust storms caused by ploughing rangeland

"Grass is the natural produce of this land, which seems to have been made on purpose to produce it; and we are not to call land poor because it will produce nothing but meat."
William Cobbett, 1832

"The rich white man, with his overconsumption of meat and his lack of generosity for poor people behaves like a veritable cannibal - an indirect cannibal. By consuming meat, which wastes the grain that could have saved them, last year we ate the children of the Sahel, Ethiopia and Bangladesh. And we continue to eat them this year with undiminished appetite."
René Dumont

RIVERS

"The dams will be in your interest. They will bring progress."

With these words, the Chief Engineer for Brazil's electricity conglomerate, Eletronorte, concluded his case for building a series of massive hydroelectric dams on the Xingu river. His audience, however, was unimpressed. It consisted of about a thousand Amazonian Indians, who had assembled at Altamira, a boom-town in northeastern Amazonia, in the first-ever mass gathering of Brazil's tribal peoples against a dam project.

The Indians rose to their feet, lifting their clubs and arrows in protest. A woman, streaked in warpaint, strode to the dais, brandishing a machete and cutting the air to emphasize her points. She brought down the machete in a swift but graceful arc, stopping just a hair's breadth from the engineer's shoulder-blade.

"We don't need electricity. Electricity won't give us food. We need the rivers to flow freely. We need our forests to hunt in. We are not poor. We are the richest people in Brazil. We are not wretched. We are Indians."

On this occasion, the protests proved successful. Under pressure from environmental groups, the World Bank announced that it would withdraw funding from the project. The Brazilian government reluctantly agreed to shelve the scheme. But the Indians know that the threat is still there. Their river may yet go the same way as hundreds of others across the globe - dammed and polluted in the name of progress.

◀ Fishing on one of the tributaries of the Amazon. Clean rivers are immensely valuable to mankind. They provide food, drinking water, a place to swim and bathe, and an easy means of travel. However, today they are few and far between, as most of our watercourses have been turned into dumping grounds for all manner of wastes and effluents.

RIVERS OF THE WORLD

Every year, over 400,000 km^3 (95,000 cubic miles) of water evaporates from the oceans into the atmosphere. Ninety per cent of that water returns to the ocean as rain, sleet or snow, but the rest falls over land. As the snow melts and the rain runs off the land, it collects in streams and rivers, cascading down mountains and into the plains on its long journey back to the sea.

When the ancient Greek philosopher Heraclitus said that one could never put one's feet in the same river twice, he captured the forever changing, yet seemingly unchanging, nature of rivers. Their waters are always on the move - sometimes violently so, sometimes at a snail's pace - rising and falling with the seasons. To survive, their plants and animals must be able to cope with such change.

The speed at which a river flows makes a critical difference to the wildlife that is found there , and the strategies they have evolved to survive. As rivers tumble down from the mountains where they rise, the sheer physical force of the water as it pours over rocks and shingle leaves little sediment on the riverbed. Here, fish such as salmon use their powerful muscles to swim against the stream. Apart from them, only those plants and animals which attach themselves to rocks can survive. In the more sluggish lower reaches of a river, the sediment carried down from the mountains provides a nutrient-rich bed in which plants can readily establish themselves. Fish are more numerous, as are the birds that prey on them, from iridescent kingfishers to ungainly pelicans.

Millions of people around the world depend on rivers for their livelihood. The fertile alluvial soils of floodplains, the rich fisheries, and the ready access to water have encouraged settlements along riverbanks for millennia. Over a million people inhabit the Senegal River basin in West Africa, tens of millions the Niger basin and hundreds of millions the Ganges basin in India. The world's greatest cities have grown up along the banks of rivers - Paris, London, Rome, New York and Bangkok. But as humanity demands more of its rivers, so their future health is in increasing jeopardy.

THE IMPACT OF DAMS

In the arid southwest of the US, where rainfall rarely exceeds 50 cm (20 inches) a year, local farmers have an expression, "a wild river is a wasted river". Their whole way of life depends

◀ Young rivers, cutting down through newly upraised landscapes, flow fast and straight. Where rivers have flowed for many years, they carve out wide valleys for themselves, developing a meandering course across the flat ground. The rich silt brought down by a river over millennia makes such land especially valuable for agriculture, as this view of the River Orme in northern France shows.

▲ At their source, most streams and rivers are clean, so plant and animal life is usually healthy. But these mountain streams are less valuable for wildlife than the placid lower reaches, due to the sheer force and speed of the water.

▼ A European kingfisher returning to its nest hole. Kingfishers need fairly quiet, unpolluted streams to survive.

on capturing the water in rivers, storing it, and rerouting it to areas - often hundreds of kilometres away - which would otherwise have none. Without dams and canals, much of the land would still be scrub, useless for agriculture. Cities such as Phoenix, Arizona, or Denver, Colorado, would cease to exist.

The southwest of the US is by no means exceptional. Over half of the world's land surface is classified as arid or semi-arid. In some areas, rainfall is so sparse as to make rainfed agriculture impossible. In others, farmers must survive for most of the year with little or no rainfall, followed by the annual monsoon, when rain falls in such torrents that much of the water simply runs off and is eventually lost to the sea. By damming rivers, water can be stored for use in the dry summer months.

Dams have the additional advantage of enabling man to tap the vast energy potential of rivers. By channelling the water in reservoirs through giant turbines, electricity is generated - and in vast quantities. The Itaipu dam in Brazil alone generates 12,000 mW, enough electricity to power New York. Dams already supply more than a fifth of the world's energy and if all the

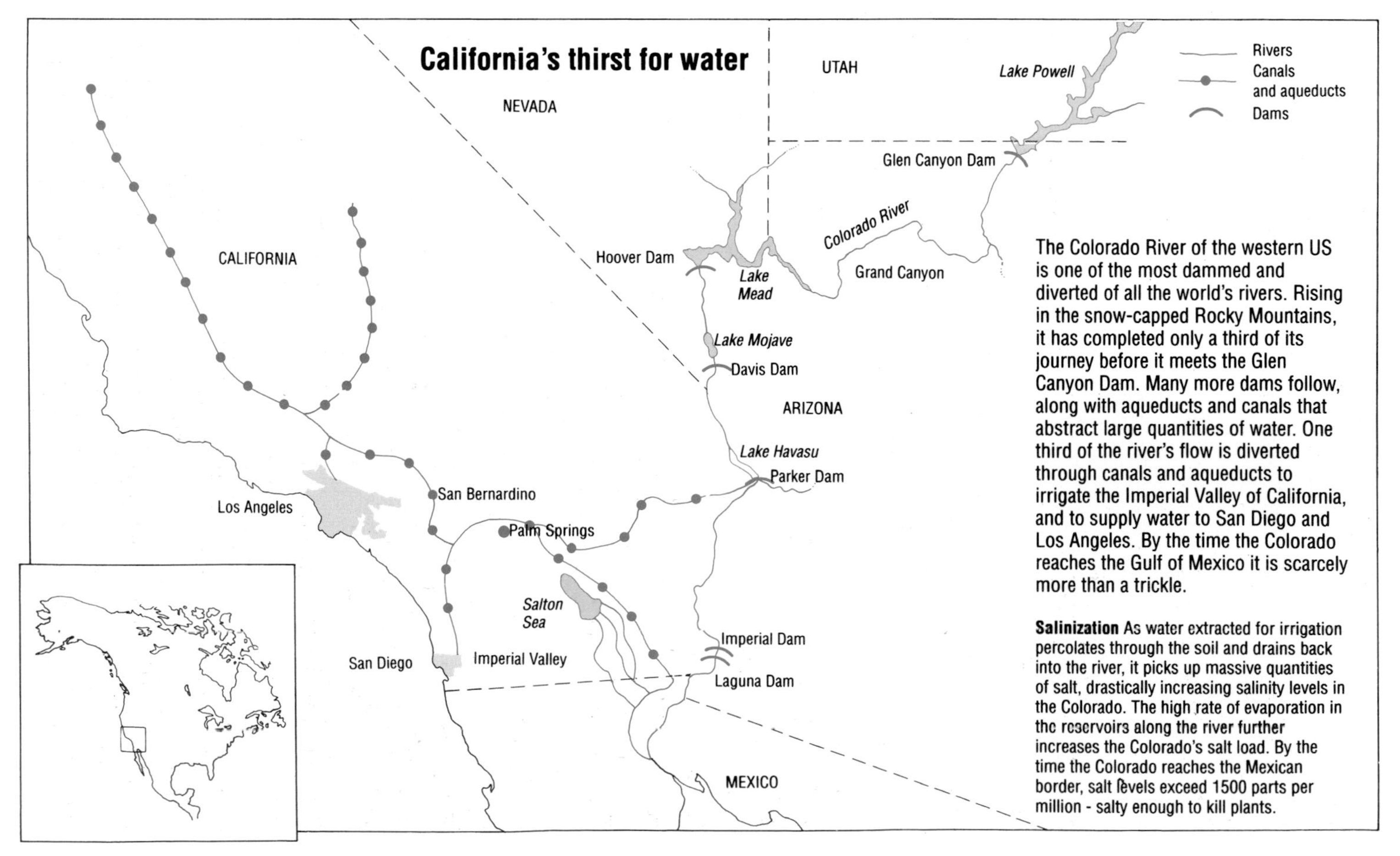

The Colorado River of the western US is one of the most dammed and diverted of all the world's rivers. Rising in the snow-capped Rocky Mountains, it has completed only a third of its journey before it meets the Glen Canyon Dam. Many more dams follow, along with aqueducts and canals that abstract large quantities of water. One third of the river's flow is diverted through canals and aqueducts to irrigate the Imperial Valley of California, and to supply water to San Diego and Los Angeles. By the time the Colorado reaches the Gulf of Mexico it is scarcely more than a trickle.

Salinization As water extracted for irrigation percolates through the soil and drains back into the river, it picks up massive quantities of salt, drastically increasing salinity levels in the Colorado. The high rate of evaporation in the reservoirs along the river further increases the Colorado's salt load. By the time the Colorado reaches the Mexican border, salt levels exceed 1500 parts per million - salty enough to kill plants.

rivers of the world were to be dammed, they could produce as much electricity as 12,000 nuclear reactors. It is small wonder that dams are seen as vital to economic development.

The technology that has brought water to the arid southwest of the US or to the drylands of Africa and India is not new. Nor is the use of water power to drive machinery. For centuries, man has built dams and watermills. Indeed, the embankments and water storage tanks that remain at such ancient capitals as Angkor Wat in Cambodia bear proud witness to the engineering skills of those who constructed them.

But advances in concrete technology and the development of vast earthmoving machines have enabled dams to be built and rivers to be manipulated on a scale that would have staggered the ancients. In California, for example, engineers have defied nature, redirecting whole river systems in order to bring water to the thirsty south. Massive 300,000-horsepower pumps siphon billions of cubic metres of water over the Tehachapi Mountains to irrigate the farmland of the Imperial Valley, south of Los Angeles. All but three of California's large rivers have been dammed, many of them more than once.

The era of large dam building began in 1935 with the construction of the massive Hoover Dam on the Colorado River in the US. Rising to over 210 m (700 feet) from the bedrock of the river, the dam was at the time the largest single engineering structure ever built. Since then, river after river in the US has been dammed, to the point where few still flow unimpeded to the sea. Some 50,000 major dams now straddle American rivers. Five dozen dams have been built on the Missouri river alone, 25 on the Tennessee. On such rivers, the transformation is so complete that hydrologists no longer refer to them as rivers but as "regulated streams".

With funding from the World Bank and the international aid agencies, the Third World too has set out to tame its rivers. India in particular has invested billions of dollars in its dam building programme, with over 1,500 dams built to date. Along the Narmada River, which flows through the states of Maharashtra, Madhya Pradesh and Gujarat, work is underway on one of the biggest dam building programmes in Asia. By the time it is finished, 30 large dams, 135 medium-sized dams and 3,000 minor dams will impound the river and its tributaries.

◀ The Glen Canyon Dam, Arizona. This is the first of many dams and diversions to confront the Colorado River as it flows towards the Gulf of Mexico.

▼ The Colorado River, cutting through the soft rock of Utah, before it encounters the obstacle course that man has created on its lower reaches.

▲ A waterway in Burma clogged by water hyacinth. This plant floats by means of air-filled leaf-stalks, and as it grows, it produces plantlets which quickly spread over the water's surface. Water hyacinth can become an immense problem wherever man interferes with a river's natural flow by building dams or weirs.

THE ADVERSE EFFECTS OF DAMS

Few technologies have had a greater impact on rivers and their peoples than large dams. For those living in the river valley immediately above a dam, that impact begins even before the flood-gates are closed. Reservoirs cover large amounts of land, and millions of people have been uprooted to make way for them. The building of the Kariba Dam on the River Zambezi, for example, displaced 57,000 people who had previously farmed the fertile floodplains on either side of the river. Although the government promised that they would be no worse off as a result of the dam, they were moved to land so marginal that for several years they depended on food aid for their survival. The Narmada complex in India threatens even greater disruption.

In some cases, local people have resisted the move. In 1989, as the floodwaters of the Kedung Ombo reservoir in Indonesia began to rise, over a thousand families took to the roofs of their houses, then to rafts, hoping that the government would relent and stop the project. It would not and the reservoir kept on rising.

As the reservoir of a dam fills, so the river ecosystem is changed forever. The rising flood-waters spell death for the many animals that live in the river valleys. Efforts to rescue wildlife before the reservoir begins to fill have proved a limited success. Although it makes good 'positive' publicity, the numbers rescued are often small, and many do not breed successfully after being moved since they lack an established territory among the existing inhabitants of their new home. The rescue operations inevitably concentrate on the largest animals. Little or nothing can be done to save the rest of the wildlife under threat, including small mammals, reptiles, insects and plants.

Fish are also affected, but not always - as might be thought - for the better. In a reservoir itself, those species that depend on free-flowing water give way to species better adapted to a lake environment. Initially, the fish tend to thrive, encouraged by the release of nutrients from the submerged soils. Local fisheries therefore benefit, but often it proves a short-lived bonanza.

As the submerged vegetation rots down, it uses up oxygen in the water. Eventually the reservoir

◀ A giant otter from the Amazon. These impressive animals, which can grow up to 2 m (6 feet) long, are endangered by hunting. Although protected in some countries, there is little chance of enforcement in most parts of the giant otter's range.

▼ Trees killed by the damming of a river in Thailand.

▲ A beaver carries a branch home, part of the food store it is amassing for the winter. Beavers are among the most extraordinary river animals, using the water only as a home and food-store, not for food. By damming streams, beavers create deep pools of water that will not freeze at the bottom in midwinter. Beavers also create small canals to transport trees to their homes.

is transformed into a stinking and stagnant morass in which few fish can survive. Invasive waterweeds, such as the beautiful but prolific water hyacinth, often spread over the water's surface, blocking out sunlight and reducing fish populations.

Downstream fisheries, too, can suffer. For example, before the Aswan Dam was built, the sardine fisheries along the eastern Mediterranean were plentiful. But when the dam deprived the coast of the nutrients in the Nile's silt, catches plummeted by 97 per cent.

For migratory fish, such as salmon and sturgeon, dams can prove particularly disastrous. In the south of the Soviet Union, sturgeon catches fell drastically after dams cut the sturgeon off from their spawning grounds in the major rivers entering the Caspian Sea.

The building of reservoirs can also lead to an increase in waterborne disease. In contrast to the flowing waters of a river, the still shallows of a reservoir provide an ideal breeding ground for the mosquitoes that carry malaria and the snails that carry bilharzia, or schistosomiasis, a group of tropical diseases which cause diarrhoea, fever

▶ **Flooding in Bangladesh. Deforestation in the Himalayas, combined with various forms of interference in the course of rivers, has greatly increased the amount of flooding.**

“Five hundred and seventy square miles of dense virgin rainforest was flooded for this reservoir. As the trees decomposed, they produced hydrogen sulphide and a stink that brought complaints from people many miles downwind. For two years workers at the dam had to wear gas masks. The worst effect of the decomposing vegetation was that it made the water more acidic and corroded the dam's expensive cooling system.”
Catherine Caufield on the Brokopondo Dam in Surinam

“The Narmada project is the latest in a series of dam projects encapsulating every imaginable social, environmental and economic folly... A massive disinformation campaign in the beneficiary state of Gujarat has convinced people that the project is their lifeline. It has ignored the fact that it is years of deforestation which has led to the severe water crisis, and that the most obvious solution to the problem lies in afforestation.”
***Bittu Sahgal, editor of India's* Sanctuary Magazine**

and anaemia. Following the building of the Aswan Dam, the rate of schistosomiasis trebled among the local population. In some villages, nearly everyone was affected.

The impact of dams does not stop with the dam itself. In those river valleys where farmers rely on the annual flood to irrigate their crops and bring nutrient-rich silt to fertilize their land, dams can have disastrous consequences. Held back by the dam, neither the annual flood nor its precious silt reaches the floodplains downstream, jeopardizing the livelihoods of thousands of farmers.

The reduced silt-load of the river also results in the bed of rivers gradually being eroded, since the eroded sediment is no longer replaced by new silt as it would be normally.

Silt trapped behind dams often leads to their premature closure. The silt reduces the capacity of reservoirs and hence their useful lives. The problem is particularly severe in the tropics, especially in those areas that have been heavily deforested. In China, the Laoying Dam silted up before it had produced a single megawatt of electricity. Even where silt is not a problem, weeds and rotting debris clog up the turbines of many new dams, putting them out of action.

TAMING THE FLOOD

For those cultures which practise seasonal irrigation, the annual flood is an occasion to be welcomed. But for those who have built their homes on reclaimed wetlands *(see pp.150-163)* or who farm drained land, floods spell devastated villages and ruined harvests.

During the last 30 years, flooding has been on the increase, particularly in the tropics. Deforestation in the watersheds of rivers is a major cause of the increase. Without trees and vegetation to soak up the monsoon rains, the water cascades down into the valleys below. To prevent flooding - and to make navigation easier - riverbeds are dredged, river courses straightened and rivers turned into little more than concrete waterways. Dams are also built to store the floodwaters for slow release later in the year.

This strategy has encouraged people to settle permanently on floodplains in the belief that they will be safe from the ravages of even the wildest rivers. Inevitably, when floods occur, the damage is all the more extensive.

The tragic irony is that the 'channelization' of rivers and streams fails to prevent the problem of flooding, and in many cases exacerbates it. In India, nearly $1 billion was spent on building

◀ Intensively cultivated farmland alongside a river: an apparently tranquil scene, but one which may conceal immense unseen damage to river life from agricultural run-off. Fertilizers, particularly nitrates and phosphates, provide excessive nourishment for algae in the water, and these proliferate, depleting the oxygen supply and killing fish. Insecticides and fungicides poison river animals more directly. 'Top predators' such as otters suffer the most because their body tissues accumulate persistent pesticides such as dieldrin. Even though some of the worst pesticides have been banned in many countries, dead otters with visible signs of poisoning are still found.

▶▶ A North American stream, choked with refuse. Metals and plastics, which are not biodegradable, will remain in the stream indefinitely, unless active steps are taken to remove them.

▶▼ A riverside shanty town in the Philippines. The river is used as a latrine, as a source of drinking water, and for bathing, resulting in the rapid spread of waterborne diseases.

“The spring and the rivulet, the brook, the river, and the lake, seem to give life to Nature, and were indeed regarded by our ancestors as living entities themselves. Water is beautiful in the morning mist, in the broad lake, in the glancing stream, in the river pool...beautiful in all its varied moods. The refreshing power of water upon the earth is scarcely greater than that which it exercises on the mind of man.”

Sir John Lubbock, 1898

embankments and canals between 1953 and 1979, yet the damage done by floods continues to increase year by year. Embankments certainly prevent a river from bursting its banks and so depositing its silt on the floodplain around it, but the result is that the silt accumulates in the river, making the riverbed rise. The height of the embankments must therefore be raised to keep up with the river. Year by year, the river rises threateningly above the surrounding land, so that when a breach occurs - as invariably they do - the result is calamitous.

WATERBORNE WASTE

For thousands of years, humanity has consigned its wastes to rivers. Where settlements are small, and the quantities of wastes limited, the harm done is minimal. Provided the wastes can decompose naturally, and there is enough water to wash them away, it takes little time before they are broken down and rendered harmless. But as cities have grown and industry has expanded, the burden and toxicity of wastes has massively increased, overwhelming the capacity of many rivers to cleanse themselves.

Untreated sewage has turned whole stretches of water into stinking cesspools, so depleting the water of oxygen that fish can no longer survive. In the slums of the Third World, open drains, clogged with human excrement, have spread such age-old scourges as cholera, causing thousands of deaths. To sewage has been added a cocktail of highly toxic and persistent chemicals, created by industry. Chemical fertilizers and pesticides, washed off the land, have further polluted rivers and streams. Fish have been killed in their thousands, poisoned or suffocated, and drinking water sources for millions of people have been contaminated to the point where they are no longer fit for human consumption.

The rivers of the developed world fare little better. Travelling east from Brussels to the border where Belgium meets Holland and West Germany, one enters an area dubbed by industry as the 'Euregio Rhine-Meuse'. Consisting of a triangular wedge stretching from the port of Rotterdam in the west, to Dortmund and the Ruhr Valley in the east, and to Luxembourg in the south, the Euregio is one of the most densely populated and heavily industrialized areas in the world. It is also one of the most polluted.

Since the middle of the last century, one heavy industry after another has set up on the banks of the rivers and canals that criss-cross the area. Coalfields in the Ruhr Valley and in Holland's

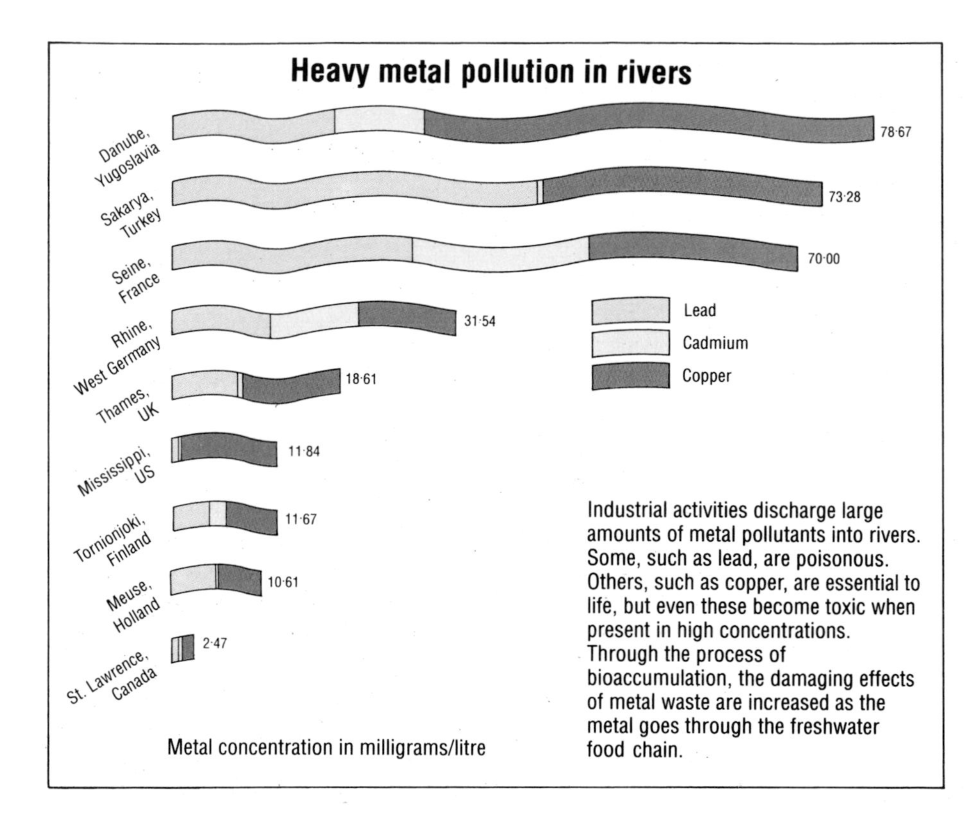

Industrial activities discharge large amounts of metal pollutants into rivers. Some, such as lead, are poisonous. Others, such as copper, are essential to life, but even these become toxic when present in high concentrations. Through the process of bioaccumulation, the damaging effects of metal waste are increased as the metal goes through the freshwater food chain.

"Polluted first by Czechoslovakia, the Odra River is biologically comatose even before it crosses the border. And the Vistula, Poland's main artery, is so chemically active for 80 per cent of its length that it is unfit even for industrial use: it corrodes the machinery. Around 10,000 factories sluice their effluent into it, unfiltered, and half the 800 communities along its banks have no sewage treatment plants."
Peter Martin

"Die meisten Flüsse sind Kloaken"
(Most of the rivers are sewers)
Headline in Berliner Morgenpost, *an East German newspaper*

Limberg province - a spur of land sandwiched between Belgium and West Germany - powered the expansion. The coal mines, steel works, bottling plants, papermills, factories and mineral works along the Ruhr, the Rhine and the Meuse have been joined by chemical plants and, more recently, nuclear power stations. Their wastes have been dumped wherever it was most convenient - into the atmosphere, into holes in the ground and, above all, into local rivers.

CRISIS ON THE RHINE

The Rhine typifies the problem of waterborne pollution. Over a fifth of all the chemicals produced every year in the world are manufactured along its banks. At Basle in Switzerland, the Sandoz chemical plant produces pesticides. Further downriver, BASF, Hoechst and Bayer manufacture fertilizers, pharmaceuticals and other chemicals. For years, their wastes - like those from the numerous other factories that cram the river's banks - have simply been poured into the river.

In the mid-1980s, industry along the Rhine was discharging over 2,000 different chemical compounds a day, for the most part without any controls on their disposal. By the time its waters enter the North Sea, the Rhine spills out over 100 tonnes of toxic heavy metals and 30 tonnes of toxic chemicals a day.

The Meuse, which flows through France and Belgium before entering Holland to the south of the Rhine, is even filthier. Running through the heart of the Euregio, levels of its many pollutants - from human faeces to cadmium, zinc and lead - regularly exceed EC safety limits and in many instances are higher than the levels in the Rhine. The pollution is particularly severe between Liege and Maastricht, where discharges to the river are virtually uncontrolled. Indeed, in Belgium as a whole, only 10 per cent of domestic sewage is treated before it is discharged to rivers, and the percentage of industrial waste treated is only marginally higher. The river must also cope with discharges of radioactive tritium, caesium and cobalt from the nuclear power plants that France and Belgium have recently constructed. Since 1985, levels of tritium entering the North Sea from the Meuse and other rivers have risen by 75 per cent.

Illegal discharges and discharges due to accidents are common. In 1986, more than 30 tonnes of pesticides, fungicides and chemical dyes were released into the river after fire at the Sandoz chemical plant near Basle. In just two hours, the Sandoz fire released more pollutants into the Rhine than the river normally receives in a year. For over 100 km (60 miles) below the Sandoz plant, the river was rendered lifeless, and it is only now recovering.

Following the fire, the authorities intensified the monitoring of discharges to the Rhine, revealing the true extent of accidental or illegal dumping. Within a single week, 12 major pollution incidents had been unearthed. First there was a spill at Ciba-Geigy; then 1,100 kg (2,425 lbs) of the herbicide 2,4-D were discharged from a BASF plant at Ludwigshafen; then the Hoechst plant on the Main owned up to a major leak of the solvent chlorobenzene.

THE COST OF CLEANING UP

The countries which border the North Sea are now committed to halving the volume of the most dangerous pollutants entering the sea from rivers by 1995. But many fear that even a 50 per cent reduction in pollution will not be sufficient to restore many rivers to good health. According to the Dutch government, emissions of heavy metals, pesticides and chlorinated hydrocarbons alone will have to be cut by 75-90 per cent if salmon, long since vanished from the river, are to return and commercial fisheries be restored. Such cuts cannot be achieved through better pollution control alone: changes in patterns of consumption will be critical - and that will require wider ranging changes than many countries are prepared to contemplate.

Even the agreed 50 per cent reductions are proving difficult to achieve. Many governments have been reluctant to make the investment necessary to clean up their rivers. To restore Belgium's river Scheldt, for example, will require 4 to 5 times more money than either government or industry are prepared to spend on pollution controls. Lobbying by industry has also hampered the clean-up. In West Germany, negotiations with the fertilizer and chemical industries have been so slow that new regulations on discharges are not expected to be agreed until 1993 at the earliest - far too late to reduce emissions by the 1995 deadline.

Across the North Sea, Britain has been particularly stubborn about reducing emissions from its rivers, reinforcing its reputation as the 'dirty man of Europe'. At one time, Britain held a promising track record in countering river pollution. In the 1960s, major efforts were made to reduce the amount of sewage and organic waste entering Britain's rivers, and the return of fish to the once lifeless Thames provided evidence that the campaign was working. However, in the 1970s and 1980s, the clean-up ground to a halt. Economic recession and pressure from industry led to controls being relaxed, and the quality of Britain's rivers is once more on the decline.

On the other side of the Atlantic, the reckless pollution of rivers has also taken its toll. Many rivers in the east of the US are so polluted by industrial chemicals that they can no longer be used for drinking water.

Many rivers of North America and western Europe are clean, however, by comparison with those in eastern Europe, where industry has been given a free hand to pollute almost at will.

And so the catalogue of pollution and poisoning continues, from the rivers of industrial countries to those of the Third World, where controls on discharges are often non-existent. The cost of cleaning up rivers is prodigious. But so, too, is the cost in environmental damage and ill-health if we continue to use them as running rubbish dumps. The world, as we now know, is too small a place for one habitat to be ruined without affecting many others.

"Bats ... love to frequent waters, not only for the sake of drinking, but on account of insects, which are found over them in the greatest plenty. As I was going, some years ago, pretty late, in a boat from Richmond to Sunbury, on a warm summer's evening, I saw myriads of bats between the two places: the air swarmed with them all along the Thames, so that hundreds were in sight at a time."
Gilbert White, naturalist, 1767

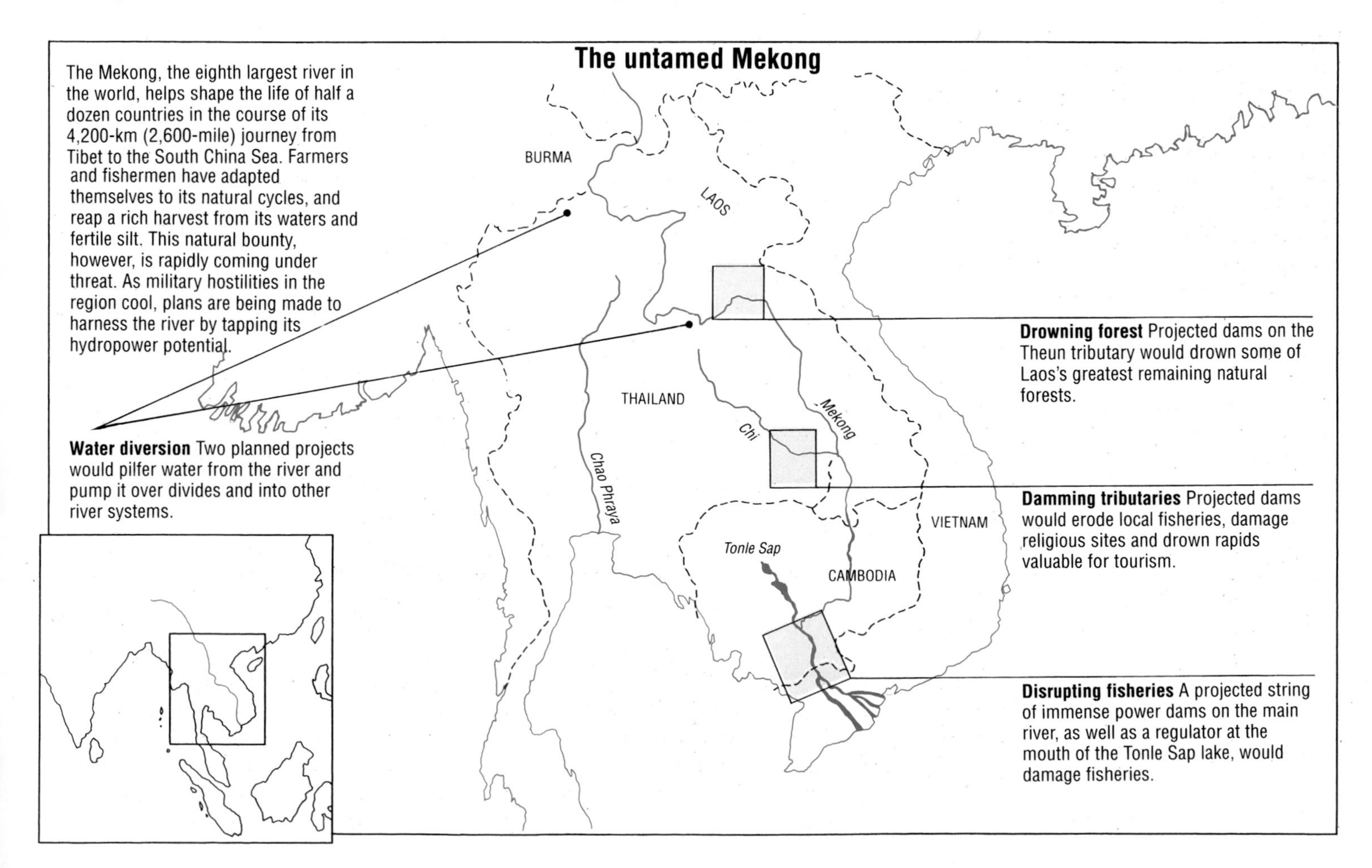

GROUNDWATER

Like many other communities in the world, Woburn, Massachusetts, is a town that relies on wells for its water supply. For the past century, it has also been home to a number of chemical plants and factories, which manufacture or process a wide range of products from glues and pesticides to dry-cleaning fluids and leather. For years, the companies dumped much of their wastes on a local site which was less than 800 m (1/2 mile) from two of the town's wells. In the late 1970s, tests revealed that water from the wells was heavily contaminated with numerous highly toxic chemicals, including trichloroethylene, chloroform and benzene - all of which are known to be carcinogenic.

The wells were closed. But the damage had already been done. Childhood leukaemia rates in Woburn are double or treble the national average, and a study by the Harvard School of Public Health has concluded unequivocally that contaminated drinking water is to blame.

WHAT IS GROUNDWATER?

Much of the rain that falls on land is taken up by vegetation, lost through evaporation or returned to the sea via streams and rivers. Where the soil is permeable, however, some of the rainwater gradually works its way downwards, filling the gaps between soil particles and then seeping into the empty pores of the underlying rock. The water thus stored in underground rock is known as groundwater.

Rocks in which every available pore has been filled with water are said to be saturated, while unsaturated rocks are those in which the pores can still absorb more water. The dividing line between the saturated and unsaturated rock is called the water-table. In some areas, this is only a few millimetres below the surface - indeed, it may reach the surface. In other areas the water-table is several metres deep.

◀ A wind-operated pump taps into groundwater in Australia, providing water for livestock. This traditional method only removes water to a depth of about 3 m (10 feet) and does not normally deplete reserves.

◀ Blauloch Spring in the Swabian Alps, West Germany. One of the largest freshwater springs in western Europe, this 'blue hole' supplies water from deep below the ground. Once clean and unpolluted, such water supplies are increasingly tainted by pesticides and chemical spillages.

“Of all our natural resources, water has become the most precious... In an age when man has forgotten his origins and is blind even to his most essential needs for survival, water along with other resources has become the victim of his indifference.”
Rachel Carson

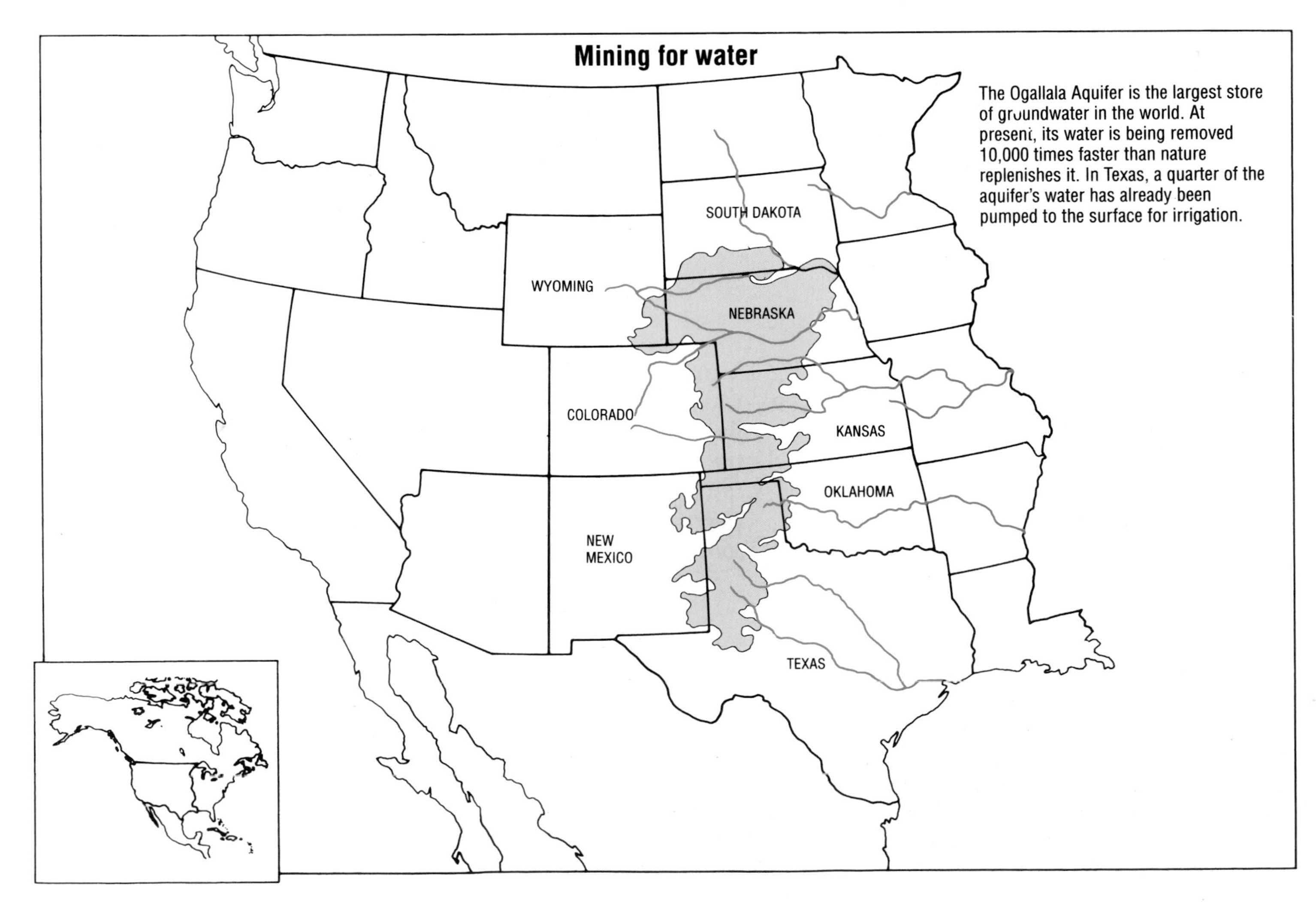

The Ogallala Aquifer is the largest store of groundwater in the world. At present, its water is being removed 10,000 times faster than nature replenishes it. In Texas, a quarter of the aquifer's water has already been pumped to the surface for irrigation.

Wells reach down to the water-table so that water can be brought to the surface. Where the volume of groundwater stored in underlying rock is great enough to be exploited in this way, the rock is termed an aquifer.

Water travels slowly on its long underground journey. Depending on the depth of the water-table and the permeability of the rock, rainwater passing through an aquifer may take many years to emerge in a spring, or in a river or well. The groundwater beneath central London, for example, is 20,000 years old. In many deep sedimentary deposits of silt and sand, the flow of groundwater is so slow as to be almost imperceptible, the water moving just a few metres every thousand years. By contrast, rocks like limestone, which are frequently heavily fissured, permit the easy flow of groundwater. Nonetheless, it may still take five years for groundwater to pass through a limestone aquifer.

It is estimated that between 40 and 60 million km^3 (9.5 to 14.3 million cubic miles) of water are stored underground, although only a fraction of that volume is recoverable. Some aquifers are enormous. The largest in the world is the Ogallala Aquifer which stretches from southern South Dakota to northwest Texas, covering an area three times larger than New York State.

For years, it was assumed that groundwater was virtually lifeless, primarily because it is cut off from supplies of oxygen. This is certainly the case with deep-lying aquifers, but the shallow groundwater beneath and alongside rivers is now known to support complex communities of small aquatic animals, from stonefly larvae to blind shrimps and primitive worms. This subterranean ecosystem may play a vital part in maintaining the health of rivers, serving as a refuge for creatures during times of drought, and supplying food for animals that live within the river. Stonefly larvae, for example, emerge from the groundwater to mature along the banks of rivers, providing food for salmon and other fish.

POLLUTION FROM THE PAST

Despite its importance as a water source, little care has been taken to prevent the pollution of groundwater. Much is now severely contaminated with agrochemicals and toxic wastes. Because the movement of pollutants through soils is so slow, the contamination we are picking up today results from the downward drainage of chemicals that were dumped or sprayed years ago. In effect, it is the first wave of pollution: worse is yet to come as our past catches up with us.

Once contaminated, groundwater is notoriously difficult - or sometimes impossible - to clean. The cool, dark, virtually lifeless nature of deep underground aquifers allows contaminants to be stored for hundreds or thousands of years. They cannot evaporate and, without bacteria, they cannot be biologically broken down. In Norfolk, England, groundwater contaminated with whale-oil in 1815 still contained residual toxic compounds when wells were dug there in 1950 - almost 150 years later. Indeed, in most cases, contaminated aquifers must be written off as future drinking-water supplies.

AGROCHEMICALS UNDERGROUND

In the pursuit of higher yields, farmers throughout the globe have applied millions of tonnes of artificial fertilizers and pesticides to the world's arable lands. In many areas, these chemicals are now polluting groundwater. In the US, some 50 million Americans are potentially at risk from pesticide-contaminated groundwater which is used for drinking. Two of the most pervasive of these pesticides, alachlor and atrazine, are known to cause cancer in laboratory animals, and are probably human carcinogens.

Nitrates are also a major contaminant of groundwater. In large doses they can cause illness in very young babies, and they are suspected of contributing to stomach cancer. The major source of nitrates in drinking water is artificial fertilizer, the use of which has doubled worldwide over the past 20 years. Leaking septic tanks, silage effluent and farmyard manure are also sources of nitrate pollution.

Nitrate levels in groundwater are steadily increasing. In Britain the number of drinking-water wells which exceeded recommended nitrate limits more than doubled between 1970 and 1987. In the Brie area of France, nitrate levels in the Petite Traconne spring, one of the many sources supplying Paris with drinking-water, have trebled over the past 30 years and regularly exceed official safety limits.

HIDDEN THREATS FROM TOXIC DUMPS

Every year, millions of tonnes of waste, from highly acid tars and cancer-causing solvents to pesticides and domestic rubbish, are generated by our modern industrial society. For the most part, this waste is dumped in landfill sites, a method of disposal which, at its crudest, consists of little more than tipping or pouring the waste into holes in the ground.

“...a group of Amsterdam municipal workers stumbled across some suspicious-looking drums during routine work... A little spadework uncovered no fewer than 10,000 drums filled with toxic waste. Many of the drums, containing residues from the production of pesticides, had rusted through and surrendered their contents to the mercies of percolating groundwaters... Meanwhile, reports of new finds became a daily occurrence. One began to wonder if there was anywhere where toxic waste had not been dumped... Within six months the inventory of illegal dumps had topped 4000... Clearly, the environmental control for the past decade was less than water-tight. But surprising as it may seem, Holland is undoubtedly one of the better countries when it comes to environmental protection. The extent of the toxic waste problem has come to light because the Dutch were plucky enough to ‘fish around in the drains’. And that raises the question of what might lie waiting to be discovered elsewhere. The lessons seem to be clear enough: seek and ye shall find.”

Graham Bennett, Consultant to the Institute for European Environmental Policy

“What I sometimes feel like doing, is enclosing a little packet of waste material with every one of the products we sell, just to remind people that they can't have one without the other.”

Chemical company waste disposal officer

Hundreds of thousands of such sites, many of them abandoned, now pockmark the landscape of industrialized countries. As rainwater percolates down through the waste in landfill sites, it picks up pollutants on the way, producing a highly toxic liquid known as leachate. It has been estimated that a typical dump, just 7 ha (17 acres) in size, produces over 20 million l (4.5 million gallons) of leachate a year, and will continue to do so for perhaps 50 years, long after the dump has been closed and its surface covered with soil.

Depending on what has been tipped into the landfill, leachate may be contaminated with heavy metals such as cadmium and lead, solvents, ammonia, phenols, PCBs, cyanide and numerous other chemical compounds. Leachate from domestic landfills can be just as polluting as that from chemical waste dumps. In particular, the prodigious quantity of organic waste in domestic dumps - food scraps, old newspapers and the like - produces a leachate so rich in nutrients that it can cause algal blooms if it leaks into streams or rivers. When the algae die, their decomposition rapidly removes oxygen from the water, killing fish and other animals as well as plants. Discarded household chemicals and consumer goods add toxic chemicals to the already dangerous mixture.

The vast majority of landfills operate on the principle that “dilution is the solution to pollution”. On most sites, no attempt is made to contain leachate within the site. On the contrary, it is encouraged to percolate downwards into underlying soil. The justification for this practice is that pollutants within the leachate will be rendered harmless, either by bacteria breaking them down, or because they become attached to soil particles and thus immobilized. These processes are known collectively as attenuation. To an extent this does happen, but the capacity of soils to attenuate pollutants has been greatly overestimated, with the result that much groundwater is now contaminated with a witch's brew of highly toxic chemicals.

In the US, some 160,000 old chemical waste dumps are thought to pose a threat to groundwater. At one illegal dump in Kentucky, known as the ‘Valley of the Drums’, 100,000 drums of chemical waste were piled on top of each other and abandoned, their deadly contents spilling out into a local creek. Ten thousand sites are now classified as being in need of urgent remedial action, and the total cost is expected to exceed $100 billion. According to the US Environmental Protection Agency, at least 1 per cent of all usable groundwater in the US is now contaminated by toxic waste. Others put the figure as high as 20 per cent. In Michigan, for example, officials now admit that they face a virtual explosion of contaminated aquifers. In New York, half the landfills in the state are now known to be leaking.

Europe has fared no better. In West Germany, 35,000 problem sites exist, while in Denmark, 2,000 are thought to have contaminated groundwater. In Britain, the Department of Environment acknowledges that 10 per cent of aquifers could be contaminated with by industrial solvents, many of them carcinogens, exceeding safety levels set by the World Health Organization (WHO). Some 1,300 dumps have been identified as a threat to groundwater. Traces of trichloroethylene, a carcinogenic solvent, have been found in 61 out of 168 aquifers used for drinking water in Scotland.

A CHEMICAL TIME-BOMB

Concern over the dangers posed by landfill sites has led several countries - notably the US, West Germany and Holland - to tighten the rules governing disposal of wastes in landfill sites. In the US, for example, all new landfills are required to have a lining of clay or some other impermeable material in order to prevent leachate from seeping into groundwater. But such containment sites have not always proved effective. Some chemicals - particularly industrial solvents - rapidly destroy the liners, reducing even thick layers of clay to sieves. Indeed, the Environmental Protection Agency now admits that all liners eventually leak. The US has now banned the land disposal of liquid wastes, and landfilling with untreated wastes is soon to be phased out.

Most countries, however, remain firmly committed to landfill. Moreover, despite the danger of groundwater pollution, unlined sites remain the rule rather than the exception. Britain in particular has strongly resisted tighter control on landfill: more than half of the country's domestic waste tips and almost all its toxic waste tips are unlined. The lax controls over landfill have encouraged countries with more stringent regulations to export their wastes to Britain - wastes which they would not be permitted to bury in their own countries.

In France, it has been estimated that more than three-quarters of the poisonous wastes that are generated every year are either exported to countries where waste disposal is less strictly controlled, or are simply dumped in old quarries, gravel pits, ravines and fields, and even in ordinary municipal rubbish tips. Dumps contaminated with poisons such as pesticides and solvents have been cleaned up simply by covering them with soil. At many known sites, and doubtless at many more waiting to be discovered, toxic chemicals are leaking into groundwater. After 700 tonnes of waste containing the highly dangerous insecticide lindane were dumped into a gravel pit near Colmar on the German border, locally produced milk and potatoes were found to contain lindane's active ingredient, a chemical known as HCH. The levels of HCH were 9,000 times higher than the WHO recommended limit.

Many developing countries have become dumping grounds for wastes from industrialized countries, 20 million tonnes of waste being shipped to the Third World every year. At one notorious site in Nigeria, 3,500 tonnes of toxic chemicals were found to be leaking from over 8,000 rusting and corroding drums, poisoning both soil and groundwater.

◀ Slurry escaping from silage-making bags on a farm. Like the slurry from intensively reared animals, this liquid is rich in nitrates which are now contaminating groundwater as well as streams and ponds.

▼ Drums of chemical waste awaiting disposal. Tipped into waste dumps - whether legally or illegally - such drums are as ineffectual as paper bags in retaining the deadly wastes within. Rain rusts them from the outside, the corrosive chemicals eat them away inside, and within a few months or years the contents escape.

WHEN THE WELLS RUN DRY

Every year, more and more water is needed, particularly for industry and irrigated agriculture. This had led to massive overexploitation of groundwater in many parts of the world. As a result, wells are drying up, villagers are left without water and irrigated land is being forced out of production.

In India, groundwater levels in many states have declined by 5-10 m (16-32 feet). In the state of Maharashtra, some 23,000 villages are now without water while in Gujarat, the figure is 64,500 villages. The decline is due in large part to the 'Green Revolution' *(see p.109)* which encouraged farmers to adopt new hybrid varieties of wheat and rice that require far more water than traditional varieties. This has massively increased the area under irrigation, from about 50 million ha (120 million acres) after the Second World War to 250 million ha (620 million acres) today. The introduction of water-intensive export crops such as sugar-cane has also put enormous strain on water supplies, sugar-cane consuming 10 times more water than wheat. Widespread planting of eucalyptus under the Indian government's Social Forestry Programme has also played a part. Eucalyptus draws up large quantities of water from the soil and wherever it has been planted the water-table has fallen dramatically.

The destruction of traditional methods of storing rainwater has exacerbated the problem of groundwater depletion. For hundreds of years, Indian villagers stored rainwater and run-off from streams in large tanks. One tank was connected to the next through a series of channels, ensuring that the overflow from tanks on higher ground was collected in those lower down. The water stored in the tanks replenished groundwater, and was used for irrigation.

The introduction of commercial cash crops, however, dramatically changed farming patterns throughout rural India. In particular, agriculture ceased to be a communal activity and the co-operation necessary to keep the tanks maintained began to break down. Many tanks have now silted up, with the result that rainwater is lost as run-off and groundwater is no longer replenished from this source. Meanwhile the extraction of groundwater has increased.

The fall in water-tables is proving ruinous for the bulk of India's farmers. Traditional methods of extracting groundwater for irrigation only work if the water-table is higher than 7 m (23 feet) beneath the surface. Below that depth, expensive tube wells and mechanical pumps are needed. Consequently, only the richer farmers can continue to irrigate their land, the poor farmers being reduced to the status of landless labourers or migrant workers. Many end up in the slums of the nearest city.

Groundwater depletion is not a problem restricted to the Third World. In the US, one-fifth of all irrigated land is irrigated by groundwater extracted from the Ogallala Aquifer, a body of water equal to that in Lake Huron. It took nearly half a million years for that groundwater to accumulate: at current rates of extraction it will be depleted within 25-50 years.

As aquifers become increasingly robbed of water, so the land above them starts to compact. The result is widespread land subsidence, gaping holes appearing in some areas. In California, over 13,000 km^2 (5,000 square miles) of the San Joaquin Valley are affected. Around Mexico City, where groundwater has been exploited for drinking and for industry, some land has subsided by 9 m (30 feet) or more.

Calls by environmentalists to limit the abstraction of groundwater in the US have fallen on deaf ears. In the San Joaquin Valley, one of the major areas of irrigated farming in the US, the rate of groundwater pumping now exceeds replenishment by more than 2.3 trillion litres (500 billion gallons) a year, and by the end of the century that figure is expected to double. Much of the land is farmed by large corporations - Exxon, Getty Oil and Texaco to name but a few - and lobbying has ensured that virtually no controls exist over the amount of groundwater they are allowed to extract.

In the American southwest, those states dependent on the Ogallala Aquifer have taken a deliberate policy decision to continue present rates of extraction, insisting that groundwater left in the ground is water wasted. The prospect of the Ogallala running dry does not disturb them: by then, they argue, alternative water supplies will be available as a result of river diversion schemes. Such diversion schemes will cause tremendous ecological damage, however, and are also likely to prove too expensive to build. The chances are that, sooner or later, the arid southwest of the US will simply run out of water. Here, as elsewhere, the pollution and waste of water, our most precious resource, is one of the most striking examples of how industries and governments put short-term gain before our long-term survival.

▲ Irrigated farming on a grand scale. Throughout the world, groundwater is being steadily depleted by such schemes. When the water runs out, whole areas will have to be abandoned.

◀ A traditional irrigation system in West Africa. Using the muscle-power of man, ass or bullock alone, water can be raised in considerable volumes. However the amount that can be obtained from below 7 m (23 feet) is very limited. This has prevented groundwater from being over-exploited in the past.

▼ An irrigated garden with date palms in the mountain area of Oman. Date palms have deep roots which can tap into aquifers up to 10 m (33 feet) down. This allows the trees to grow without irrigation, but can have serious consequences if there are too many palms. During the past century, thousands of new palms have been planted along the edge of the Sahara in Tunisia, causing the water-table to drop by 5 cm (2 inches) a year.

WETLANDS and MANGROVES

Throughout the world, in innumerable different cultures, there have been myths and legends about cranes. Their sonorous, haunting calls, their elaborate dance displays and courtship ceremonies, and their elegant silhouettes in flight, have inspired humanity for many thousands of years. These most graceful of birds depend almost entirely on wetlands for their survival.

Sadly, cranes are today one of the world's most endangered families of birds. The whooping crane of North America hovers on the brink of extinction, while other species, including the Siberian and Japanese cranes, are almost as rare. The precarious state of these magnificent birds bears witness to the massive destruction of wetlands around the world in a way that no statistics ever could. For often wetlands are small, unremarked areas, geographically too insignificant for their destruction to be recorded. Only the more extensive areas, such as the threatened Okavango Delta in Africa, or the Florida Everglades, receive much public attention. Meanwhile smaller marshes, ponds and swamps vanish in the name of agriculture, construction or other forms of development, or are poisoned by pollutants, leaving birds like cranes with no refuge.

FENS, CARRS AND SWAMPS

The English language is rich in words for different types of wetland, an indication of their great diversity. A mire is an area of land that is waterlogged but contains no open water; these are found particularly in cooler regions of the world. Wetlands which owe their watery state to groundwater (*see p.143*) are known as fens, while those fed primarily by rain are called bogs. Wetlands can be covered by scrub or trees, a type of habitat known as carr in Britain, but as swamp in North America.

◀ The Okavango Delta in Botswana - a haven for wetland wildlife on the edges of the Kalahari Desert. In this inland delta, the waters of the Okavango River flow around countless reed-flanked islands before sinking into the sandy soil.

Here the unsettled soil beneath the trees causes their trunks to lean this way and that, producing 'drunken forests'. Along coasts and estuaries, wetlands with brackish water occur. These include saltmarsh, with its hummocks of compact salt-tolerant herbs, and the dense stands of reedswamp, where the tall, feathery seed heads of reeds sway with the wind from the sea.

Tropical areas have their own characteristic wetlands, reflecting the influence of temperature. The best known of these is the mangrove swamp *(see p. 160)*, which flourishes along estuaries and in swamplands intimately connected with the sea. Papyrus swamps, dominated by the tall papyrus reed, are characteristic of the edges of slow-flowing rivers and other marshy areas in Central Africa.

Several factors determine the type of plants that dominate a wetland. Temperature and salinity (the amount of salt) are important, but most crucial of all is the level of the water. Does it cover the soil or lie just below it? And does it drop markedly during the dry summer months, or does it keep close to the surface? Some wetlands, including reedswamps and the swamp-cypress forests of the Everglades, have a permanent though shallow cover of water throughout the year. As an adaptation to such conditions, swamp cypresses send up knobbly aerial roots to obtain oxygen from the air, this being scarce in the stagnant water below.

Other wetlands are submerged completely for part of the year but then dry out. The varzeas, which support a rich forest in the whitewater floodplain regions of Amazonia, are wetlands of this type, as are the igapo forests that grow in the blackwater floodplains of the same region. Igapos are distinguished from the varzeas by the absence of silt in the water and consequent lack of fertility. Their dark brown colour is due to tannins in the acidic water, these being derived from leaves that have not decomposed completely. The tannins, in turn, inhibit decomposition.

WETLAND WILDLIFE

Where land is permanently waterlogged and the water static, peat may build up, radically changing the nature of the habitat. Peat is a brown, usually acidic material, made up of the compressed remains of plants which have not decomposed. In wetlands, bacteria and fungi may have difficulty in breaking down plant remains, because the oxygen in the water is quickly used up, and without oxygen most decomposers cannot function.

The steady accumulation of peat raises the level of land, which eventually makes it less wet and more suitable for other plants. Thus the vegetation may gradually change, a process known as succession *(see p. 59)*. Peat formation also leads to a lack of fertility, because the nutrients from past generations of plants are not returned to the soil. Under such circumstances, some plants have evolved unusual means of satisfying their nutritional needs. Insect-eating plants, such as the pitcher plants (*Sarracenia* species), the Venus fly-trap (*Dionaea muscipula*) and sundews

▲ Swamp cypresses, festooned with Spanish moss, in the bayous of Louisiana. Like mangroves *(see page 160)*, swamp cypresses have special roots which help them to 'breathe' in waterlogged ground. These trees are typical of the bayou country - an extensive area of wetland that has formed where the Mississippi meanders its way to the sea.

► An eastern long-necked turtle from Queensland, Australia, peers through a blanket of waterweed. Australia's wetlands are particularly rich in reptiles, including turtles, crocodiles and aquatic snakes.

▼ A flock of egrets feasting on stranded fish in the Pantanal, Brazil. The Pantanal is a seasonal wetland covering an area of over 12 million ha (30 million acres). Every year, rains swell the rivers that flow through the Pantanal, flooding the flat ground. Gradually it dries out, forcing many fish into smaller and smaller pools where they make an easy meal for birds.

▲ An aerial view of the Camargue, the extensive wetland in France which has formed where the River Rhône reaches the Mediterranean. The Camargue's soil, like that of most deltas, is very fertile, and much of it has been drained for agriculture. The remaining areas of the Camargue that have not been drained are internationally important as a refuge for birds, particularly flamingoes.

(*Drosera* species) are all wetland plants which supplement the mineral intake of their roots with precious nitrogen extracted from the bodies of insect prey.

By contrast, those wetlands which are flooded annually, or fed by flowing streams or tides, are often extremely fertile, being continually topped up with nutrient-rich silt by the floodwaters. In the tropics, this high fertility combines with abundant moisture, sunlight and warmth to create the most productive natural habitats on Earth. In the papyrus swamps of Uganda, the rate of biological productivity is twice that of tropical rainforest, and some temperate wetlands also support luxuriant growth.

Rather than being locked into a massive stand of trees, the sugars made by wetland plants and algae during photosynthesis pass quickly through the food chain to support an abundance of animal life. The dense flocks of birds feeding in wetlands are largely there because of the generous supplies of worms, shellfish and other animals in the water. The Camargue in France, for example, provides a unique environment for flamingoes and avocets, while on the Austrian-Hungarian border, as many as 10,000 geese may be in view at any one time during the winter, in the marshlands of the Neusiedler reserve. Nor are birds the only beneficiaries of the biological wealth of wetlands. The Okefenokee Swamp in Georgia, with its swamp-cypress forests and flooded meadows, contains a rich fauna of turtles, snakes and alligators. In India, wetlands alongside the Brahmaputra contain rhinoceroses, tigers, marsh deer, elephants, crocodiles and water buffaloes, as well as wood storks, ibises and cranes.

THE VALUE OF WETLANDS

Because of their high productivity and their role as stepping stones for migratory species, particularly birds, the world's wetlands contribute in untold ways to maintaining the health of numerous ecosystems. Indeed, the continual movement of animals into wetlands and then beyond in their migrations, acts as an effective means of dispersing the rich accumulation of nutrients into less productive areas.

◀ Okefenokee Swamp, in the state of Georgia, forms part of a great band of low-lying and waterlogged ground that stretches down the east coast of the US. The surface waters of Okefenokee Swamp lie over a thick layer of woody peat, which has been created from the accumulated remains of swamp cypresses. The swamp covers an area of over 1,650 km^2 (640 square miles), much of which is a protected wildlife refuge.

▼ Fish-eating birds in the Everglades, Florida. The three species shown here all fish in different ways. The great egret, in the centre, stabs fish with its beak and then swallows them. The brown pelicans, which flank the egret, trap fish in a scoop formed by elastic skin that joins the two halves of their lower beak. The double-crested cormorant, in the background, catches fish by pursuing them underwater.

Consequently, the destruction of wetlands has wide repercussions. Almost the entire population of North Sea herring is dependent, at some period during its life cycle, on the coastal wetland associated with the Wadden Sea, now seriously threatened by pollution. Indeed, it is believed that two-thirds of the world's fisheries are directly dependent on the fertility of coastal wetlands. In this respect, wetlands are particularly important in the Third World as food sources. In Africa, for example, one-fifth of all the animal protein consumed by local people comes from fish dependent on wetlands.

By quickly absorbing water, which is then released gradually over time, wetlands, particularly marshes and swamps, also play a vital role in preventing flooding and recharging local streams and groundwaters. Moreover, wetlands are extremely efficient water purifiers - so much so that one ecologist has referred to Africa's papyrus swamps as "natural septic tanks". Nutrients, such as nitrates and phosphates, are quickly taken up by plants and thus removed from the surrounding water, preventing them from causing a damaging overgrowth of algae. One wetland on Long Island, New York, was found to remove 90 per cent of the nitrogen compounds in the water flowing into it.

Wetland plants also absorb many other pollutants - pesticides and heavy metals for example - although there is a limit to the pollution that a wetland can tolerate before its cleansing properties are overwhelmed. There is also the danger that animals higher up the food chain may be poisoned, owing to the accumulation of pollutants in the animals they feed upon *(see p.33)*. Economists in the US have calculated that the water treatment and other functions carried out for free by wetlands are worth some $400,000 per ha ($160,000 per acre) - the amount it would cost to perform the same functions using man-made treatment plants.

LIVING WITH WETLANDS

In many parts of the world, people have learned to exploit the natural productivity of wetlands without destroying the ecosystem. In and around Lake Xochimilco in Mexico, farmers still practice an age-old method of wetland agriculture, first developed by the Aztecs, but now under threat from drainage schemes and the expansion of Mexico City. The farmers pile up layer upon layer of vegetation and silt to build rectangular garden plots, known as chinampas, raised several centimetres above the water level. The organically-rich silt from the bottom of the lake provides ample nutrients and the chinampas are renowned for their productivity. Until the 1950s, the chinampas produced almost all of Mexico City's vegetables.

Today, the chinampas cover a fraction of the area they did under the Aztecs: only 200 ha (about 500 acres) compared with 20,000 ha (about 50,000 acres) that existed when the Conquistadors first invaded Mexico. Indeed, the future of the chinampas looks bleak. Much of Lake Xochimilco has been drained, both for flood control and to provide land for urban expansion. Meanwhile, water levels in the lake are falling owing to deforestation in the catchment area and the use of groundwater to supply Mexico City's population. To add insult to injury, the discharge of raw sewage into the lake is rendering crops unfit to eat.

Other traditional forms of wetland agriculture are also under threat. For thousands of years, the annual floodwaters of the River Nile brought water to peasant farmers of the Nile Delta. As the river began to rise in the late summer, vil-

▲ Swamp cypresses in the Everglades, Florida, with epiphytic plants growing on their trunks. The Everglades National Park covers over 5,000km² (2,000 square miles) of swamp and lakes at the southern tip of Florida. The park attracts many tourists every year, but the annual influx of so many visitors creates problems for the region's wildlife because it increases the demand for water. Without abundant water, the Everglades could not survive.

▲► Rows of tomato plants growing on floating gardens on Inle Lake, Burma. The combination of a tropical climate and abundant water makes this sort of agriculture highly productive. Unlike drainage for dry-land farming, it causes little damage to the wetland habitat.

◄ A coastal marsh at Lantau Island, Hong Kong, earmarked as a potential site for an international airport. Throughout the world, flat coastal wetlands like this are vulnerable to many forms of development, from airports and industrial complexes to docks and housing.

lagers would channel the floodwaters into large basins, releasing it onto their fields where it would be allowed to stand for 40 days or more, saturating the parched soils. Once the floodwaters had receded, the excess water from the fields was returned to the Nile, allowing the land to produce its rich harvest. The floodwaters deposited millions of tonnes of natural fertilizer in the form of silt onto the land, in addition to washing out salts that had accumulated in the soil over the previous year. It was an eminently sustainable system and one which supported the great Pharaonic civilizations of ancient Egypt. Today, however, the annual flood is checked by a series of dams along the Nile, in particular the Aswan Dam. The river's silt never reaches the fields of the delta but accumulates behind the dams, clogging up reservoirs. To make up for the lost fertility, Egyptian farmers must apply artificial fertilizers, at great cost. To make matters worse, the salts are no longer flushed from the soil, causing much land to become salinized *(see pp.111-112)*. Some suggest that the delta's fertility will only be restored by allowing the Nile its annual flood.

Other seasonal irrigation systems in wetland areas - from Senegal to Southeast Asia - have been similarly undermined, and still more are in jeopardy as new dams are built *(see pp.130-136)*. Happily, however, a few governments have recognized the importance of maintaining seasonal floods. In Botswana, the Department of Water Affairs recently abandoned plans to use the waters of the Okavango Delta for irrigating new areas. Formed by the River Okavango as it spills over the Kalahari Desert, the Okavango Delta - or swamp - has been described as "Africa's largest and most beautiful oasis". Instead, they have chosen to build upon traditional methods of flood irrigation, called *molapo* farming.

In the Sudd, the vast swamps that stretch across southern Sudan, tribal pastoralists such as the Nuer and the Dinka have developed a migratory way of life that is in perfect harmony with the floodplains. At the end of the rains, the grasslands which surround the floodplains are burned, leading to a flush of growth. As the grasslands dry out, and the floodwaters begin to recede, the people and their vast herds of cattle move to the permanent floodplains nearest to the rivers that meander across the swamp. Here rich *toich* grass flourishes even in the dry season, providing ample grazing for the cattle. With the return of the rains, it is time to move again, first to the grasslands that border the floodplains and then to the higher ground. During the rainy season, agriculture is practised.

Since the mid-1970s, a 350-km (217-mile) canal through the Sudd - the Jonglei Canal - has been under construction, although work is at present halted owing to the war in the southern Sudan. The canal is intended to increase the flow of water into the Nile, much of it at present evaporating as it crosses the Sudd. The canal, which is three-quarters finished, would fatally disrupt the traditional pattern of life in the Sudd and would do severe damage to its unique wetland ecosystem.

ROBBED OF WATER

Wherever wetlands are deprived of water for a significant length of time, their future is under threat. Vast areas have already been drained to reclaim the land for agriculture or urban expansion. The alluvial forests of the Rhine, for example, are less than a tenth of their extent prior to the Second World War, while the US has drained more than half of its wetlands, primarily for agriculture.

Much of the drainage has been encouraged by government subsidies or by loans from the international development banks. In 1984, 30,000 ha (74,000 acres) of tropical wetlands in South Sumatra were drained under a World Bank resettlement scheme, and in 1983 the World Bank loaned billions of dollars to Mexico for a massive drainage scheme in the state of Chiapas. Nearly a fifth of the wetlands that have been identified as internationally important in Central and South America are threatened by direct drainage for farming or ranching.

As reclaimed land is built upon, or brought into agriculture, there is inevitably pressure to protect it from the floods that previously nurtured the wetland. Such flood control programmes can pose as great a threat to remaining wetland habitats as the original reclamation schemes. Since the 1920s, thousands of hectares of the Florida Everglades have been drained, transforming the swamp into some of the richest agricultural land in the US. To protect that land, a vast and intricate network of canals, levees, tidegates and floodgates have been built, severely disrupting the flow of water into the remaining swamps and marshes. By the late 1960s, the area was so deprived of water that it looked possible that the entire Everglades National Park would dry out. Although the park is now guaranteed nearly 400 million m^3 (315,000 acre-feet) of water a year, the water is frequently released into the park with little regard for nature's cycles. In 1983, the floodgates were opened at precisely the time that the swamp's alligators were nesting, normally a dry period: a third of the alligator nests were destroyed by the deluge.

▲ Mechanized peat extraction in Pollagh Bog, Ireland. Peat in 'raised bogs' such as this can be as much as 10 m (33 feet) deep, and is built up over thousands of years through the accumulation of dead plant remains. Many of Ireland's finest raised bogs have been all but stripped of their peat. At one time, peat was dug by hand, but today mechanical cutters slice through it, turning it over to dry out in the air.

The growing demand for water - for industry, agriculture and domestic use - poses a further threat to wetlands, as more and more water is extracted from them. One area particularly under threat is the Okavango Delta. To the south of the delta, at Orapa in Botswana, lies a rich seam of diamond deposits. To supply water to the diamond mines, one of the Okavango's rivers, the Boro, has already been dredged in order to increase its flow, causing large areas of the Okavango to dry out. But the Boro is still an unreliable source of water and there are plans to further increase its flow to safeguard the mines and the growing city of Maun against water shortages. A larger area of the lower delta, including the feeding grounds of waterbirds, is likely to be affected. Fish movements and breeding grounds would also be disrupted. Further threats to the Okavango come from a livestock scheme, backed by the World Bank, which threatens large areas with drainage.

WETLAND POLLUTION

Drainage and flood-control programmes have long been a threat to wetlands. Chemical pollution is a more modern threat. In the Okavango, for example, a campaign to eradicate the tsetse fly has caused large areas to be sprayed with a lethal cocktail of pesticides. The long-term effects on wildlife are not known. Other wetlands are threatened by fertilizer and pesticide run-off from agricultural land, and from the discharge of industrial waste.

One area where wildlife has been severely affected through pollution is the Kesterton National Wildlife Refuge and its surrounding marshlands in California. Until the mid-1980s, contaminated drainage water from the irrigated farmland of California's Central Valley was pumped into 12 evaporation ponds within the reserve. The ponds now contain high levels of selenium, salt, pesticides and other toxic chemicals. Selenium contamination has been blamed for the mass poisoning of wildfowl in the refuge. Deaths and deformities among wildfowl in adjacent wetlands have also been reported.

DIGGING UP PEAT

Peat, cut into brick-sized blocks, burns well when dried and has been used for centuries as a source of fuel, particularly by small farmers: until recently, there was scarcely a household in the west of Ireland that did not rely on peat for heating and cooking. But the demands of powering national electricity grids are now causing massive damage to peatlands.

Worldwide, the tonnage of peat dug for fuel doubled in the 30 years between 1950 and 1980, some countries such as Finland burning 90 per cent of the peat cut. In the tropics, many countries where fuelwood is in short supply are being encouraged to exploit their peat resources, which in some instances (China and Indonesia, for example) are considerable.

Peatlands are also being lost as a result of the boom in horticulture, the peat being used as a rich source of fibrous matter (humus) which improves the soil or which can be used in potting compost. Ireland is a major exporter of peat,

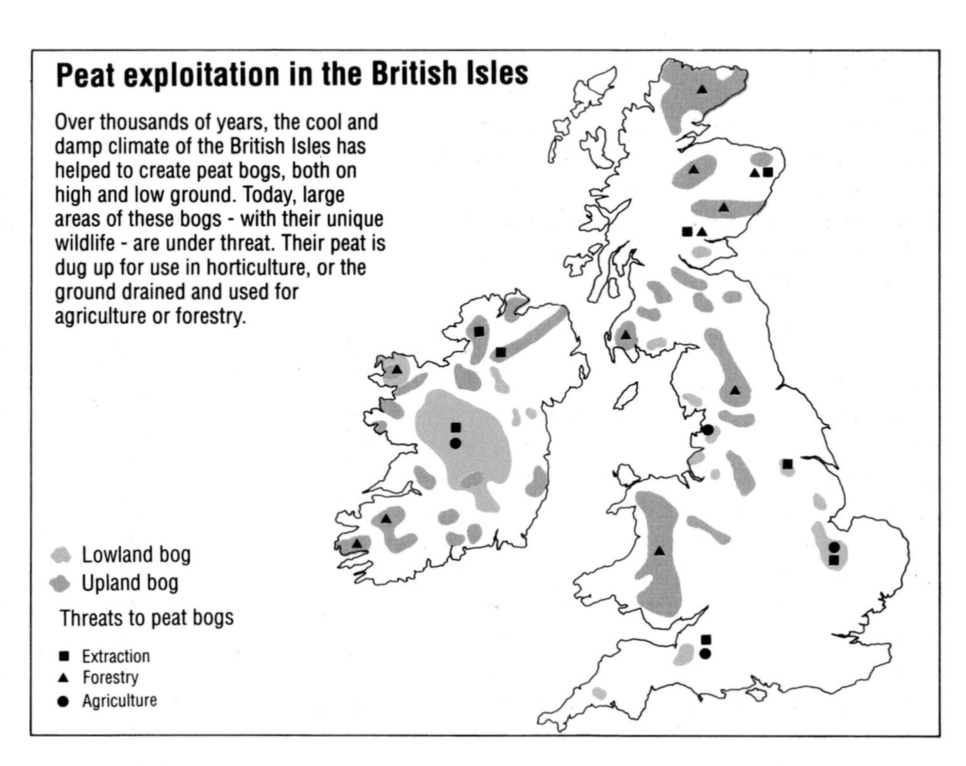

Peat exploitation in the British Isles

Over thousands of years, the cool and damp climate of the British Isles has helped to create peat bogs, both on high and low ground. Today, large areas of these bogs - with their unique wildlife - are under threat. Their peat is dug up for use in horticulture, or the ground drained and used for agriculture or forestry.

and even sites designated as 'Areas of Special Scientific Interest' have been dug up to satisfy the export market. Compost, which can be made from fruit and vegetable wastes, dead leaves, straw and other organic matter, could well serve as a substitute for peat. Yet the vast majority of the organic waste that is generated in peat-importing countries is simply allowed to rot in rubbish dumps, indiscriminately mixed with plastics, metal, glass and other types of waste which will not rot down.

At a time of fears over global warming *(see pp.40-53)*, the loss of peatlands is of particular concern. Active peat bogs are efficient 'fixers' of carbon, locking it up in the soil. As peatland is dug up or dried out, its store of carbon is released into the atmosphere, adding to global warming. It is a double blow, since the peatland no longer exists as a sink for carbon.

CONSERVING WETLANDS

For many years, the value of wetlands remained unrecognized. In 1971, however, the urgent need to protect the remaining wetlands of the world led to the setting up of the Ramsar Convention, with the express aim of conserving the most important wetland sites. Wetlands are the only ecosystems to have their own international convention, a measure of the severe threat posed by their worldwide destruction.

But despite the good intentions, Ramsar has been hampered by lack of funds and political

"Wetlands are wastelands; that, at least, is the traditional view. Words like marsh, swamp, bog and fen imply little more than dampness, disease, difficulty and danger... Nothing could be further from the truth. Far from being wastelands, they are among the most fertile and productive ecosystems in the world."

Dr Edward Maltby, University of Exeter

will. In 1988, almost two decades after it was drawn up, it still had no permanent secretariat and no regular source of funding. Of the 357 wetland sites that were supposedly protected under the convention - covering an area the size of Wales or just 2.5 per cent of all wetlands in the world - at least 20 were listed by the World Wide Fund for Nature as threatened. Moreover, many countries with major wetlands (Botswana and Indonesia, for example) are not signatories to the convention.

Without international agreement, it is unlikely that wetlands can survive. The dynamic nature of wetlands - with small lakes giving way to reedswamps, then to marsh, then carr woodland, and finally dry land through the process of ecological succession - means that they are constantly changing. Before the widespread impact of man, new wetlands would have regularly opened up - due to the changing flow of a river for example - replacing those lost by succession or changes in drainage patterns. Animals, such as migratory birds, or breeding frogs and dragonflies, would have switched to new wetlands as old ones disappeared. In our modern, human-dominated world, however, where rivers and coastlines are constantly being moulded to suit our whims, the natural recruitment of new wetlands is lost, depriving wildlife of a vital habitat and local people of an often critical source of food. It is time to put an end to the destruction - and to take active care of the remaining wetlands of the world, for they are among our most precious natural resources.

MANGROVE SWAMPS

Cienaga Grande, Colombia's 'Great Swamp' region, has for centuries provided some of the richest fishing grounds in the entire country. Until about 20 years ago, local people could make a reasonable living from the harvest of fish and shrimps. Even today, a cursory look at the swamp could lead one to think that all was well. The fishermen are still there, plying their nets from their flat-bottomed boats that glide silently over the still waters. But, year by year, the swamp is deteriorating. The mangroves are being logged and cattle ranching is causing more and more water to be taken from the swamps. The delicate natural balance between the ebb and flow of saltwater and freshwater has been disrupted, and the Cienaga Grande is slowly being transformed into a putrefying morass.

Mangrove swamps cover some 15.8 million ha (39 million acres) worldwide, the largest concentrations being in tropical Asia. The name 'mangrove' covers any species of tree that can live partly submerged in the relatively salty environment of coastal swamps, some species being more salt-tolerant than others. About 20 different families from 12 orders are considered mangroves. Among them, *Rhizophora*, with their reddish bark and breathing roots, are probably the best known and the most ubiquitous of mangrove species.

Mangroves do not require a salty environment for growth. On the contrary, they often do better when the salt content is low, but under those conditions they tend to be ousted by other species of tree that grow more vigorously. Their ability to tolerate salty water is the secret of their success, for it allows them to survive where other trees cannot. They cope with the briny environment around their roots in various ways. In some species, salt is removed through being excreted from special glands, or is isolated in inactive tissues within the plant.

The shifting silt bed of mangrove swamps is lacking in oxygen and provides little in the way of solid ground. To overcome this, mangroves

▶ Mangroves growing on the coast of Queensland, Australia. At low tide, the mangroves' pneumatophores, or 'breathing roots', protrude like miniature spires from the damp sand. The tree in the centre is surrounded by a network of stilt roots, which anchor it in the shallows. Mangroves spread both by sending up shoots from their roots, and also by seed. In some species of mangrove, the seeds begin to germinate while still attached to the tree. Each one produces a long, sword-shaped root, which anchors the seedling in the mud when it eventually falls from the parent tree.

"The pressure on an area can grow progressively greater and, on the surface, everything is alright, until you reach a certain point after which the decline is extremely abrupt. We don't know how close we are to that limit, but there are a lot of naturalists who think we are very near it."

Dr Jesus Casas, Conservation Director, Coto Doñana National Park, a vast wetland area in Spain under threat from a planned leisure complex on its boundaries

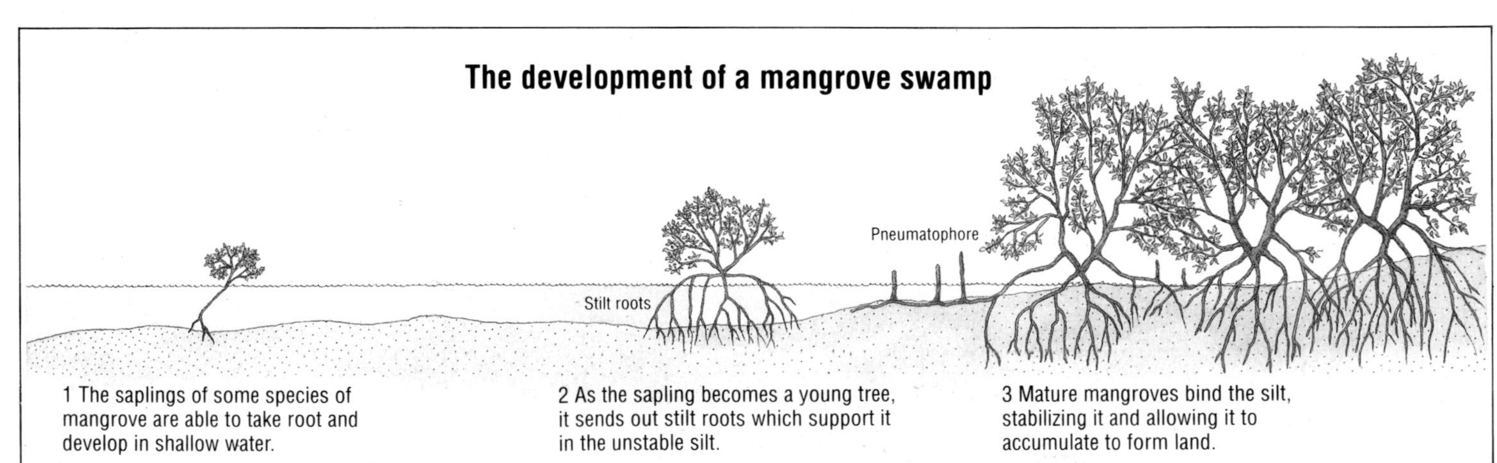

▲ A proboscis monkey with its main source of food - the leathery leaves of a mangrove tree.

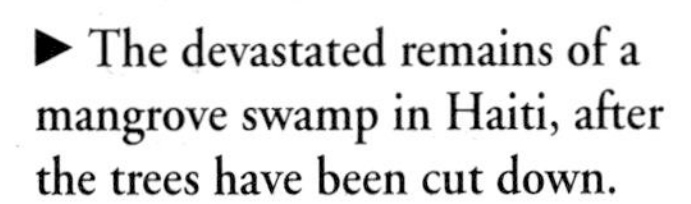

► The devastated remains of a mangrove swamp in Haiti, after the trees have been cut down.

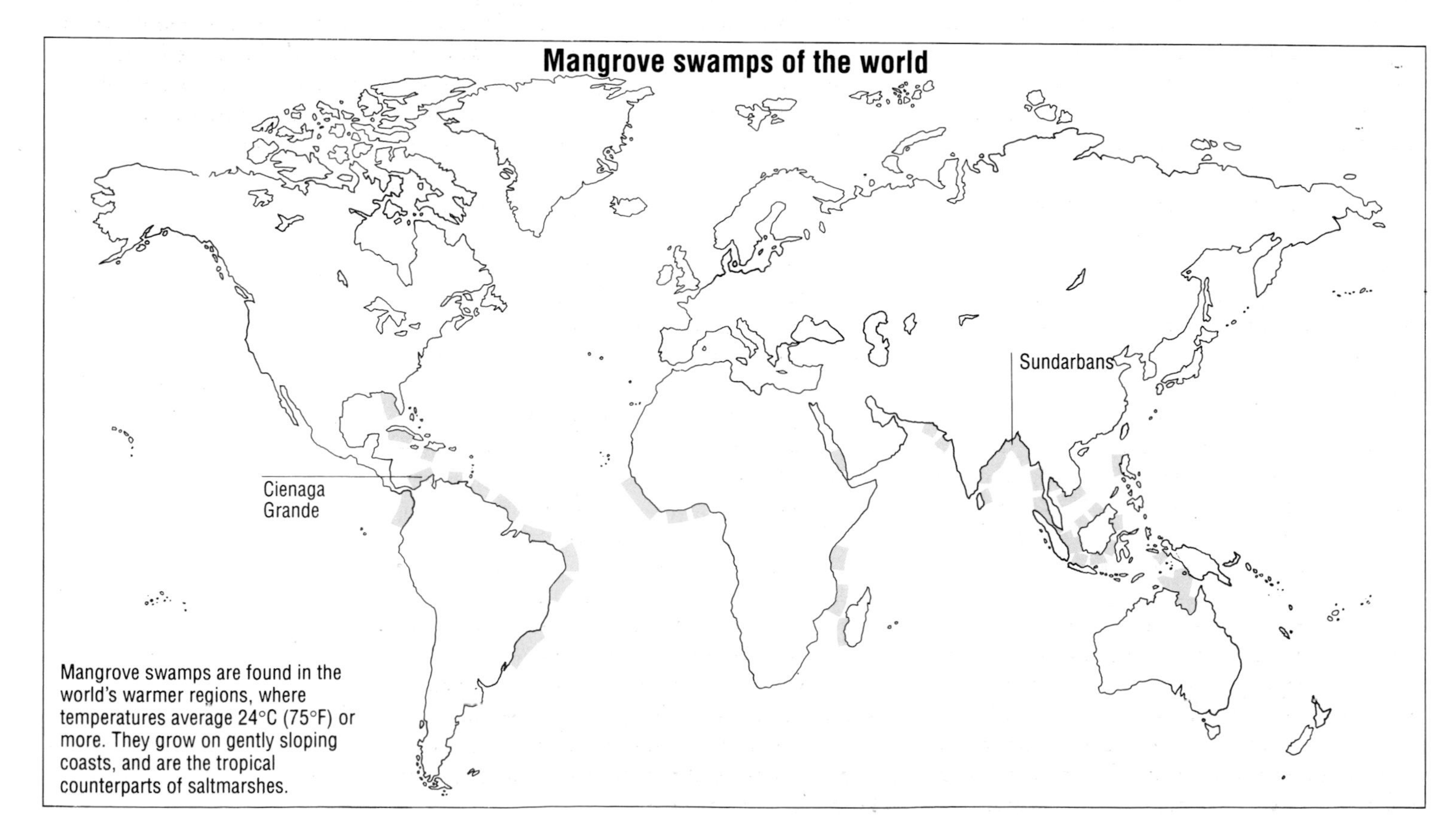

have evolved stilt roots, which act as buttresses, and pneumatophores or 'breathing roots', through which air is funnelled down to their roots which would otherwise be starved of oxygen. Mangroves also face another problem. Bacteria in the silt on which they grow decompose plant debris, deriving their oxygen from dissolved nitrates and sulphates, and converting the latter into toxic sulphides. To prevent themselves becoming poisoned, a mangrove's roots are sealed against the influx of minerals. Nutrients, including water, are taken up by fine roots above the silt layer.

In colonizing marginal coastal areas, mangroves stem the flow of tidal and estuarine waters, trapping silt and contributing to the build-up of mud banks. As the mud banks become free of water, the mangrove species are replaced by other vegetation. Consequently, over long periods of time, mangroves act as a major force in reshaping coastlines. Moreover, by creating a natural buffer against typhoons, cyclones and hurricanes, mangroves protect the coastline against erosion.

For centuries, or in some instances thousands of years, mangrove swamps have provided traditional peoples with a rich, highly productive environment. In the Cienaga Grande, for example, local people have long exploited the mangroves for wood, charcoal and tannin, the wood being used for construction of their houses and for firewood, the charcoal for tile and brick manufacture, and the tannin as a preservative for fishing tackle.

Equally important, mangroves provide a rich habitat for fish and shellfish, many of which depend on mangroves at some stage in their life cycles. In Bangladesh, where fish are a major source of protein, 80 per cent of fish catches in the Lower Ganges Delta come from the Sundarbans, the largest stretch of mangroves in the world.

Despite their importance for traditional coastal-dwelling communities, mangroves are frequently regarded as "wastelands ripe for development". Their reputation as inhospitable, disease-ridden swamps, full of lurking dangers, has encouraged their systematic destruction.

Since the mid-1960s, some countries have lost a half or more of their mangrove forests. Industrial logging has condemned thousands of hectares. In Indonesia, more than 2,000 km^2 (770 square miles) of mangrove are converted to woodchips every year, mostly for export to Japan where they are used to make rayon and pulp.

Fish and shrimp farming has also caused massive destruction. In Ecuador, 16 per cent of the total mangrove area has been cleared since the 1960s, its silt scooped out and massive ponds created for shrimp farms. In the Philippines, aquaculture development is accelerating at such a rate that few mangroves are likely to be left in a decade from now.

Dams and water development schemes *(see pp.130-136)* have also taken their toll, reducing the flow of rivers and allowing seawater to encroach further and further inland. Salt levels in many mangroves have become intolerably high, causing the trees to die. The Sundarbans have been particularly badly affected, salinity levels rising thirteen-fold after the building of the Farraka barrage on the Ganges.

Land reclamation for agriculture and urban development has caused further destruction. Pollution, too, has created problems. The coastal waters of the Straits of Malacca, one of the busiest shipping lanes in the world, are home to some of the lushest mangrove swamps on Earth. The mangroves trap the oil discharged from the tankers and other ships that ply the Straits: many are now poisoned beyond recovery.

The tragedy is that the commercial exploitation of mangroves, whether for logging or for shrimps, overlooks their real value. When mangroves are exploited traditionally, they can support at least 10 times as many people at a subsistence level than when they are exploited by commercial concerns. Moreover, the traditional use of the mangroves is one that has been shown to be sustainable on a long-term basis, unlike commercial exploitation for timber.

The United Nations, through its Development Programme and UNESCO, has now established a Regional Mangrove Project for Asia and the Pacific region. The project, begun in 1983, has 16 countries participating, the aim being to preserve still extant mangrove forests. Indonesia, in responding to the initiative, established a 200-m (650-foot) wide 'green belt' along its coastline and intensified work on the rehabilitation of degraded mangrove forests in Java and Sumatra. Two mangrove islands have also been designated nature reserves.

Despite those constructive initiatives, the Indonesian government is still committed to the conversion of 10 per cent or more of the remaining mangroves into fishponds, especially for prawns. There, as in other countries with mangroves, the notion of mangroves as 'wastelands' is taking a long time to die.

“... early mornings are noisy with the splashing of young Lechwe as they romp around, running and jumping through the shallows, leaving swirling patterns in mud that has the texture of churned grey oil paint. The Lechwe are soon surrounded by crowds of storks, ibises, egrets and Long-toed Plovers, while hosts of wagtails dance around in pursuit of mosquitoes. Carmine Bee-eaters come in after grasshoppers... At nightfall the Lechwe seek firmer ground or shallows where they are surrounded by the mumbling of crakes and piping frogs and insects. Long-tailed Nightjars chase moths where the pratincoles hunted by day, flying in and out of the laminated layers of mist and smoke that drift out from fires on higher ground. Here the Nuer and Dinka herdsmen talk and laugh, enveloped in the smoke from smouldering dung fires.”

Jonathan Kingdon on the Sudd wetlands, now threatened by the Jonglei Canal

COASTS and ESTUARIES

The fishing port of Grimsby lies on the mouth of the River Humber, one of Britain's fastest flowing rivers. Inshore fishing in the estuary was once a thriving industry. The river teemed with life. Cod and codling, skate, brill, turbot, plaice, soles, monkfish and herring were all caught in its waters. Salmon and sea-trout were also to be found on their way upriver to spawn, and shrimps, crabs, whelks, cockles and mussels were plentiful.

Today, after 40 years of industrial development, the fishing industry is all but dead, destroyed by pollution from the chemical factories that line the banks of the Humber. As in so many estuaries in the industrialized world, chemical companies were attracted to the area because the estuary provided a cheap and convenient sink for the disposal of waste.

THE NATURE OF ESTUARIES AND COASTS

An estuary is the area at the mouth of a river where its freshwater meets the saltwater of the open sea. Estuaries are ever changing their form as they ebb and flow with the tide, never quite a river and not yet the sea.

Each estuary has its own specific characteristics, depending on the size of the river system, the lie of the land at the river mouth and the type of terrain through which the river has flowed. Where the land is flat, as at the mouth of the River Thames, the estuary may meander over mudflats and saltmarshes for as far as the eye can see. If, as in the Rhône and Mississippi, large quantities of silt are brought down by the river, and there are no offshore currents to disperse it, deltas form, the silt building up to form a fan-shaped plain. In complete contrast are the deep, narrow estuaries, which lie in steep-sided valleys, known as fjords.

◀ A weather front blows in over an Irish estuary. On shallow, protected shorelines such as this, the waves lap gently, and plants such as seaweeds can get a firm hold and grow luxuriantly.

In most cases, the freshwater carried down by the river mixes with seawater within the mouth of the estuary. Rivers like the Amazon, however, carry such massive amounts of water that mixing with seawater takes place well offshore. During the dry season, or where the flow of the river is impeded by dams (*see pp.128-141*), saltwater will intrude further upstream than when the river is in full spate. The consequences of such saltwater intrusion can be severe. Agricultural lands may become so tainted with salt that they are useless for growing crops, and mangroves may die from salt poisoning. In some areas, however, the intrusion of seawater has positive advantages. In the Mekong Delta, for example, farmers rely on the rich marine sediments deposited during the low river flows of the dry season to neutralize and fertilize the delta's otherwise acidic soils.

THE LIVING SHORE

The silt which is carried down by a river and deposited as its meets the sea is rich in essential nutrients such as nitrates and phosphates. Many types of algae, from tiny single-celled phytoplankton to luxuriant masses of seaweed, prosper in the nutrient-filled water, and these in turn provide food for many animals. The abundance of food results in large populations of shellfish, crabs, shrimps and fish. Feeding on these are many types of birds - in Britain and the Netherlands, for example, mudflats at the mouths of rivers support higher concentrations of birdlife than any other habitat. Huge numbers of waders, such as oystercatchers, redshanks, dunlins and knots, feed in these estuaries, taking flight when disturbed in vast, wheeling flocks, shimmering as the sunlight catches hundreds of beating wings. Estuaries also support many types of waterfowl such as ducks and migratory geese which fly south every year to escape the bitter Arctic winter.

The coastline that links estuaries has its own ecology. The plants and animals of most seashores must be able to withstand the tremendous force of incoming waves and the pull of the outgoing tide. Rocky coasts are the most favourable habitat, because the rocks provide a firm base to which plants and animals can attach themselves.

In the tropics, few species are able to survive the great changes in temperature between low tide, when they are exposed to the full heat of the Sun, and high tide, when they are covered by water. Consequently, few plants and animals live within the tidal zone, and those that do are mostly nocturnal. Shoreline plants and animals are also scarce in coastal areas near the poles, but here it is the relentless scouring of the ice that prevents them from becoming established.

In temperate areas, however, shores abound with life. Some species, such as limpets and barnacles, attach themselves to rocks and rely on tightly closing waterproof shells to keep their bodies moist at low tide. Others, like shrimps and sea anemones, may survive in the water that remains in rock pools when the tide is out. Still

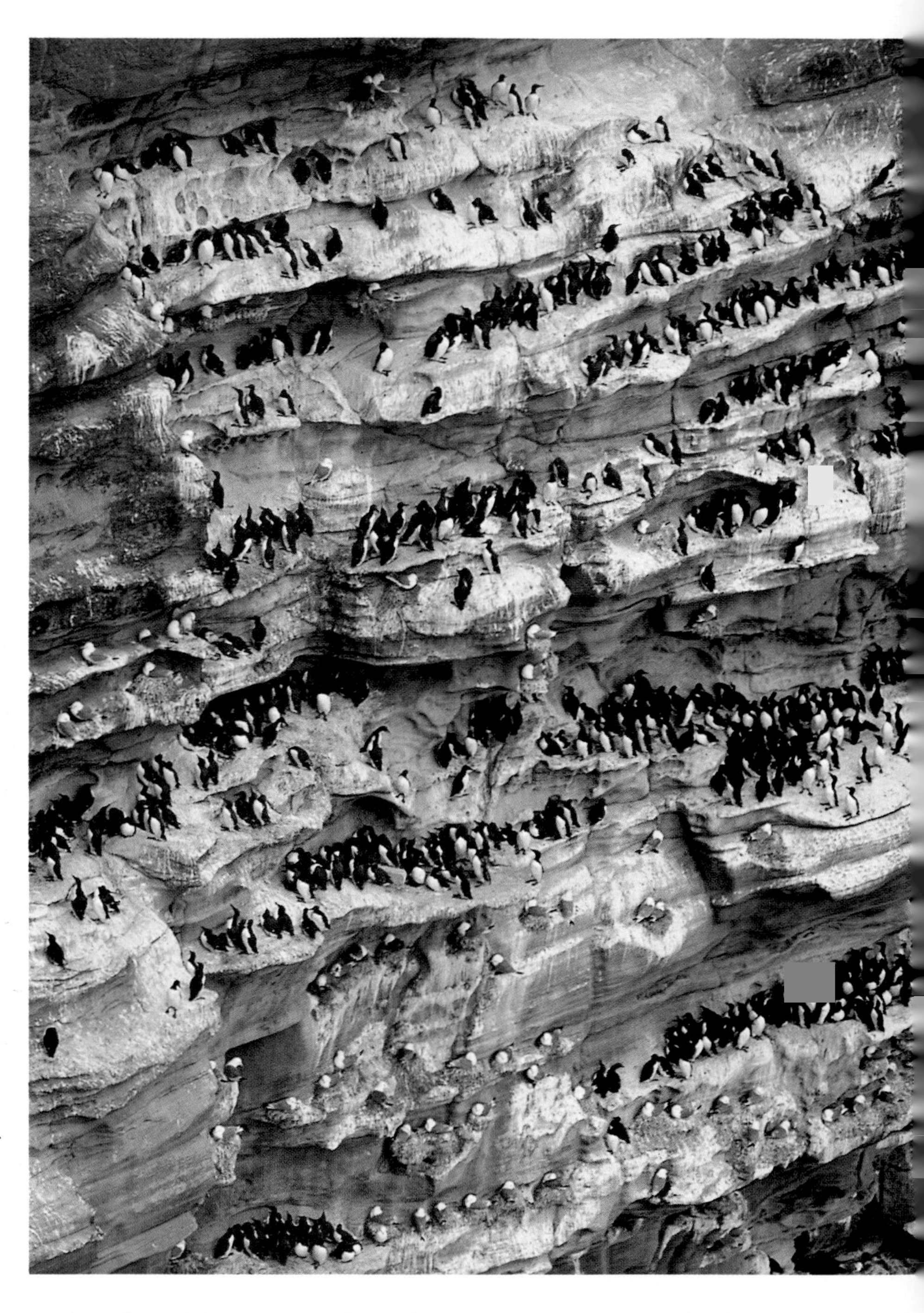

▲ A colony of guillemots, fulmars and other seabirds, clinging precariously to a cliff face on Britain's western coast. The perpetual calling of the birds can be heard far away, and the cliff face is alive with restless movement as the birds alight and take off. Such cliffs provide a safe nesting site, secure from foxes, rats and most other predators.

other species live on sandy beaches, some, such as the masked crab and burrowing starfish, digging into the wet sand when the tide is out. Others, such as cockles and razorshells, live permanently in the sand, breathing and feeding through long tubes that extend to the surface.

Inland from the shoreline, sand dunes may develop. As the dunes build up, they are colonized by tough, spiky grasses, such as sea twitch or marram grass. Such grasses help stabilize the dunes, binding the sand with their roots. As time passes, and provided the dunes are not disturbed, marram grass gives way to other vegetation, first sand couch grass and sea holly, then, eventually, small shrubs and trees.

Sand dunes play an important role in protecting the land from erosion by the sea, as well as providing ideal nesting sites for such birds as the crab plover, a native of East Africa, which burrows into the sand to build a nesting chamber. Cliff tops and cliff faces too provide nesting sites for seabirds, their inaccessibility protecting the birds from predators. Some cliffs are towering 'tenement blocks', alive with screaming seabirds

▲ Bladder wrack, kelp and other seaweeds, pounded by the full force of the waves, are supple and rubbery, with a protective gelatinous surface to minimize the damage they suffer. They have no need of roots, since they absorb all the mineral nutrients they need from the water around them, but they have an anchoring base known as a holdfast.

▲ Sand dunes on the beach of Jencoacoara Ecological Reserve, in the northeast of Brazil. Such dunes are created by the opposing forces of wind and water. Streams, eroding the soil of the hinterland, wash down huge quantities of sand and stilt. Strong onshore winds pick up this sediment from the sand bars and spits where it collects, blowing it back inland and piling it into massive dunes.

such as kittiwakes, guillemots and fulmars. Their meagre nest-sites are squeezed together on the narrow ledges of precipitous cliffs, from which birds swoop gracefully down to the sea to catch their food.

It is not only birds that benefit from coasts and estuaries. Millions of people, principally in the Third World, depend on estuarine and coastal ecosystems for their livelihood. Estuaries provide some of the richest fishing grounds in the world, supplying about 80 million tonnes of fish a year. At least three-quarters of all fished species in the eastern US spend a portion of their life-cycles in estuaries. Not surprisingly, over the centuries, fishing villages have developed along the banks of most estuaries. Agriculture too has benefited from the deep layers of silt deposited in estuaries and deltas, supplying vast quantities of nutrients to the land.

A CHAIN OF POLLUTION

The huge volume of water that passes through estuaries has encouraged people to believe that pollutants discharged into estuaries will rapidly be dispersed and diluted to harmless levels. Thousands of large industrial plants, including chemical works, oil refineries, power stations, gas terminals, iron and steel foundries, metal smelters and pulp and paper mills, have thus been built on the banks of estuaries. Where once there may have been fishing villages, today sprawling towns have built up.

The siting of industry along the banks of estuaries takes little account of the fact that freshwater and seawater do not quickly mix, and that pollutants tend to become incorporated into the silt. As a result, pesticides, heavy metals and other contaminants have been building up to toxic levels in a wide variety of estuarine species, starting in the algae and other plants at the bottom of the food chain and increasing in concentration as these are eaten by animals higher up the chain.

Many estuaries are now dangerously polluted. An increasing proportion of the fish harvested in waters such as the Thames Estuary are found to

▲ Fish carriers transport the catch from the fishing boats in Goa, India. Unpolluted coastlands are rich in fish and other food resources.

◀ Gladstone aluminium works in Queensland, Australia, the largest in the world. The chance to dispose of wastes cheaply has made estuaries and sheltered coastlands attractive to industry, resulting in serious pollution.

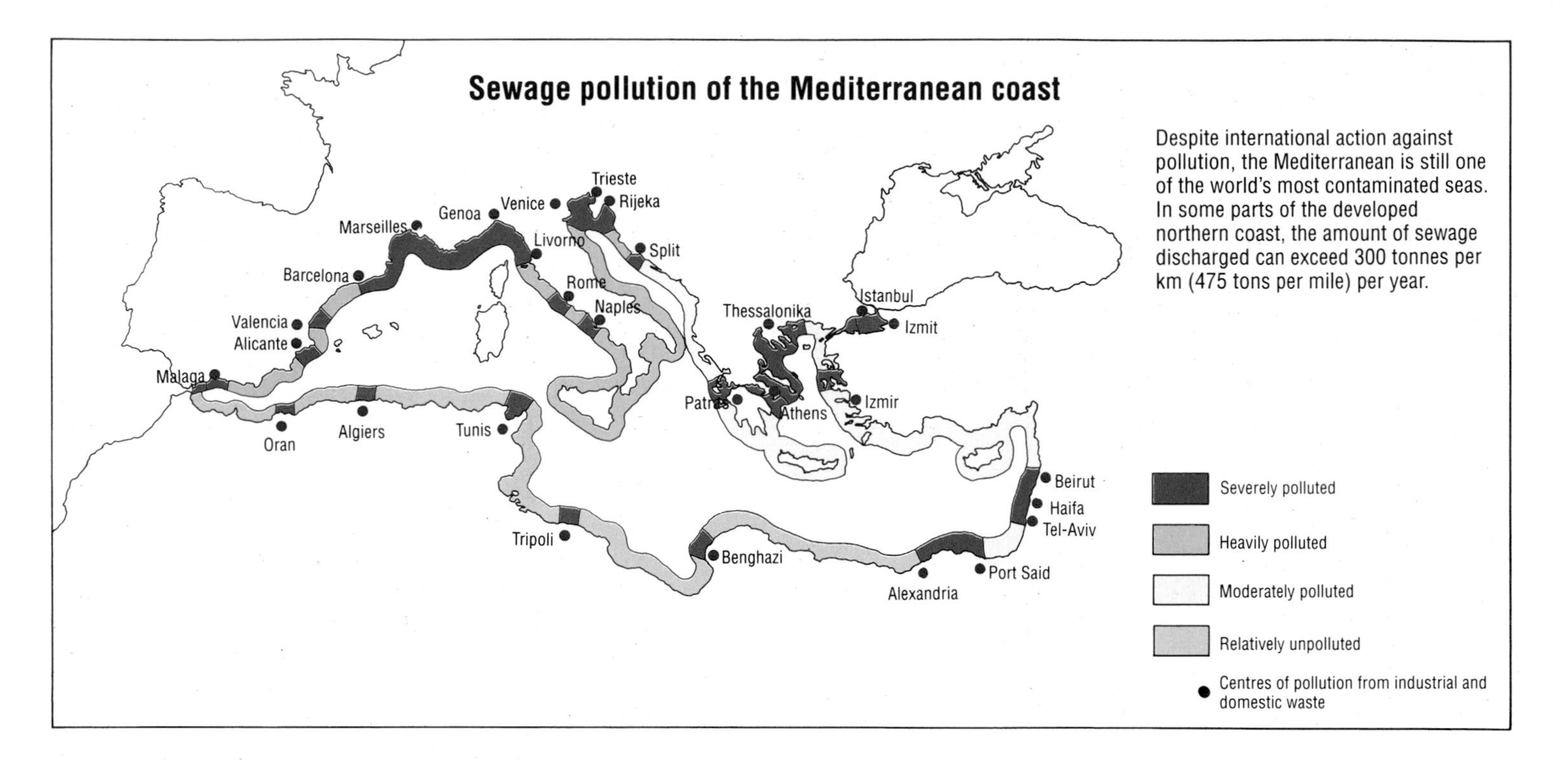

► A southern fur seal being slowly strangled by a fragment of fishing net, which cuts ever more tightly into its flesh as the animal grows. This is far from being an isolated case: every year thousands of marine animals are killed by such discarded materials.

"There is a rapture on the lonely shore,
There is society, where none intrudes,
By the deep sea and music in its roar:
I love not man the less, but Nature more."
Lord Byron

be diseased or deformed. Fish caught in Liverpool Bay, at the mouth of the River Mersey, have levels of mercury which are only marginally below the safety limits set by the European Community. The PCB levels in some fish are twice the level that would render them unfit for consumption in the US. A recent survey revealed that a quarter of the dabs (a kind of flatfish) caught in Liverpool Bay were diseased.

The Tees in northeast England, which is home to the largest concentration of chemical works outside the US, has fared little better. Large amounts of highly toxic materials, such as cyanide and ammonia, are regularly discharged into the river, which has been described as smelling "like a thousand tom cats".

The once-plentiful fisheries of Chesapeake Bay, at the mouth of the James River on the east coast of the US, have all but been destroyed through pollution. In the mid-1970s, all fishing was banned in the lower James River when high levels of kepone, a highly toxic pesticide, were found in local sediments and fish. The ban was lifted in 1981, despite evidence that fish were still contaminated with levels of kepone considered unsafe for human consumption.

As Third World countries become industrialized, so more and more of their estuaries and coasts are severely polluted. Mangroves (*see pp.160-163*) and coral reefs (*see pp.186-193*) are particularly vulnerable, many having been destroyed already by pollution.

▲ Pollution in Chesapeake Bay, on the eastern seaboard of the US. Bordered to the north by the cities of Washington and Baltimore, the bay receives large amounts of effluent from the rivers that flow into it. Pollution has affected the fish and shellfish for which the bay is famous.

The dumping of sewage into coastal waters and estuaries, together with the run-off of fertilizers and pesticides from agricultural land, has added to the pollution. Sewage and fertilizers have massively increased the amount of nutrients entering coastal waters. Algae thrive on these nutrients, causing their numbers to explode. The resulting algal blooms can produce toxins and may form putrefying masses as they rot down, taking up large quantities of oxygen. In relatively enclosed bodies of water such as the Baltic Sea, North Sea and northern Adriatic, algal blooms along the coast have caused oxygen levels in the sea to plummet, killing fish and other marine organisms. Some Swedish coastal waters are now completely devoid of marine life. In Norway, algal blooms have threatened to ruin the country's coastal fish farms and hatcheries.

Many beaches are now seriously polluted as a result of the direct discharge of sewage through outfall pipes. Waste dumped at sea and then brought ashore by the tides is also a major cause of contamination. In most cases, the waste consists of plastic bottles or tar balls from oil discharged perfectly legally at sea. In some cases, however, the wastes are more deadly. In 1989, for example, medical wastes, including used hypodermic needles, were washed ashore near New York. More deadly still, canisters of cyanide and other poisons, washed overboard or deliberately dumped at sea, are turning up on beaches with increasing frequency.

THE COAST UNDER ATTACK

As waves pound against the shore, they inevitably cause erosion. However, this is normally balanced by the building up of new beaches and sandbars as sediment is carried down by rivers and then deposited back on land by the sea. Human activities, however, often greatly accelerate coastal erosion.

Where a river has been dammed, for instance, the sediment it previously carried down to the sea is trapped in the dam's reservoir. This can cause severe erosion along the coastline. Nowhere is this demonstrated more dramatically than in Egypt. The building of the Aswan Dam *(see p.157)* has deprived the Nile Delta of an estimated 60 million tonnes of silt a year, and the delta is retreating for the first time in history.

The destruction of coral reefs is also a major cause of accelerating coastal erosion, since the reefs no longer provide a buffer against storm surges. Such storms are likely to increase dramatically as a result of global warming, changing the landscape of many coastal areas.

It is not just the advancing sea that threatens coastal ecosystems. Tourism has ruined many coastlines as hotels and villas have sprung up to cater for holidaymakers. Beaches have been polluted, sand dunes built over, and many coastal species have lost their habitats to the developers. The loggerhead turtle is now seriously endangered due to tourism on the Aegean beaches where it lays it eggs. Beach umbrellas disturb or destroy the turtle's nests, and hatchlings often cannot emerge because people walking on the beach compact the sand above. Lights from hotels and bars distract the turtles when they emerge from the water during the night to make their nests. Confused, they often lay their eggs in the water where they cannot hatch, or they may be disoriented by the lights and not find their way back to the sea, dying from exposure to the Sun the following day.

The death of loggerhead turtles on the beaches of the eastern Mediterranean may seem of little importance to those seeking to develop other coasts for tourism or industry, but they are an ominous symptom of the disregard with which we treat our estuaries and coasts. If the abuse continues, it will not be long before many other species go the way of the turtles, their habitats destroyed by bulldozers or pollution. For millions who live on estuaries and coasts, and who rely on them for their livelihoods, the results could be disastrous.

SEAS and OCEANS

Every evening, as the Sun goes down over the deep blue of the Pacific Ocean, thousands upon thousands of kilometres of almost invisible nylon netting - enough to stretch right around the equator and then double back across the Pacific again - are unfurled into the sea by boats from Japan, Taiwan and South Korea. From the sea's depths, these drift nets present an eerie sight, with hundreds of fish hanging strangely suspended in the water, apparently frozen in time and space.

Besides their intended catches of squid and tuna, these almost indestructible walls of death, each up to 65 km (40 miles) long and 12 m (40 feet) deep, trap hundreds of thousands of dolphins, seals, turtles, sharks, salmon and seabirds in an indiscriminate marine holocaust. When drift nets are lost, often due to whales becoming tangled up in them, they join hundreds of other 'ghost nets', drifting with the currents and continuing to reap their harvest, sometimes for years, before sinking to the bottom under the weight of their 'catch'. Eventually, the carcasses of the fish, marine mammals and birds trapped in the nets decompose, allowing the nets to rise to the surface again.

By strip-mining the seas, drift-netting has brought big profits to the fishing industries of Japan, Taiwan and South Korea, but financial ruin and the threat of hunger to the unfortunate fishermen and islanders elsewhere in the Pacific. In the South Pacific, the economies of the many tiny island nations depend almost entirely upon tuna, but the drift-net boats are presently taking two to three times the sustainable yield of tuna from their waters. Fisheries experts estimate that the stocks of tuna could collapse in the early 1990s.

◀ A school of dolphins in the Atlantic Ocean, off the coast of Brazil. Dolphins are believed to be highly intelligent, with brains that are as large, in proportion to their bodies, as those of humans. The cortex of a dolphin's brain - the site of intelligent thought - shows as much complexity as our own. The Ancient Greeks believed that killing a dolphin was equivalent to murdering a human being.

“While sailing a little south of the Plata on one very dark night, the sea presented a wonderful and most beautiful spectacle. There was a fresh breeze, and every part of the surface, which during the day is seen as foam, now glowed with a pale light. The vessel drove before her bows two billows of liquid phosphorus, and in her wake she was followed by a milky train. As far as the eye reached, the crest of every wave was bright...”
Charles Darwin, in The Voyage of the Beagle

▼ A comb jelly, or ctenophore, shimmers with phosphorescence. Many ocean animals generate their own light, often to attract mates and distinguish one species from another. Some lights serve as warning signals for ill-tasting animals who wish to deter predators. The predators themselves may use lights as lures, or to illuminate their food. Light-generation by living creatures is far more efficient than anything technology has devised, because it produces cold light, with no energy wasted as heat.

But it is not just overfishing that now threatens life in the oceans. Pollution, the destruction of fish nurseries, algal blooms and other threats have brought many marine ecosystems to the point of collapse. Ozone depletion could provide the final blow *(see p.48)*.

AN UNEXPLORED WORLD

The days when intrepid explorers discovered new continents, or could claim to be the first to have reached the source of this river or to have climbed that mountain, are rapidly passing. There are few places on land where humans have never set foot, and even these have largely been mapped with the aid of satellites. But the oceans are different and, to the extent that their waters cover seven-tenths of the planet, the Earth may be said to be still largely unexplored.

Drain the seas of their water and a landscape as physically spectacular as anything on Earth would be revealed. Stretching out for as far as 1,500 km (930 miles) from the shoreline of the Atlantic, and for as little as 20 km (12 miles) from the shores of the Pacific, are the continental shelves, gently sloping plains broken up by valleys and small hills. Here, the seabed consists largely of sand and gravel, or, in the tropics, of coral *(see pp.186-193)*. Where the shelf ends, the seabed plunges down massive escarpments. These are the continental slopes, their surfaces often scarred by massive gullies and canyons, scoured out during the last Ice Age and continuously widened by underwater avalanches carrying sediment down from the higher slopes at speeds of up 160 km/h (100 mph).

At the base of the slopes, the seabed begins to rise again, where foothills have formed through the accumulation of sediment. Beyond this continental rise lies the abyssal zone, the area that makes up the deep ocean floor. Although physically unexplored by humans, the abyssal zone can now be mapped by sonar with the aid of computers sensitive enough to pick up even small crevices. Sonar reveals that the ocean floor forms a vast plain, in some places rent by deep trenches. The biggest, the Marianas Trench east of the Philippines, plunges to a depth of 11,700m (38,380 feet), deeper than Mount Everest is high. In other places, it is broken by spiky mountains rising several kilometres.

Running down the middle of the Atlantic, and continuing through the Southern Ocean and up the eastern Pacific, is the longest mountain range on Earth - the Mid-Ocean Ridge. Some 800 km (500 miles) wide and 55,000 km (35,000 miles) long, its mountains soar up to 4 km (2.5 miles) above the seabed and are still 2-3 km (1.25-2 miles) below the water's surface.

Although mariners may boast of having sailed the seven seas, they have in fact sailed one ocean, for, like the atmosphere, the oceans know no physical boundaries. Driven by energy from the Sun, by wind and by the rotation of the Earth, the ocean is continually on the move. Powerful surface currents, sometimes described as rivers in the ocean, carry vast volumes of water from the poles to the tropics and back again. Rising in the Gulf of Mexico, the Gulf Stream surges up the eastern coast of North America at speeds of up to 5 knots, carrying some 55 million m^3 (2 billion cubic feet) of water per second across the North Atlantic to northern Europe. On the other side of the globe, the Kuroshio current sweeps warm water from the Philippines to Japan, Hawaii and the west coast of America.

Beneath the surface, equally powerful currents can create massive underwater cataracts as seawater tumbles down the sides of ridges and mountains. On the bed of the Denmark Strait, off Greenland, lies a deep-sea waterfall that dwarfs anything on land, with 5 million m^3 (175 million cubic feet) of water a second cascading down a slope that descends for some 3.5 km (2 miles). By comparison, the tallest waterfall on land, the Angel Falls in Venezuela, is only 1 km (0.6 miles) high, while the largest river on Earth, the Amazon, carries ‘only’ 200,000 m^3 (7 million cubic feet) of water a second.

THE MARINE FOOD CHAIN

At the base of the marine food chain are billions of free-floating, microscopic plants known as phytoplankton *(see p.45)*. A single cubic metre of seawater contains as many as 200,000 such

organisms. Like plants on land, phytoplankton rely on sunlight to convert carbon dioxide, mineral salts and other nutrients into food, and are thus only found in the top 100 m (330 feet) of the sea, photosynthesis being impossible at lower depths. Where surface waters are particularly rich in nutrients, as at the Arctic and Antarctic convergence zones, phytoplankton 'blooms' occur. Off the Antarctic, such blooms frequently extend up to 400 km (250 miles) into the Ross Sea, providing a major food source for krill, the tiny animals that are critical to the Antarctic food chain *(see p.221)*.

Above phytoplankton in the food chain are zooplankton, varying in size from single-celled organisms to jellyfish, which either graze phytoplankton or are carnivorous, preying on other zooplankton. Like phytoplankton, zooplankton are largely dependent on the movement of the sea for their mobility. Feeding on them are a variety of predators, from the massive baleen whales to shoaling fish like herring or mackerel. Small fish in turn provide food for larger fish, such as tuna, and marine mammals such as seals, themselves often serving as prey for sharks. Dead and decaying organisms, meanwhile, fall to the ocean floor to feed scavengers, such as crabs, sea-urchins and sea-cucumbers. Other scavengers - sharks and some shrimps, for example - are adept at intercepting dead and decaying organisms before they reach the seabed.

Although the ocean is one ecosystem, its plants and animals vary considerably from region to region depending on the temperature

▲ A humpback whale leaps from the cold waters of the Arctic Ocean where it feeds in summer on shoals of tiny fish. There are three separate populations around the Arctic, and five or six around the Antarctic. Before winter sets in, the humpbacks migrate to the warm waters of the tropics to breed. It is here that they sing their haunting songs which carry for hundreds of kilometres through the deep layers of the ocean.

◀ Animals that have never known light before are revealed by a deep-water research vessel. The food chain of this strange animal community is based on hydrogen sulphide, spewing from volcanic vents, which is turned into useful food by bacteria. The warmth from the vents offsets the chill of the deep ocean, allowing life to flourish in this precarious oasis of the abyss.

▼ A British ship dumps sewage sludge in the North Sea.

and depth of the water, the saltiness, and the availability of nutrients. Life flourishes even in the deepest ocean, with some 2,000 species of deep-sea fish and as many invertebrates now identified. In the very depths of the trenches on the seabed, 11,000 m (36,000 feet) below the surface, remarkable communities of worms, clams and blind white crabs feed on bacteria which have the unusual capacity to metabolize hydrogen sulphide from volcanic vents.

Life in the deep ocean is sparse compared to that along coasts, where the water is shallow enough to allow sunlight to penetrate to the seabed, and where nutrients washed down from the land provide a plentiful source of food. Here fish can be found in abundance. Although they constitute just one per cent of the total area of the ocean, coastal ecosystems - principally wetlands and mangroves *(see pp.150-163)*, estuaries *(see pp.164-171)*, and coral reefs *(see pp.186-193)* - are where the vast bulk of marine organisms are concentrated. Mangroves, for example, are some 20 times more productive than the open ocean, and reefs and estuaries up to 18 times more productive. Yet it is precisely these coastal regions where the impact of man is most evident and most destructive.

DEPTHS OF IGNORANCE

Much of our knowledge of the forces that drive the oceans, of their geography and of the myriad forms of life that inhabit their depths is relatively recent, and even today there are major areas of uncertainty and ignorance. One hundred and twenty years ago, those who sailed the oceans knew next to nothing of the vast mountain ranges below them: the Mid-Ocean Ridge, for example, was not discovered until 1872. It was not until the 1960s, when transistors replaced vacuum tubes, that oceanographers were able to study deep ocean cataracts and currents.

As our knowledge of the oceans increases, so many widely accepted theories turn out to be myths. Until the early 1980s, it was generally assumed that conditions on the seabed of the deep oceans - at depths of 4,800 m (15,750 feet) below sea level - were utterly calm. However, researchers at the Woods Hole Oceanographic Institute in the US have now discovered that far from being calm, the deep oceans experience violent underwater storms which scour the sea floor, transporting large volumes of silt and sediment from one area to another.

Because of their size, and the vast quantities of water they contain, the world's oceans have long

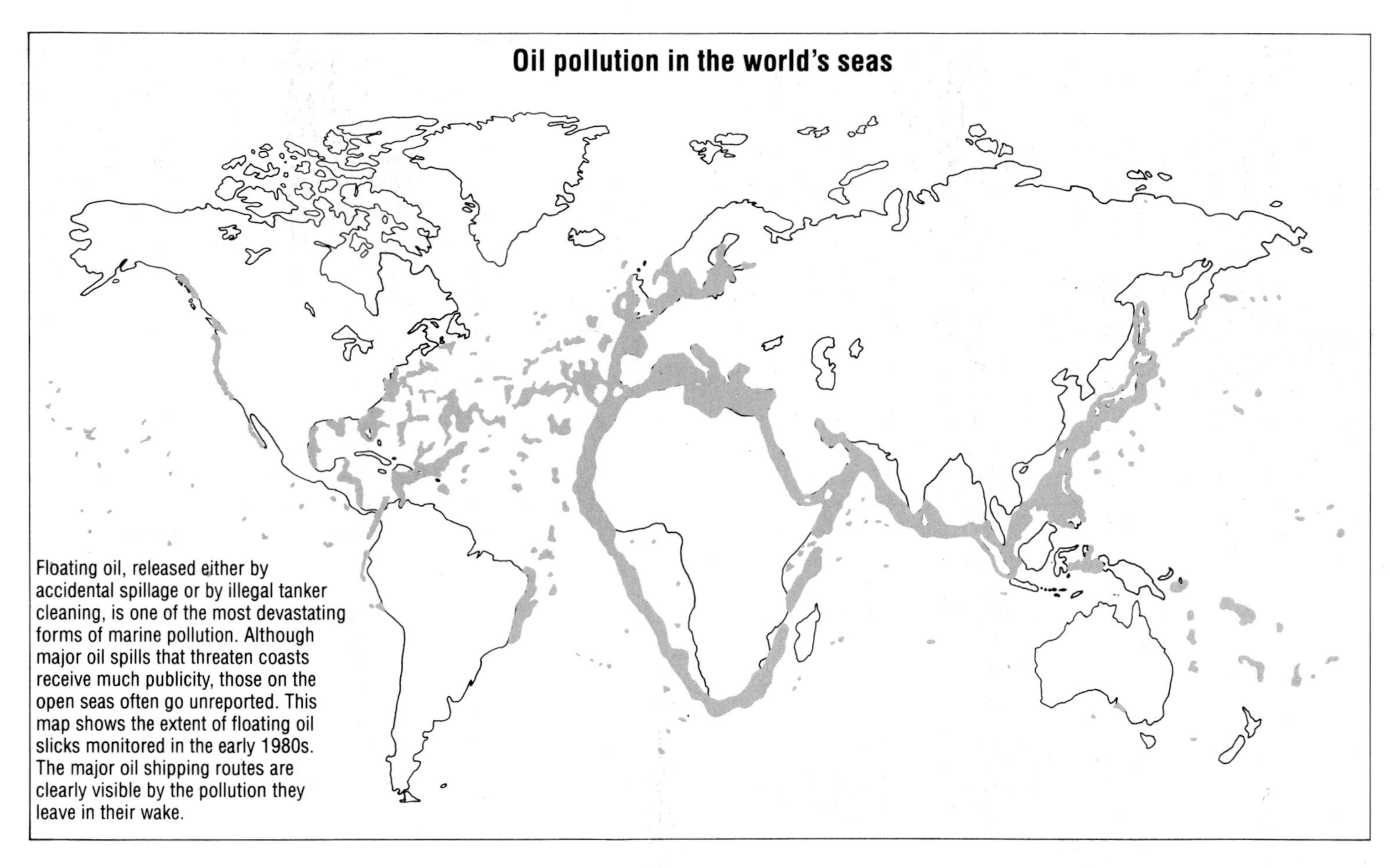

Floating oil, released either by accidental spillage or by illegal tanker cleaning, is one of the most devastating forms of marine pollution. Although major oil spills that threaten coasts receive much publicity, those on the open seas often go unreported. This map shows the extent of floating oil slicks monitored in the early 1980s. The major oil shipping routes are clearly visible by the pollution they leave in their wake.

been regarded as having an almost infinite capacity to absorb wastes, the assumption being that the wastes are rapidly diluted and dispersed to levels where they pose no threat to the environment or human health.

In many areas, the reverse is proving to be true. Inshore currents often trap wastes along the shoreline, where they may accumulate in sediments. In others, wastes are swept offshore only to be carried to other coastal areas, a problem that is particularly acute in seas such as the Baltic and the Mediterranean, which are almost completely surrounded by land. Moreover, as pollutants are taken up by marine organisms, so they accumulate in the food chain, often reaching toxic levels *(see p.33)*. As a result, vast stretches of coastal waters have been rendered all but lifeless through pollution, threatening wildlife and destroying fisheries. Several seas are now in a critical condition.

The Baltic is one of the most polluted seas in the world. Wastes from factories, farms and private households have turned 100,000 km^2 (40,000 square miles) of the sea into an ecological wasteland, devoid of life below depths of 80 m (250 feet). The damage is particularly severe in the narrow straits between Denmark and Sweden and in the bay off Gdansk, where thousands of tonnes of phenols, oil, lead, zinc and other pollutants are carried into the sea by Poland's main river, the Vistula. High levels of DDT have been blamed for thinning the eggshells of such seabirds as razorbills, guillemots and white-tailed eagles, while PCBs have been blamed for a crash in the population of grey seals. Algal blooms are common in many coastal waters, encouraged by the million tonnes of nitrogen and 48,000 tonnes of phosphorus that the sea receives every year. Thousands of fish have been killed, poisoned by the algae.

The Mediterranean, like the Baltic, is virtually an enclosed sea, and it is almost as sick. Because of its high rates of evaporation and its very slow-moving waters (it takes a hundred years for the sea's waters to renew themselves as they flow through the Straits of Gibraltar), pollutants tend to accumulate without degrading.

But it is not only enclosed seas that are now under threat. The North Sea is also showing signs of biological collapse. Worst affected are the coastal areas of Holland, West Germany and Denmark. Here, strong inshore currents have

"After the USSR, Poland is the Baltic's worst enemy. Via the Vistula, 90,000 tons of nitrogen pour into it annually, plus 5,000 tons of phosphorus, 80 of mercury and unmeasured quantities of cadmium, zinc, lead, copper, phenol and chlorinated hydrocarbons. Baltic fish are so scarce that the Polish fleet can catch no more than 20 per cent of its quota, and the last eels - taken three years ago - were apparently part-marinated."
Peter Martin

The disappearing Aral Sea

The health of the great inland seas of the world is in many cases even worse than that of the oceans. In the Soviet Union, the over-extraction of water for irrigation is causing the Aral Sea - once the fourth largest lake in the world - to disappear. The area of the Aral shrunk by 40 per cent between 1960 and 1989. Fishing grounds have been replaced by salt-encrusted deserts, with huge trawlers rusting on the dunes. Salt from the dried-out bed of the Aral has been blown as far away as the Arctic. In 1990, an agreement was signed between the USSR and the United Nations Environment Programme for international action to save the Aral Sea. But, with the Soviet economy in crisis, the massive sums of money required to stop the further devastation of the region are unlikely to be forthcoming.

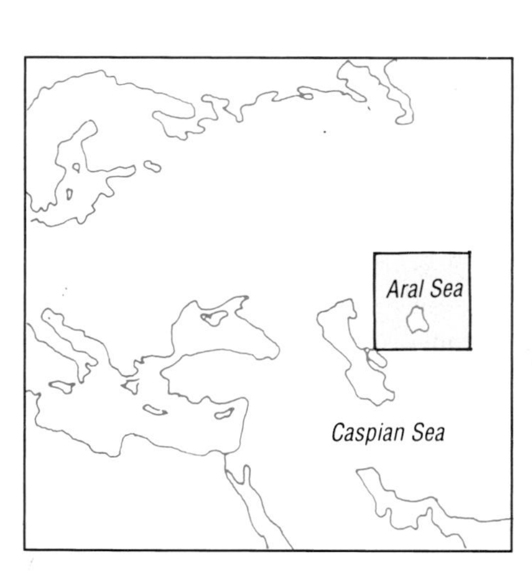

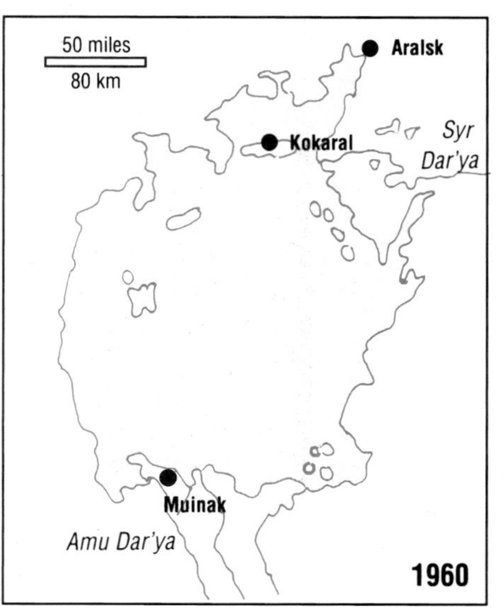

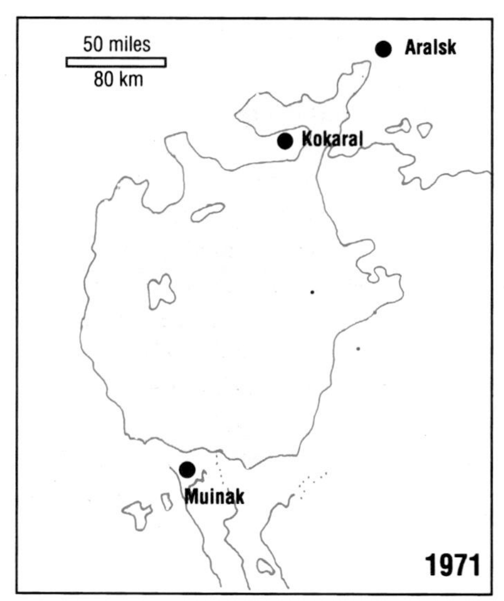

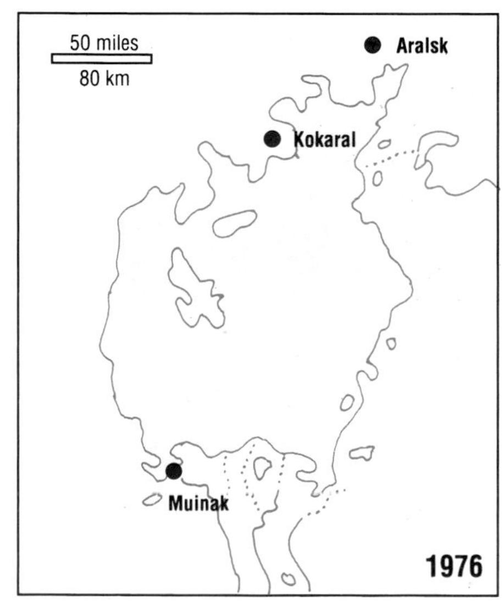

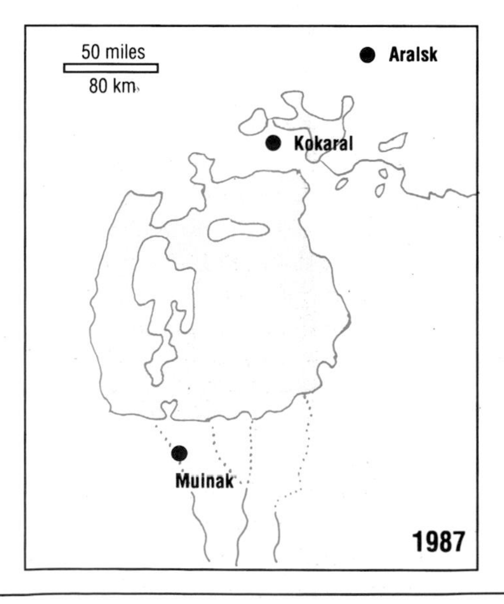

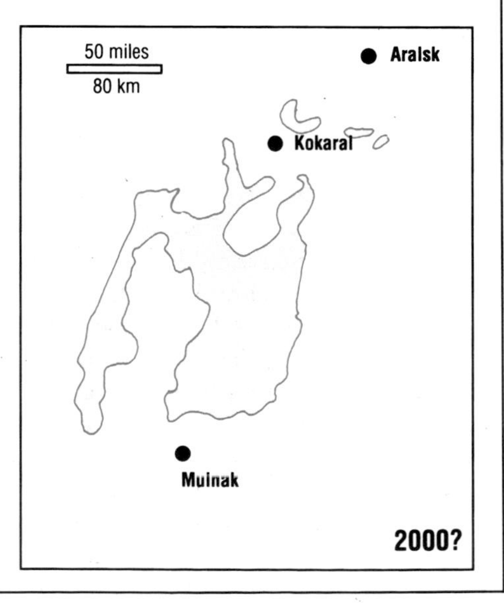

trapped tonnes of pollutants carried into the sea by the major rivers of northern Europe - the Elbe, Rhine, Meuse and Scheldt - causing them to accumulate to dangerously high levels in the Wadden Sea and the German Bight. Adding to this fatal burden are pollutants driven across the North Sea from Britain by westerly winds.

More than 1,000 species of birds, fish, seals and invertebrates rely on the Wadden Sea and 40 per cent of North Sea fish spawn in its waters. Its mudflats are also a major stopping point for migratory birds. In the German Bight, deformities in fish are abnormally high due to pollution by pesticides and wastes from the paint industry, 40 per cent of dab showing sores, tumours and lesions in their skins. High levels of PCBs have been held responsible for reproductive failures among harbour seals in the Wadden Sea and have been linked to the epidemic that destroyed whole seal colonies throughout the North Sea during the late 1980s *(see p.33)*. Porpoises and killer whales, once common in the Wadden Sea, are now a rare sight, while once-thriving fisheries have collapsed, as have populations of several seabirds.

Even coastal areas as yet unaffected by industrialization have not escaped the effects of pollution. Ninety-eight per cent of PCBs in the sea enter the ocean via the atmosphere, borne up into the air through evaporation and often transported for thousands of kilometres by the wind. Even the high Arctic and other remote regions are now contaminated. PCBs, which can suppress the immune system and cause reproductive failure in marine mammals, have been found in striped dolphins, melon-headed whales, Dall's porpoises and other cetaceans living in the open oceans. In some animals, the PCB concentrations are far above those that are found in toxic industrial waste.

Every year on land, large quantities of PCBs are disposed of, often by encasing them in drums which are then buried. However, this does not get rid of the PCBs. It merely contains them for as long as it takes the drums to disintegrate. Should the chemicals then leak into groundwater *(see pp.142-149)*, and make their way into the oceans, the results could be catastrophic. One marine biologist, Professor Joseph Cummins, has predicted that they could cause the extinction of a wide range of marine mammals, if not all.

Growing public concern over marine pollution has led to a series of conventions aimed at cleaning up the seas. In 1987, the European states bordering the North Sea agreed on a number of measures, including an end to the dumping of harmful wastes by 1989 and a halt to the incineration of wastes at sea by 1994. They further agreed that, by the end of the century, they would halve inputs of nutrients such as nitrates and phosphates, and also substances that are toxic, persistent or liable to build up in food chains *(see p.33)*.

However, no clear programme has yet emerged for achieving those goals. Some countries - notably Holland and Denmark - took strong measures to comply with the 1990 deadline for halting toxic waste dumping, to the extent of closing down offending factories. But Britain unilaterally extended the deadline until 1992, and even then reserved the right to issue 'one-off' licences, allowing it to continue past that date. Sewage dumping by Britain - the only country still continuing the practice - will not be halted until 1998. To cloud the picture further, different countries have put different interpretations on the agreement. Denmark has singled out 50 chemicals for special controls, Britain 23 and Norway 13. By 1990, Belgium still had not produced a list.

Even if all the measures agreed are eventually carried out, half of the chemicals currently entering the North Sea will still be permitted to flow into it, adding to the already critical pollution load. Environmental groups are almost unanimous in calling for zero inputs of toxic wastes by the end of the century if the North Sea is to recover its health. But a complete clean-up will never be possible: much of the radioactive waste in the sea, for example, will remain active for thousands of years.

THE IMPACT OF MODERN FISHING

For about a decade, the world's annual fish catch has exceeded what the UN Food and Agriculture Organization considers to be the maximum yield sustainable in the long term. The old adage 'there's plenty more fish in the sea', no longer applies in this age of modern fishing technology.

In 1986, the harvest from the world's oceans was around 108 million tonnes, 8 million tonnes above the maximum sustainable yield and increasing steadily by around a million tonnes a year. It is now only a matter of time before catches start to plummet.

This has already happened for many species of marine life, most notably the great whales *(see pp.222-223)*. There are numerous examples of commercial fisheries which have similarly collapsed due to overexploitation in the last few decades, disrupting marine ecosystems and destroying the livelihoods of fishermen and the communities in which they live.

The common skate, for example, is thought to be extinct in the Irish Sea because of overfishing in the 1970s. In the North Atlantic, the herring has been fished to the verge of extinction, while stocks of the Atlantic cod, haddock and capelin have been severely depleted. In the South Atlantic, pilchard catches have seriously declined and in the Pacific, even before the devastation caused by the widespread use of drift nets, the anchovy, salmon, halibut, king crab and Pacific-Ocean perch had all been overfished.

Overfishing does not only affect fish. For example, in the 1980s, stocks of capelin, a type of sardine which is used in animal feed and commercial fats, were so drastically reduced by Norwegian fishermen that all capelin fishing was banned indefinitely in 1988. The damage, however, had already been done. The cod which normally ate the capelin now ate their own young instead. With their waters depleted of both capelin and cod, hundreds of thousands of starving harp seals invaded the Norwegian coast in search of food, depleting coastal fish stocks and destroying salmon farms in the fjords. In February 1987, 60,000 seals were accidently caught and drowned in Norwegian fishing nets. Tight-knit coastal communities, which have

◀ French fishermen at Belle Ile. Throughout the world, small local fishing boats such as these are losing their livelihood as the fishing fleets of the industrialized nations plunder their traditional fishing grounds.

“Oceanic fishing today is oceanic anarchy.”
Jacques Cousteau

lived off the seas for generations, are now breaking up as their young people leave for the cities to look for employment, the sea no longer able to yield enough fish to provide them with a livelihood.

On the other side of the Atlantic, dwindling fish resources, especially of northern cod, threaten the disappearance of many of the fishing communities on Canada's eastern seaboard. The Newfoundland economy is heavily dependent on the fleets of trawlers which are now being sold off, and the shore-based fish-processing factories which are closing down, putting thousands out of work.

THE IMPACT OF MODERNIZATION

The driving forces behind overfishing in the developed world, and increasingly also in the developing world, are improvements in fishing technologies. In Britain, for example, grants and loans from the government and the EC have encouraged fishermen to build bigger and bigger boats, and to use nets with smaller and smaller mesh sizes, which take even very young fish. Grants also allow them to buy ever more sophisticated sonar equipment with which to locate the dwindling shoals. After investing in such expensive equipment, fishermen have to bring in huge catches to pay back their loans. They thus become ensnared in a vicious circle, having to catch more and more of a decreasing resource to

▲ Overfishing of the anchoveta shoals is just one threat facing these fishermen in Ecuador. Fish stocks here depend on an upwelling of cold water that brings nutrients to the surface. Every three to eight years, an abnormally warm current, *El Niño*, arrives at Christmas-time. The upwelling then ceases, the shoals of fish vanish, thousands of seabirds starve and the fishing boats lie idle. *El Niño* begins with a small change in atmospheric pressure and sea temperature thousands of kilometres away, off Australia. If a small natural change can have such a devastating effect on one fishery, what might global warming do for world fish supplies?

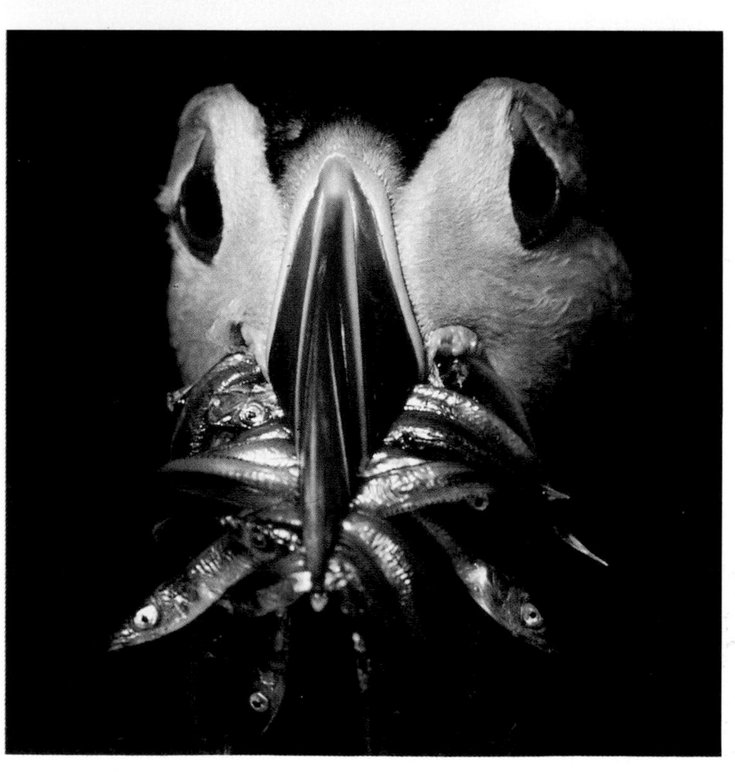

◀ A puffin carrying sand eels for its growing chicks. In 1989, almost all of the 48,000 puffin chicks on one island in the Shetlands starved to death. The same devastating losses occurred elsewhere in Scotland and parts of Norway, affecting terns, kittiwakes and other seabirds as well. These terrible losses are largely due to the massive overfishing of sand eels, which are used primarily to make animal feed for mink farms. It was the sixth year in succession that these seabirds had failed to raise chicks. Catching small fish, such as sand eels, capelin or anchoveta, simply to make foodstuffs for animals, frequently has catastrophic effects on other wildlife.

stave off bankruptcy. Even though the fishermen may be well aware of the consequences of overfishing on their long-term livelihoods, they have little choice but to continue if they are to survive in the short term. Although governments may realize the need for conservation measures, they are usually reluctant to impose strict quotas on fishing communities undergoing hard times.

In the North Sea, the EC has been progressively cutting down the fish quotas set for individual nations' fleets, but their proposals have been consistently watered down by the various national governments, worried about the political fallout from job losses in the fishing industry. Not surprisingly, stocks of haddock, cod, herring, whiting and other commercial species in the North Sea continue to fall rapidly. The hundreds of thousands who depend upon the North Sea fleets for their jobs face a bleak future.

▲ A trawler fishing at night. Fishing is now a highly intensive industry in which work continues round the clock, in order to maintain catches and pay for ever more expensive machinery.

▶ A tuna-fish canning plant reflects the enormous scale of today's fishing industry. The oceans cannot sustain such massive production indefinitely, and there are already signs that fish stocks are being exhausted.

In the Third World, too, it is the modernization of fishing methods which is spelling disaster for fishermen and the fish stocks they rely upon. Fishing communities in Malaysia, which is surrounded by one of the richest fishing grounds in southeast Asia, have seen a steady deterioration in their situation since the introduction of modern trawlers in the early 1960s. Although the total fish catch remained constant through the 1970s and 1980s, the proportion of the catch that was made up by 'trash' fish, useless for human consumption, increased markedly. In 1980, about 80 per cent of the fish caught off the west coast of Malaysia were 'trash', most of which were young fish of commercially valuable species, too small for anything other than fertilizer or animal feed.

Around the world, more than 100 million people make their living from the sea. The great majority of these people are traditional or 'artisanal' fishermen in the Third World, who use small canoes or sailing vessels such as the dhow, and take their catches with handmade lines, spears or nets. In the developed world, overfishing may mean loss of livelihood, but for artisanal fishermen, it can mean starvation for them and their families.

Along India's 5,600-km (3,500-mile) coastline, fish is the main source of animal protein for tens of millions of poor. But it is now becoming increasingly expensive as mechanized fishing boats, owned by big business and multinationals, scoop up fish and prawns for sale in the cities and on the international market. Shrimp fishing has proved exceptionally damaging as it involves dragging huge, heavy nets along the seabed, taking up eggs and immature fish from other species, stirring up sediment, and driving

**“In my father's time there were hundreds and hundreds of boats in Newlyn, but very different boats than there are today. The fishing was different too... they've gone mad now, quite mad. They spent all those millions on the new quay and just look out there, look at all those big, expensive boats tied up and yet...no fish. First they overfished the pilchards, then the herrings - the German boats, trawlers as big as the Icelandic ones, came down to Plymouth and destroyed the herring, then the mackerel. Everyone told the Government you couldn't keep taking mackerel like that, the big trawlers and the factory ships and so on, but they kept saying their scientists said there was plenty out there... After the pilchards they thought it didn't matter as there was herring and when the herring was all gone there was still the mackerel.
I suppose they think that after the mackerel there'll be something else but if there is, I don't know what it is. It's live for today or at the most the week, with no thought for the future.”**

Joe Tonkin, Cornish fisherman

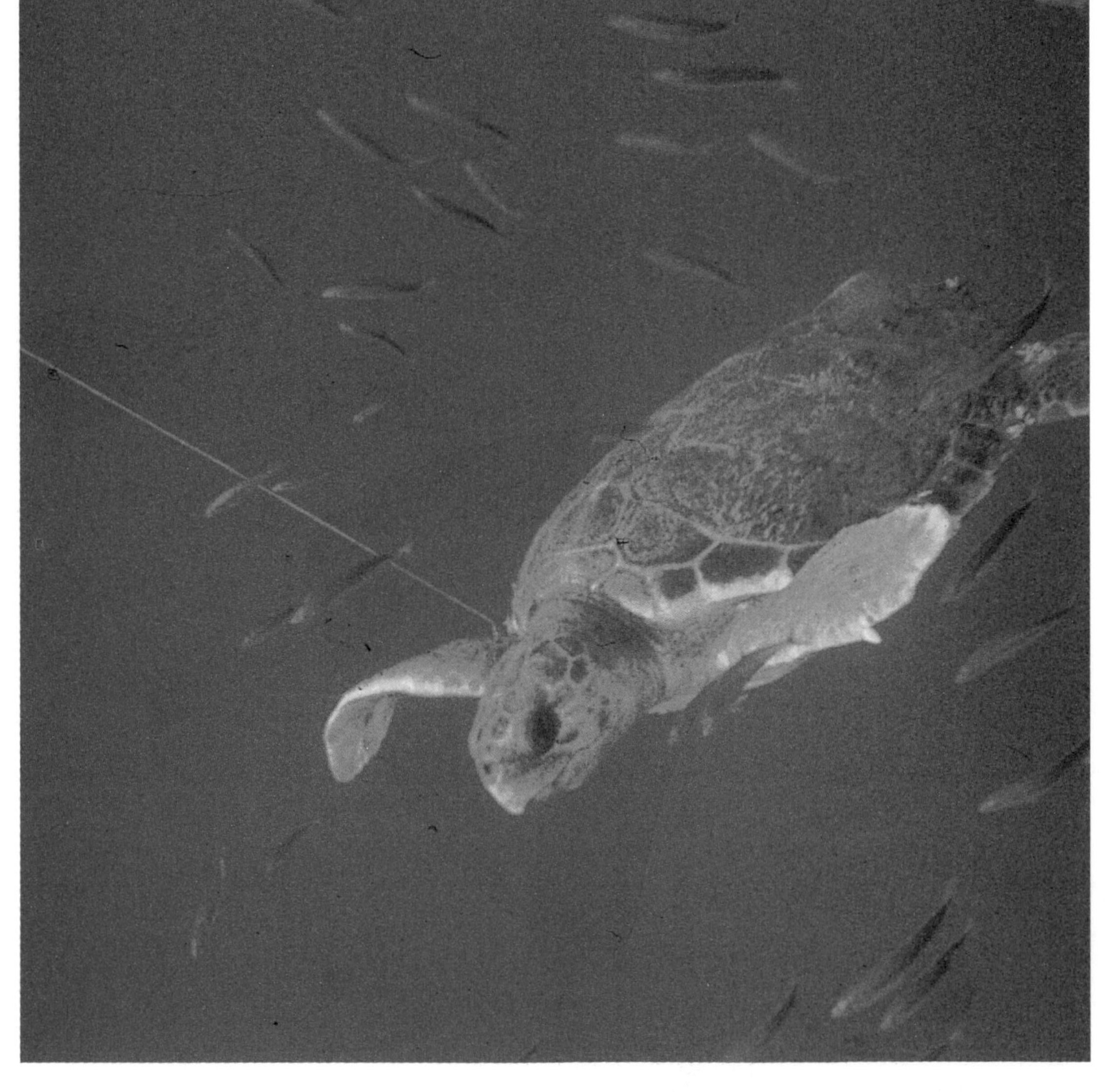

◀ A turtle caught in a longline from a fishing boat. Like whales and dolphins, these endangered reptiles must come to the surface to breathe air. Snared underwater by a fishing line or net, they quickly drown.

▶ Indonesian fishermen throwing a cart net. Their traditional fishing grounds are being destroyed by large trawlers.

▼ A skipjack tuna caught in a type of net known as the 'wall of death' for its disastrous effects on ocean wildlife. After considerable international pressure, Japan has agreed to a UN resolution banning such nets in the South Pacific.

off shoals of fish as they head for the coast for feeding and spawning.

In Indonesia, artisanal fishermen have destroyed nets and boats in protest against the use of mechanized vessels, and attacked the crews of trawlers who came too close to traditional fishing grounds. The 64,000 Thai village fishermen take only 30 per cent of the catch from Thailand's waters, the rest going to the trawlers which, in terms of numbers of boats, make up only 15 per cent of the fleet.

FISHERIES AND FOOD

The consequences of overfishing and pollution on world food supplies are likely to be severe. In many low-income countries and island nations,

"Not even the sophisticated sonar of a dolphin can always detect the danger. So dolphins swim into the net. Whales, seals and sea turtles swim into it. Attracted by the trapped fish, seabirds dive into it and drown. Some victims of the nets may struggle through, only to die painfully, weeks or months later. Most marine creatures stay snared, writhing and dying. In the morning the net is lifted from the water and you can see the body count..."

Greenpeace report on Japanese driftnet fishing in the Pacific

fish are the principal source of animal protein. Fish provide almost a quarter of humanity's animal protein, although only two-thirds of the world's commercial catch is eaten directly by humans, the rest going to feed pigs and chickens in intensive rearing systems, a tragically wasteful use of natural resources.

Many food and development specialists hope that aquaculture, the farming of fish and shellfish, will make up the losses from ocean fisheries, but ironically fish farming has often destroyed natural fisheries. In Ecuador, for example, shrimp farming has led to the destruction of extensive stretches of mangrove swamps on which the fish depend. Fish farms also cause pollution problems, with antibiotics and waste products fouling the water, and their produce is often too expensive for the poor.

The simple truth is that the open seas can no longer be looked on as a limitless source of food or a bottomless dustbin. The seas are the cradle of life on our planet, a habitat in which living things have flourished and diversified for four billion years. As Thoreau remarked: "We do not associate the idea of antiquity with the ocean, nor wonder how it looked a thousand years ago, as we do of the land, for it was equally wild and unfathomable always. The ocean is a wilderness reaching round the globe..." Unless pollution and over-exploitation are brought to a rapid halt, that wilderness, with its immeasurable riches, is in danger of turning into a desert.

CORAL REEFS

Every year, during the monsoon rains, violent storms and raging seas batter the coast of Sri Lanka. Along those stretches where the island's protective coral reef remains intact, the seas do little damage. But where the coral has gone - mined for its limestone - the sea drives ashore in all its elemental fury, uprooting coconut groves, tearing up entire stretches of beach, washing away roads and railway lines, and destroying houses. So extensive was the damage in 1986 that a government minister warned that if emergency measures were not taken immediately, "there would be no coast to protect".

It was the Dutch who first began to mine Sri Lanka's coral reefs, using the tough limestone to build their forts and houses and even the pavements of their towns. The Sri Lankans wisely continued to build in mud. But independence brought a rush to modernize the country, and with it an unprecedented building boom. One massive housing scheme followed another, the aim being to provide European-style concrete housing for over a million people. Coral supplied the limestone. By 1987, 10,000 tonnes of coral were being mined a year - and the fate of large areas of Sri Lanka's coral reef was effectively sealed.

Covering some 600,000 km^2 (230,000 square miles) of the globe, coral reefs thrive in warm, shallow waters, the bulk of them lying in the tropics. Some corals are found in colder waters, but they do not build reefs.

◀ A sea fan coral branching into the waters surrounding a reef. Reefs are composed of many different species of corals. Some corals are adapted to withstand being pounded by waves. Others - like the sea fan - are more delicate, and can only survive away from the sea's turbulent surface, or in the calmer waters of lagoons.

The world's coral reefs

Coral reefs can only develop where the water is clear and where the water temperature does not fall below 20°C (68°F). Although reefs grow only slowly - sometimes just a few millimetres a year - the skeletons of dead corals build up over millennia to form vast underwater ramparts. The largest single reef, the Great Barrier Reef off Australia's east coast, is over 2,000 km (1,240 miles) long and covers an area of over 200,000 km^2 (80,000 square miles).

Northern and southern limits of reef-building corals

Areas with the greatest diversity of reef-building corals

“Deeper down are miniature blue trees with white blossoms. These are the real coral, the semiprecious *Corallium rubrum* in brittle limestone fantasies of form. For centuries coral was commercially dredged in the Mediterranean with ‘coral crosses’, a type of wooden drag that smashed down the trees and recovered a few branches. The once-thick trees on the floor that may have taken hundreds of years to grow, are no more.”
Jacques Cousteau

▼◀ A school of fish patrols the surface of a reef at dusk. Although life around a reef may look spectacular during the day, its true abundance is only revealed at night. Under the cover of darkness, countless kinds of fish and other animals emerge from crevices to feed. Some eat the coral itself, while others feed on planktonic animals or plants, or prey on the fish that do so.

There are three main types of coral reef, shaped to a large extent by the underlying geology. At the end of the last Ice Age, as the atmosphere warmed up, so sea levels rose. Some reefs kept pace with the sea, forming fringing reefs close to the shore. Others, known as barrier reefs, became separated from the land by deep channels formed as the seabed subsided. A third category, atolls, are reefs which have formed on the rim of a sunken volcano, the crater forming a lagoon. The world's greatest bank of corals is to be found in the Great Barrier Reef off the eastern coast of Australia.

BUILDING A REEF

The process through which reefs are created provides a remarkable example of mutualism, the development of mutually beneficial relationships between living organisms. The basic building-block of coral is a small organism known as a coral polyp. Living inside the cells of the polyp are countless tiny algae, known as zooxanthellae, which utilize minerals in the polyp's wastes. Unlike the polyps, zooxanthellae are photosynthesizers and are thus able to make sugars using the energy in sunlight. They convert the materials containing nitrogen, phosphorus and also carbon dioxide that the polyps excrete into oxygen and rich organic matter, which in turn nourish and support the polyps. Other algae grow outside the polyps on the surface of the reef, as grass would grow on land, providing plentiful food for fish and other animals.

The result is an extremely efficient method of mineral recycling, and one that enables corals to survive in waters that are poor in nutrients. Thanks to the activities of the algae, the productivity of corals is very high - they have been described as the marine equivalent of tropical rainforests. Like rainforests, they are among the most diverse ecosystems in the world, a typical reef containing as many as 3,000 species.

To protect themselves against predators, the polyps build up calcium-containing skeletons in which they live and feed. As the coral dies, so the reef grows, the empty skeletons being solidified by the action of bacteria and algae. Some of the algae also have protective shells which add to the process.

Coral reefs are a maze of crevices, cracks and caves that provide shelter for fish, and make them some of the most abundant fish nurseries in the world. Traditionally, reefs have thus provided a major source of food for local peoples - not just fish, but crabs, octopuses and turtles as well. Corals have also been used as building materials, as jewels and as a source of medicines.

By laying down limestone skeletons, polyps remove carbon dioxide from the atmosphere and thus play a major part in lowering the temperature of the Earth's surface *(see p.20)*. According to scientists such as James Lovelock, they may also may play a role in regulating the salt balance of the oceans. Given that the seas receive copious quantities of mineral salts every year through run-off from the land, it has always been a mystery how salt levels in the sea have remained so constant. Coral reefs may provide part of the answer. As they build up over thousands of years, they form lagoons and closed seas, effectively creating natural evaporation basins. Exposed to the Sun, these basins eventually dry out, leaving behind enormous salt deposits. Large quantities of salt are thus removed from the seas. Significantly, all the major salt deposits on land are contained within limestone barriers - which suggests that this process has been going on for millennia.

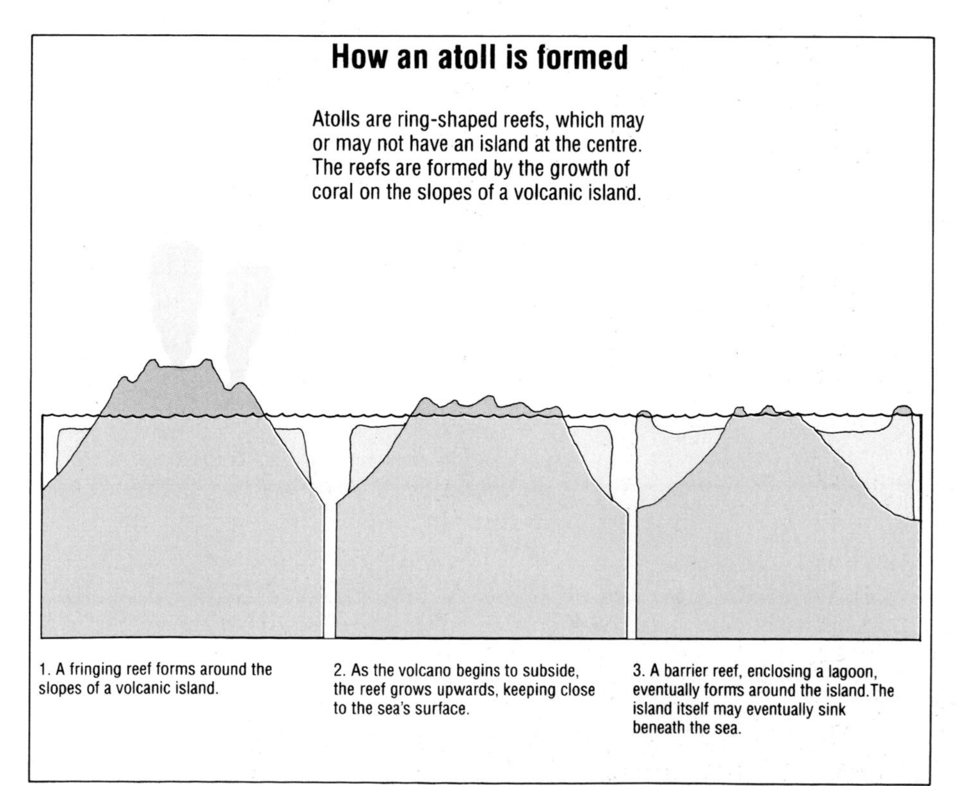

THE DESTRUCTION OF CORAL REEFS

Reefs are particularly vulnerable to any disturbance that might stir up sediment. One reason is that zooxanthellae cannot function in turbid waters; another that the sediment smothers and chokes the coral. As a consequence, the coral

▼ A fisherman in Martinique with shells and coral that will be sold to tourists. Souvenirs like these may be attractive reminders of a holiday in the tropics, but their collection causes widespread damage to coral reefs. In an attempt to preserve coral reefs and their wildlife, some countries have set up marine parks where collecting and spear-fishing are banned. The reefs inside these parks often make a striking contrast to those outside them.

► Coral mined from a reef in Indonesia. Using crowbars or explosive charges, coral miners can remove the results of hundreds of years of coral growth in a few hours. In places like Indonesia, reefs protect the shore against erosion by the sea. Mining the reefs breaches the coral wall, exposing the shore to the ocean's onslaught.

reefs of the world are under increasing threat from man. A recent survey revealed reef damage in 93 out of the 109 countries in the world with significant areas of coral reef.

Many reefs have been mined for their limestone, while others have simply been hacked to pieces. Some have been blown apart by fishermen using dynamite, the dead or stunned fish being netted as they rise to the surface. The short-sighted and wasteful practice of dynamite fishing has destroyed some of the finest reefs in Kenya, Tanzania and Mauritius.

Commercial coral- and shell-collecting has also taken its toll - imports of ornamental corals into the US rose seven-fold between 1960 and 1988 - as tourists, armed with crowbars, hammers and spears, have pillaged the coral for holiday mementoes. Anchors and chains are also damaging to corals.

Perhaps the most violently destructive of man's activities in coral reefs, however, is France's continued testing of nuclear warheads on the Polynesian atoll of Mururoa. Already some 100 devices have been tested and the atoll is gradually sinking into the sea. The French military are contemplating moving their activities to another Pacific island, Fangataufa.

◀ The crown-of-thorns starfish, which lives on coral polyps. Some marine biologists believe that man's activities are helping this potentially destructive animal to proliferate, causing the destruction of large areas of coral.

▼ Beach erosion of the Indonesian island of Bali. Following the destruction of offshore coral, storms have carried away the soil, toppling palms growing near the shore.

Deforestation, especially of hills near to coasts and of mangroves, has had a devastating effect on corals, the eroded soil and vegetation smothering many reefs. Logging in the watershed of Bascuit Bay, one of the finest fishing grounds in the the Philippines, has doubled the amount of sediment being washed downstream. Five per cent of the coral reefs in the bay are now dead.

Corals are also vulnerable to sewage and run-off of chemicals, both fertilizers and pesticides. The discharge of fertilizers and nutrient-rich sewage has led to damaging algal blooms, while chemical pollution has caused the outright poisoning of some reefs. It is also suspected, though not as yet proven, that human activities may be playing a part in causing the proliferation of the crown-of-thorns starfish *Acanthaster planci*, which is a major predator of coral polyps. The starfish is at present confined to the Red Sea and the Indian and Pacific Oceans, where it has ravaged many reefs, but there are fears that it may enter the Caribbean through the Panama Canal. It is not known if the infestations by the crown-of-thorns starfish are the result of the changes we are making to the reefs, or whether they are part of a natural cycle. Overfished reefs seem to be particularly susceptible to attack.

Changes in climate are certain to affect corals. If the tropical oceans get warmer, the zooxanthellae will disappear, since they cannot tolerate waters above 30°C (86°F). As they die, this causes the reef to bleach. In 1987 and again in 1989, mass bleaching of reefs occurred throughout the Caribbean. Bleached corals fail to grow properly and will only pick up again when recolonized by algae. Should global warming occur on a significant scale, the world's tropical coral reefs could well be condemned to extinction.

On the other hand, coral reefs could extend their range under warmer conditions, moving into subtropical zones. But they may also be 'drowned' if sea levels rise faster than the corals can grow. Some of the faster-growing corals, such as antler coral, are the most susceptible to damage from storms, which are set to become both stronger and more frequent as the temperature of the oceans rises.

STOPPING THE ROT

If used wisely, coral reefs could provide enormous benefits for mankind - as they have for centuries. One estimate puts the potential yield of fish and shellfish from coral reefs at 9 million tonnes a year. Fortunately, some governments are beginning to recognize the importance of coral reefs. Some 300 protected areas have now been listed in 65 countries, and a further 600 reefs have been recommended for protection. More encouraging still, organizations such as the World Wide Fund for Nature have now recognized the wisdom of traditional management practices, and in several parks they are actively seeking to put the management of reefs back in the hands of local communities.

But designating the reefs as national parks will not be enough. The root causes of the destruction must also be tackled. And that will prove the hardest task of all.

◄ The deeply sculpted island of Moorea, in French Polynesia, surrounded by its fringing reef. In this aerial view, the protecting effect of the reef can be clearly seen. The reef's seaward edge faces the full force of the Pacific breakers, but behind it, the lagoon's shallow waters are calm. Over millions of years, islands like this are gradually eroded to create atolls ***(see p. 189)*****.**

ISLANDS

The island of Socotra lies at the mouth of the Gulf of Aden, where the Indian Ocean meets the Arabian Sea. Part of South Yemen, it has some of the most exotic plants in the world. There are euphorbias with towering leafless stems, each growing up from a single root to form a dense mass of green columns, like giant candles. There are massive tree-like succulents whose flowers are red and moist, like figs cut open. Strangest of all is the cucumber tree, the only member of the melon and cucumber family to have shrugged off their usual lax, vine-like form to become sturdy and upright, with a fleshy, swollen trunk sprouting tufts of leaves and flowers at its apex.

Like 85 other plant species, the cucumber tree is 'endemic' to Socotra, meaning that it is found nowhere else. It is also on the verge of extinction. Another 130 of the island's endemic plants also face an uncertain future. Over the last two thousand years, since the arrival of man, Socotra has been nibbled bare by generations of goats, sheep and camels. Once home to endemic giant lizards, crocodiles and many different types of snake, it now has few of its native animals. What were formerly well-watered uplands, with rivers and green pastures, are dry, barren rock. Only on the steepest slopes, which intercept a little rain and provide some refuge from the ever-hungry goats, can the few remaining native plants survive.

◀ Giant tortoises have survived only on the Galapagos Islands, in the Pacific, and on Aldabra, in the Indian Ocean. Three other populations, on Mauritius, Rodriguez and the Seychelles, became extinct before 1900, slaughtered by whalers for food. As Charles Darwin noted, each Galapagos island has its own distinctive race of tortoise, identified by the intricate patterns of the shell.

▲▲ Guadeloupe, one of the few Caribbean islands whose native vegetation has survived.

▲▶ Sifakas on Madagascar, one of the 26 living species of lemur, fated to join the 12 already extinct, unless the desecration of their island home is halted.

▲ New Zealand's magnificent flightless takahe, threatened by introduced deer ***(see p.31).***

The story of Socotra is far from unusual. Scattered all over the world's seas and oceans are remote islands where man has done untold damage to the original plant and animal life. Much of this destruction began hundreds, or even thousands of years ago, as colonists or sailors arrived. Of the 94 birds known to have become extinct since 1600, 85 have been island species. They include the dodo of Mauritius, slaughtered by sailors for food, the great auks - the 'penguins of the North Atlantic' - plundered for their oil, and the moas of New Zealand, wiped out by Maori hunters. Many more species have suffered the depredations of animals such as rats, cats, goats, pigs and donkeys that arrived with man, particularly the European colonists. Rats eat the eggs and young of ground-nesting birds, such as petrels, flightless rails or New Zealand's endemic parrots, which have no defences against such predators because they have evolved without them.

Goats and other grazing animals destroy the native vegetation on which birds depend. In just 20 years, rabbits wiped out three endemic bird species on the Hawaiian island of Laysan. Introduced wallabies almost did the same for the Hawaiian goose, or néné, which has survived only through captive breeding.

Non-native or 'alien' plants can often be just as devastating in their effects, as the spread of gorse in parts of New Zealand shows *(see p.92)*. Increasing numbers of tourists in the Galapagos unwittingly bring the seeds of alien plants, stuck to their shoes, and invasive species are spreading from beside the islands' pathways to compete with the native vegetation.

THE EFFECTS OF ISOLATION

Island species owe their vulnerability to millennia of isolation, and to the fact that some plants and animals find it easier to reach islands than others. Among the animals, birds, bats and insects have the obvious advantage of flight, while cold-blooded reptiles can survive a fairly long sea voyage, floating on logs, without needing food. Long sea voyages are far more difficult for land mammals, and they have colonized only the more accessible islands, leaving most without mammalian predators.

Animals evolving in such circumstances do not need the defences that they would elsewhere. Island birds, for example, often lose their powers of flight, take to nesting on the ground, and can be remarkably tame. The dodo became a byword for stupidity because it was unafraid of man - fear had been unnecessary in its island world. As Charles Darwin explored the Galapagos islands, birds perched within arm's reach to take a closer look.

Isolation has also produced eccentricity and uniqueness among many island life-forms, and the more remote the islands, the more individual its inhabitants. On Hawaii, which emerged from the sea five million years ago as bare volcanic lava, new bird species may have arrived less than once every 250,000 years. Faced with untenanted niches, some of the species that did arrive expanded and diversified, producing a flurry of new species to fill the vacancies. Such evolutionary expansion is called 'adaptive radiation' and it generates striking arrays of closely related species with very different diets, such as the Hawaiian honeycreepers or the Galapagos finches.

Sometimes islands act as 'Noah's arks' for animals that once roamed widely, but which have since died out, due to competition from newer and more robust forms. If, by chance, the newer forms fail to reach an island, the original species can survive there. The lemurs are still alive for us to see today because monkeys and apes did not reach Madagascar. In a similar way, isolation preserved the egg-laying mammals in Australasia, giant tortoises on Aldabra and the Galapagos, and an ancient type of lizard, the tuatara, in New Zealand.

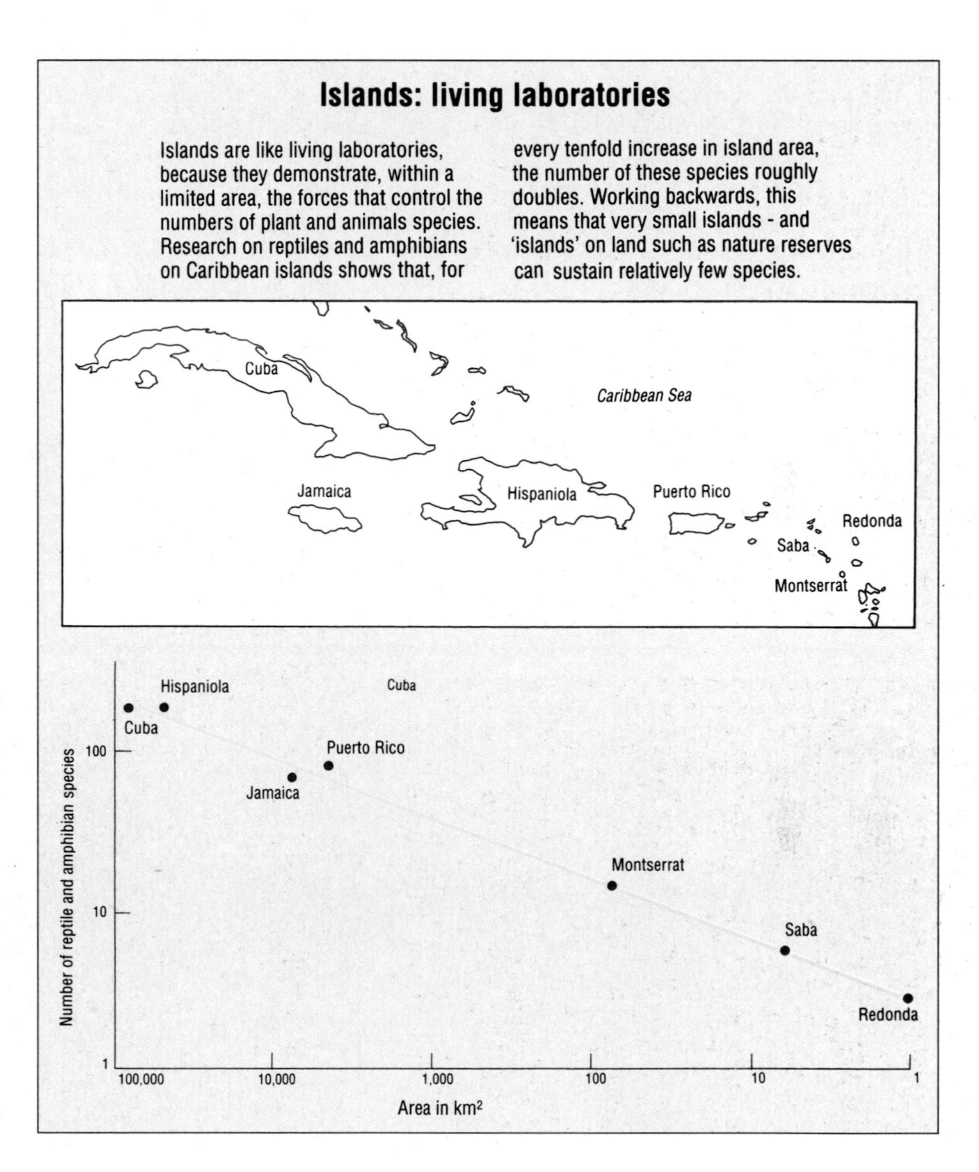

Adapted from 'Threats to Biodiversity', by Edward O. Wilson.

THREATS TO ISLAND WILDLIFE

Despite the clear messages of the past, non-native animals and plants are still being introduced in some parts of the world. Feral cats escaping in remote parts Australia do great damage to native marsupials, and mink, introduced into Britain by fur farms, threaten the few remaining populations of native otters. In New Zealand, the only hope for some species, such as the kakapo, is to move them to tiny islands that have never known rats, stoats or weasels.

"Such, then, is the face of Rona, a mere speck in the northern ocean; but such an island is much more than that biologically. Rather it is a metropolis for widely different animals through the seasons of the year, and a most important halting place for others."

Frank Fraser Darling

Habitat destruction on islands is far more damaging than on the mainland, because the habitat is already small and any further reduction can reduce populations to non-viable levels. Deforestation on Hawaii has already wiped out many of its endemic birds, and the situation facing the lemurs, tenrecs and other unique species on Madagascar is extremely urgent. Islands often have strategic importance and this too puts them at risk. But for emergency campaigning by conservationists in the 1960s, Aldabra would now be a US Air Force base, and it is doubtful if its giant tortoises and unusual vegetation would have survived.

The 'out-of-sight-out-of-mind' attitude to islands is another grave threat to their future. Dumping convicts and unwanted dictators on them is no longer considered acceptable, but the same approach is evident in US attempts to use Johnson Island in the Pacific for dumping its wartime chemical wastes. The continued testing of nuclear weapons in the South Pacific by France, which jeopardizes the health of Pacific islanders, not just for the present, but for hundreds of years to come, is environmental imperialism of the worst possible kind.

For some islands, the most serious threat is spelled out by apparently unconnected events far away: the extravagant use of fossil fuels in the developed nations, and the headlong destruction of the tropical rainforests. Even a small rise in sea levels, triggered by global warming *(see pp.40-53)*, will engulf low-lying islands such as the Maldives, Tuvalu and Tonga.

► Debris from nuclear tests on Christmas Island.

►► New Caledonia's vegetation includes the bizarre *Araucaria* trees, an ancient race of conifers which have developed into many endemic species on this and neighbouring islands. Beyond the trees, mining operations have stripped the land of these unique trees - and all other plants.

▼ The seemingly pristine environment of French Polynesia is still being poisoned by underground nuclear testing.

MOUNTAINS

For the Ok people of the highlands of central New Guinea, Mt. Fubilan was a sacred mountain, sitting on top of the land of the dead. In the late 1960s they were persuaded to lease their mountain to a mining company. To the utter astonishment of the Ok, the company began systematically to scoop away the peak of Fubilan. Within the next two decades, the 2,000-m (6,560-foot) peak will have ceased to exist. In order to exploit Mt. Fubilan's reserves of copper and gold, the mining company intends to remove the sacred mountain altogether. When the mine is finally exhausted, all that will be left is a hole in the ground, 1,200 m (3,900 feet) deep.

THE NATURAL HISTORY OF MOUNTAINS

With its covering of lush, tropical vegetation, Mt. Fubilan is a particularly rich example of a mountain ecosystem. Mountainous areas cover around a third of the world's land surface. Because of their height, they have a climate which is different from the surrounding landmass, and their own distinctive varieties of plants and animals. Climbing a mountain is like making a journey towards one of the poles, with the temperature getting colder and colder, and the types of plants and animals changing as one gets higher. A climb of 200 m (650 feet) generally results in the temperature dropping by more than 1°C (nearly 2°F).

High areas, especially when they are in the path of warm moist winds from the ocean, receive large amounts of rain and, at higher altitudes, snow. Because the air passing over a mountain ridge has already lost most of its moisture, the regions on the leeward side tend to be much drier and are described as being in a 'rain shadow'.

◀ Mont Etale in the French Alps. Although they may still appear clean and unspoiled, motor traffic has made the Alps the most polluted mountain range in the world.

Mountains contain a number of quite distinct vegetation zones. The species of plants and animals that occur, and the altitudes of the boundaries between the zones, vary widely between mountain ranges in different parts of the world, but the sequence of zones is broadly similar. A striking example is provided by Mt. Colombus and Mt. Bolivar in Colombia's Sierra Nevada range. Just 42 km (26 miles) from the coast, these 5,775-m (19,000-foot) peaks are the highest next to the sea anywhere in the world. Perhaps nowhere else in the world are virtually all the major ecological zones so clearly defined. On the shores at the foot of the mountains, turtles breed and caimans slide into the still waters of mangrove swamps. The coastal plain with its thorny scrub and cactus has a warm, steamy climate. Above 200 m (650 feet), humid tropical forest begins. This is rich in many different species of plants and animals including palms, rubber plants, cedars, carob trees, massive lianas hanging like twisted ropes and an incredible variety of snakes, birds and butterflies. Here we find the jaguar stealthily hunting the tapir, and the occasional ocelot. Eventually the humid forest makes way for the cold rains of the cloud forest, where wax palms, laurel, alder and myrtle grow. Above 3,000 m (9,850 feet), dwarf trees and golden grasses are found in the rain-swept tundra and moorland. Higher still, not even dwarf trees and shrubs can grow, and the landscape is an empty alpine grassland, which is finally superseded by the barren icy wastes of the two peaks.

At high altitudes, where the summer temperatures never rise high enough to melt all the snow, this may compact into ice and form a

▲ A glacier reaches down from the Mont Blanc massif. One likely effect of global warming will be the melting of glaciers.

◀ A North American mountain lion, or cougar. Mountain ranges often provide a refuge for large mammals that cannot survive elsewhere due to human intrusion. But for the cougar, relentless persecution by hunters has made it scarce even in remote mountain areas.

▲ Transhumance - moving herds between lowland pastures in winter and upland meadows in summer - is a very successful way of life in mountain regions, and one that has been practised for thousands of years.

◀A herd of female bighorn sheep, with their young, in Wyoming. Sheep and goats are supremely well adapted to mountain life, being sure-footed and hardy. Over-hunting has reduced the bighorn sheep to very low numbers.

glacier. Glaciers are constantly changing as new snow falls and old ice melts. As the Earth's climate is presently getting warmer, most glaciers are now shrinking - in the Alps, many glaciers have lost up to a third of their area in the last century. The meltwater from glaciers forms the sources of the Rhône, the Ganges and many other great rivers.

MOUNTAIN PEOPLES

Some 500 million people, 10 per cent of the world's population, live in mountain regions. This number includes a high percentage of tribal peoples, such as the hill tribes of Thailand, Laos, Burma and southwest China, and the many Indian tribes of the Andes, all of whose rich cultures are increasingly under threat. Only the most inhospitable mountains and plateaus are completely uninhabited.

People have been living off the land in mountain areas for many thousands of years. In the Alps, the practice of transhumance probably has its origins in pre-Celtic times. Under this system, cattle are moved in springtime from their lowland pastures up to the high alpine meadows where they graze on the new grass during the short summer growing season. On the slopes of Mount Kilimanjaro in Tanzania, the traditional irrigation system of the Chagga people amazed early European travellers with its efficiency and complexity. The system consists of a network of irrigation furrows that collect water from the mountain streams and transport it over long distances to the fields below.

▲ Mountain slopes in the Philippines, with terraces for rice-growing created over 2,000 years ago and carefully maintained by generations of farmers.

► Tibetan traders with their yaks, crossing the Tesi Lapcha Pass in Nepal. Most of the mountain ranges of the world were colonized thousands of years ago, and are still inhabited by the same hardy, resourceful peoples, whose traditional way of life is the only viable form of subsistence in such harsh conditions. For the Tibetans, yaks are the key to survival, providing transport, meat, wool, milk, butter and dung for the fire.

The greatest problem which mountain farmers face is soil erosion. When steeply sloping land is cleared for agriculture, heavy rain washes away great quantities soil and causes landslides. Downstream, the soil eroded from the hillsides silts up reservoirs and rivers. Without trees to soak up the rainfall and release it slowly over the year, heavy rain in the mountains leads to severe flooding in the valleys below.

The most common and effective method of preventing the loss of valuable soil is to reshape the mountain slopes into a series of small terraces. In the Cordillera mountains of the Philippines, an incredibly complex and efficient system of irrigated rice terraces has been used for over 2,000 years. The local indigenous peoples, known collectively as Igorots or 'people of the mountains', have chiselled their terraces from the rocky slopes, in some places forming a 1,000-m (3,300-foot) stairway from the riverbed to the Cordillera peaks. In the high Andes of Peru, traces can still be seen of the Incas' terraced fields and irrigation systems which, until the coming of the Conquistadors, supported a huge mountain empire.

Although it is capable of sustaining high yields without expensive chemical inputs or modern equipment, traditional terrace farming is under threat worldwide. Terraces, especially when they are irrigated, require a great deal of labour for their upkeep. Traditionally this labour would be provided free and would form a vital part of the community's life. The farmers' beliefs and festivals would be closely tied in with the annual cycle of planting, harvesting and repairing the terrace walls.

The migration of young people to cities, the transfer of communally-owned lands into private hands, and the introduction of wage labour have all undermined community ties and therefore also the farming systems which have protected the mountains for millennia. Throughout the world, mountain farmers are now being encouraged to tear down their terraces as they are unsuitable for tractors and other agricultural machinery which cannot operate on the narrow, contour-hugging fields.

The problem has been compounded by the displacement of peasants from fertile valley lands to make way for cash crop plantations. In El Salvador, plantations now take up half of the total farming area in the country - and almost all of the best land. The only land available to displaced peasants is on the steep rocky slopes of the mountains. Erosion on these slopes is so

▲ Wheat growing on the floor of a steep-sided valley in Peru. In mountain regions like this, land for crops is won and maintained by hard work, and careful cultivation is essential to ensure that the thin soil does not wash away. Changes in the rural economy often mean that marginal land like this falls into disuse as people move away to towns and cities.

severe that many of the peasants must move on every year. The mountain slopes are streaked with long scars of rust-red soil which is washed away into the Pacific, forming a red plume in the azure sea.

Erosion is so severe in some mountain areas that it is resulting in desertification, the final stage in the degradation of the land. The effect of the disastrous Ethiopian drought of 1984 was greatly worsened by the deforestation and resulting erosion of the once-fertile Ethiopian highlands, which now lose more than 16 billion tonnes of soil a year.

THE DESTRUCTION OF THE ALPS

The popular European image of mountains is of pure, untouched, snow-capped, ageless, rocky peaks above verdant slopes of pinewood and meadow, far removed from the pollution and overcrowding of the towns and cities below. The truth is not so romantic. Acid rain, caused by road traffic and faraway industries, is having a horrendous effect upon the highland forests of Europe and North America.

The Alps, Europe's most important mountain and hydrological system, are the world's most heavily polluted mountains. They stretch through seven countries - Austria, France, West Germany, Italy, Liechtenstein, Switzerland and Yugoslavia - and are the source of four of Europe's greatest rivers, the Rhine, the Rhône, the Danube and the Po. Although Switzerland and Austria, the countries which cover most of

▲ Deforestation in Nepal. Loss of tree cover leaves the precipitous slopes with no defence against soil erosion, which is severe in some parts of the Himalayas. Tourism and overpopulation contribute to the problem, but the replacement of the original forest by plantations, is also a factor in some areas. This removes the forest understorey, which once provided abundant firewood, and leaves the mountain people with no choice but to cut down trees.

◀ The sparse vegetation of the Himalayas is increasingly under pressure, as more is used for timber, fuel and food. Here fodder is being collected for domestic animals.

▶ Val d'Isère in France. Skiing developments, with their roads, hotels and ski-lifts, have damaged many mountain regions.

the alpine region, have some of the world's strictest environmental regulations, the relentless development of the mountains, and the pollution which these countries receive from their neighbours, has cancelled out any gains which these regulations may have achieved.

Straddling Europe's main north-south trade routes, more traffic passes through the Alps than any other mountain range. Each weekend, vehicles on Switzerland's St. Gotthard Pass deposit 30 tonnes of nitrogen oxides, 25 tonnes of hydrocarbons and 75 kg (165 lbs) of lead into the atmosphere, bringing death to the trees. More than 50 per cent of alpine trees are sick, and 15 per cent are dead or dying. Without trees to hold snow in place, avalanches increase in frequency and severity.

The skiing industry, which sells itself on the Alps' pristine beauty, has also devastated many areas. The 50 million people who ski in the Alps each year require roads, parking lots, ski-lifts and cable cars to get to the slopes, and apartments, shops, bars, restaurants and amenities such as sewage disposal and water, when they get there. In Switzerland alone, 250,000 mountain chalets and hotels cover the equivalent of 16,000 football pitches. At La Plagne in the French Alps, the site of the bobsleigh run for the 1992 Winter Olympics, a ravine has been cleared of trees and bulldozed, and a refrigeration plant for the track ice, viewing platforms and concrete walkways are being constructed. A four-lane motorway is being driven through the Tarantaise valley in anticipation of the massive increase in traffic which the French Winter Olympics will bring.

Although there have been protests, environmentalists will not stop the Winter Olympics. But global warming caused by the greenhouse effect might. Emergency state aid was given to French ski resorts in 1989-90 due to the third season in four years with late snow. The lack of snow is proving disastrous as far as the skiing industry is concerned, and they now must consider the possibility that the Alps are undergoing a permanent change in climate.

The traditional alpine landscape of pristine meadows, streams and pinewoods, and the communities and farming practices which maintained it, are rapidly becoming things of the past. Likewise, many plant and animal species such as the chamois are under severe threat. More than half the species listed by the Council of Europe as disappearing or gravely threatened, have their habitats in mountain regions. According to the Bellerive Foundation, a conservation organization headed by the Prince Sadruddin Aga Khan, the abandonment or neglect of alpine pastures is a major factor contributing to the steady increase in avalanches and other natural disasters in the region.

At a conference organized by the Bellerive Foundation in 1989, scientists predicted that if present trends continue, within 15 years there will be few trees, little snow and massive soil erosion in the Alps, and the mountains will be on their way to becoming a rocky desert.

▼ Lurcher's Gulley in the Cairngorms, Scotland, the proposed site of a new skiing centre. Its peaceful, unspoilt landscape and its importance to wildlife have made the proposal highly controversial.

DESERTS

The Negev in the south of Israel is a true desert. It has been so since the dawn of recorded history: Abraham crossed it on his way to Egypt from his home in Hebron through Beersheba - the place of the seven wells. With an average 100 mm (4 inches) of rain per year and a scorching Sun in midsummer, it would seem that nothing permanent could ever survive. Yet the Negev, like many deserts, is alive with plants and animals, all with extraordinary strategies for exploiting and conserving the little moisture that there is.

When the rains come, tumbling out of the sky, the water runs across the sloping terrain in sheets, gathering in the dry riverbeds - the wadis - and forming swirling torrents that eat away at the ground, eroding vast chunks of soil and carrying all away with them. In just a few hours the water may have vanished, except for damp spots along the bottom of the wadis. But, almost miraculously, small fragile plants are soon pushing up between the crevices, plants like *Asteriscus pygmaeus*, a tiny member of the daisy family, which completes its life cycle in just two weeks. Such plants have evolved ingenious mechanisms for adapting to the uncertain rainfall, letting just a few fruits ripen at a time. Even these will not germinate until the rains have been sufficiently heavy.

◀ Sand sunflowers paint the deserts of Utah with vivid yellow after a rare rainstorm. Deserts harbour the dormant grains of life beneath their parched surface, waiting for such ephemeral moisture to revive them. Within hours of the rains, seeds have germinated, insect eggs have hatched and toads have broken out of their subterranean cocoons. In a frenzy of growth and reproduction, they live out their lives in a matter of weeks or days, before the rainfall has evaporated.

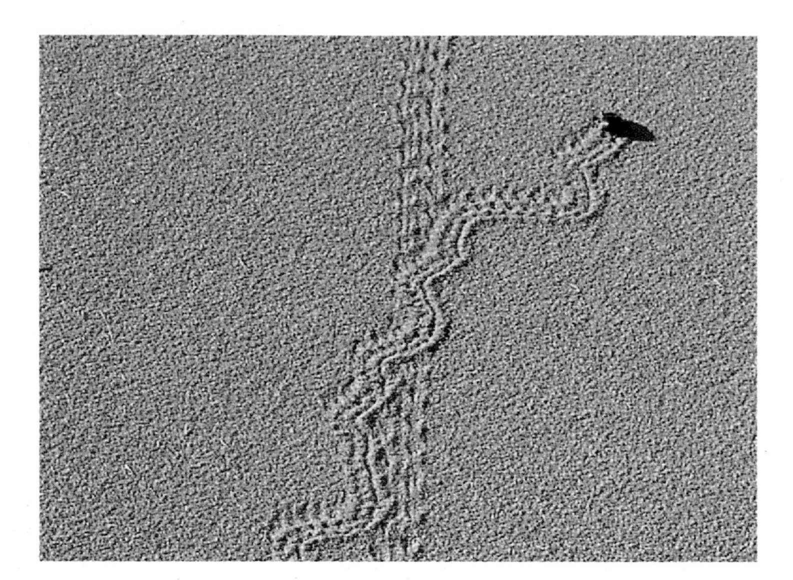

◀ Trails made by beetles as they labour up the face of a sand dune. Most desert insects escape the worst heat of the day by burrowing underground.

▼ The Sonoran Desert of Arizona. The 'rustling' of these stately cacti for sale to landscape gardeners has become a threat to their survival.

These ephemeral annuals do not, to paraphrase the words of Gray's *Elegy*, "waste their sweetness on the desert air", because their large scented flowers attract the scarce insects that have adapted to desert conditions. The perennial plants, on the other hand, have to keep a hold on life during the long hot spells. Trees and shrubs, like the acacias, the tamarisk, the creosote bush and the mesquite of the southern states of North America, develop tap roots that can reach groundwater at depths of 50 m (165 feet). The geophytes, literally 'earth plants', spend the majority of their lives underground in the form of apparently lifeless corms or bulbs, producing aerial shoots only after heavy rains. Other plants have water-bearing succulent leaves and stems with thick, waxy cuticles.

▼ Wild asses in the Great Rann of Kutch Desert, in northwest India. In the open desert large mammals make easy targets: many are endangered by hunting.

Many desert perennials, including trees like the acacias and cacti, have a covering of spines to help keep away would-be grazers. There are over 2,000 species of cacti - all of which are endemic to the Americas - and their characteristic knobbly shapes are well adapted to arid, sunny conditions. The majestic saguaro cactus, for example, presents the minimum surface to the midday Sun, but the maximum surface on its sides when the sun is rising or sinking.

Almost all plants draw up water through their roots. Some of this water is combined with carbon dioxide to make sugars through the process of photosynthesis, and some is lost through tiny pores, called stomata, which are scattered over the leaves and stems. In common with some other fleshy or 'succulent' plants, cacti have evolved a special way of minimizing this water loss. Their stomata open only after dark, unlike those of most plants, which can also open up during the day. In the cool of the desert night, very little water evaporates and escapes through a cactus's stomata. The cactus can absorb the carbon dioxide it needs for photosynthesis, storing it until daytime when it can be used. It is the special ability to store carbon dioxide that makes this strategy possible.

ADAPTING TO EXTREMES

Many animals survive in the desert by retreating under stones or burrowing in the sand during the day, emerging only at night. In sandy deserts, animals such as the golden mole of the Namib, or the marsupial mole of Australia, find their prey by 'swimming' through the sand, rarely emerging into the full glare of the Sun. The small, slender lizards known as skinks can also move about in this manner, and many have tiny limbs which can be folded away at the side of the body during sand-swimming, to make the body more streamlined.

Some of the smallest desert animals, such as spiders and insects, can derive their daily requirements of water from dew, or simply from the food they eat. Small mammals, too, such as the tiny fennec fox and the sand cat, also need very little water. Larger mammals are expert at surviving long periods of water shortage. Gazelles can lose up to 20 per cent of their body weight in water, whereas non-adapted animals such as humans would perish after losing 12 to 13 per cent. By grazing at night, gazelles 'drink' the dew absorbed in the leaves of *Disperma*, which in the daytime have only 3 per cent water, compared with 40 per cent at night.

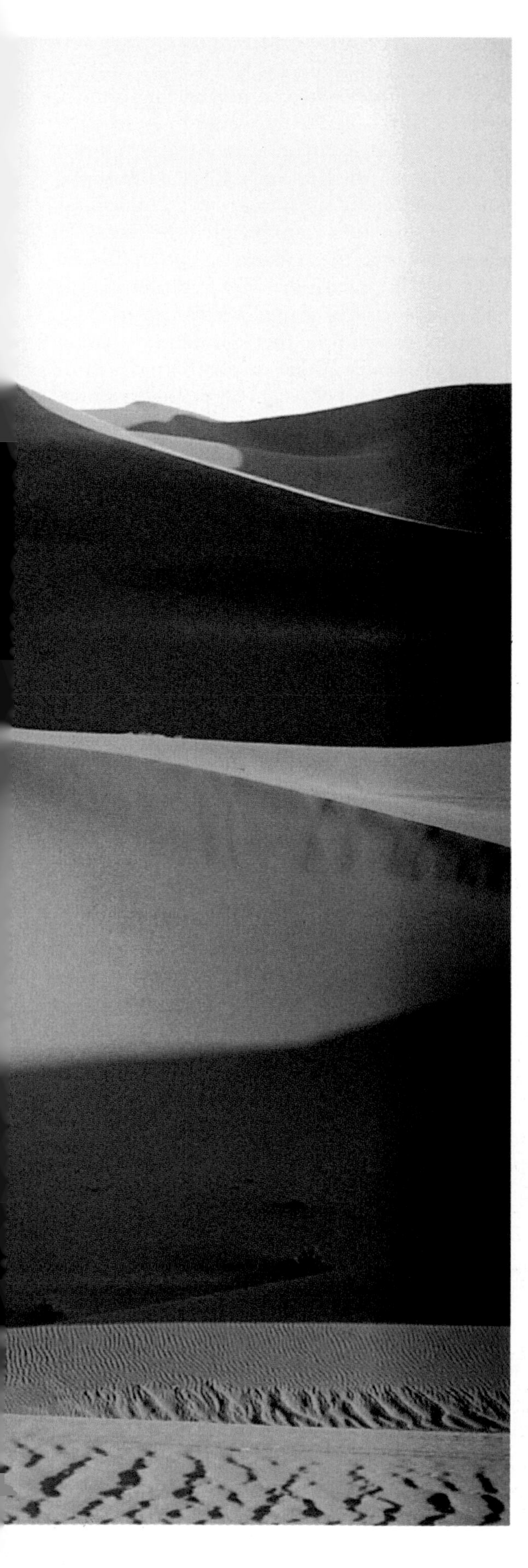

Other adaptations prevent overheating. The oversized ears of the fennec fox serve not just to capture sound but also to cool the animal. Those animals which cannot wholly avoid being exposed to the Sun often have light-coloured coats which reflect radiation. Mammals such as Grant's gazelle allow their body temperatures to climb as high as 46°C (115°F) during the day and then to cool to 34°C (93°F) in the night - a range that would kill most other species. The main threat from high temperatures is damage to the brain, and the gazelle overcomes this problem by keeping its brain 3°C (5.5°F) cooler than the rest of the body. This is achieved by routeing blood, destined for the brain, through a network of tiny blood vessels close to the nose, where it is cooled by evaporation of water.

Even man has discovered how to survive in deserts, chiefly by exploiting the camel, which can transport his tents and chattels across the sands, and which provides meat, milk and leather. The camel does not store water in its hump, as goes the popular myth, but conserves it through a heat-exchanger in its nose. This uses the relatively cool inhaled air to extract the moisture from hot, vapour-laden air being exhaled from the lungs. By bringing the two into close

◀ The forbidding beauty of sand dunes in the Namib Desert of southern Africa.

▼ When rain comes in deserts, it may do no more than skim the surface, leaving a thin layer of mud.

▼▼ The Pinnacle Desert in Nambung National Park, Australia. The pinnacles, which are made of chalk, mark the ancient sites of trees and bushes.

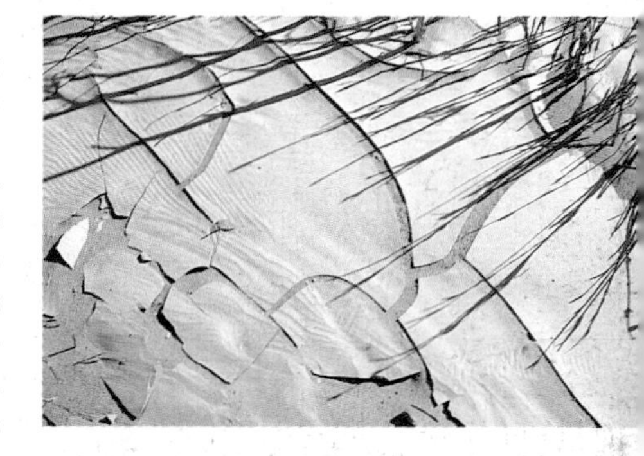

contact in the airways, the exhaled air can be induced to unload its burden of water. Once this has condensed, it can be resorbed by the body. In addition, the camel's kidneys, like those of desert rodents such as the kangaroo rat, pocket mouse and golden hamster, have 'countercurrent multipliers' which take the water out of the urine and leave it highly concentrated. The camel stores energy in the fatty tissue of its hump which, with its apex of heavy fur, insulates the part most exposed to the Sun.

Deserts may be the hottest places on the Earth's surface during the heat of the day, but at night they can become bitterly cold. The clear night skies and the light colour of sand and limestone rocks make deserts radiate so much heat that it is even possible to wake up in a desert and see the ground covered in frost. Indeed, deserts are heat sinks: they lose more heat over the course of the year than radiates down to them from the Sun. The difference is made up by heat transported in the air from the humid tropics. Deserts therefore play an important role in the circulation of the atmosphere from the equator to the higher latitudes.

Desert animals have to adapt to cold just as much as to heat. Invertebrates, as well as cold-blooded vertebrates such as lizards, position themselves in such a way as to benefit from the warmth of early morning Sun.

A FRAGILE HABITAT

In the late nineteenth century, an Englishman, E. H. Palmer, was attempting to follow the ancient camel route up from King Solomon's mines close to the Red Sea at Eilat to Beersheba, when he stumbled upon an ancient, uninhabited city. Later investigation revealed the city to be Avdat, home to the Nabateans, who in Biblical times had also carved the city of Petra out of the red sandstone rocks of the Jordanian desert.

Looking down from the high vantage point of Avdat, Palmer remarked on the pattern of mounds and lines of stones on the hills around. During the 1960s an Israeli botanist, Michael Evenari, discovered the significance of the stones. Through experiments he was able to show that they acted as run-off channels, guiding the flood waters from the scarce, unpredictable rains, into low-lying 'fields'. The run-off could therefore be multiplied up to 10 times in these special areas, providing the equivalent of 60 cm (24 inches) of rain a year. Archaeologists now had the answer to the puzzle of records from the Nabateans showing that they had grown wheat, barley, pulses, almonds and even grapes in one of the world's driest deserts. During his lifetime, Evenari established a desert farm based on the same run-off principles as employed by the Nabateans.

▼ Nomadic Bedouin, their tents dwarfed by the expanses of the Saudi Arabian desert. Traditionally, such pastoralists survive by being thrifty with the desert's resources and with those of their animal herds - almost every last hair and sinew of an animal is put to good use. However, here the presence of cars shows that the traditional way of life is changing.

► Semi-desert on the Arabian peninsula. Where seasonal streams flow, bushes and even trees can survive. But there are too few plant roots to bind the soil effectively, and little organic matter, so wind and rain easily erode the soil. Such habitats are extremely vulnerable to overgrazing, especially by goats.

▶ The Atacama Desert in Chile. Plants have different ways of coping with the arid conditions. Cacti store water in their swollen stems, and have sharp spines to keep marauding animals away from their precious moisture. Small thorn bushes have deep roots to tap underground water and leathery, unappetizing leaves.

▶▶An aloe tree in the Namib Desert. Like the Atacama, this is a coastal desert, receiving almost no rainfall, but imbibing tiny drops of moisture from sea fogs.

“...we were struck by a devastating thunderstorm one evening in northern Chad... Within a few minutes, the entire landscape, brilliantly illuminated by the continual lightning, was inundated... we seemed to be floating in a vast sea, the ripples from the wind giving the appearance of rapid currents. When the rain stopped abruptly after six hours, we literally had to shout to be heard above the fantastic chorus of croaking toads and stridulating insects.”
Professor John Cloudesley Thompson

In the midst of such dryness, there is sometimes water in abundance. Underlying many deserts are vast reservoirs of groundwater *(see pp.142-149)*, much of it laid down 20,000 of years ago, when climatic conditions were different. The northern Sahara, in Tunisia and Algeria, for instance, is estimated to contain 60 trillion m^3 (over 2,000 trillion cubic feet) of water lying at depths of 1,500 to 2,000 m (5,000 to 6,500 feet). This water lies under 600,000 km^2 (230,000 square miles) of sedimentary basins.

With assistance from aid agencies, countries such as Tunisia are hoping to exploit these reserves to help make the desert bloom. The cost of having to pump water up from such depths is extremely high, as high in fact as using a desalination plant. Nevertheless, by 1988 Tunisia was extracting 500 million m^3 (17.5 billion cubic feet) per year - enough water to irrigate 18,000 ha (45,000 acres). At Al-Kufrah in Libya, where 50,000 ha (120,000 acres) are under irrigation, the water-table is believed to have sunk by 35 m (115 feet) in 40 years.

In general the world's deserts - those extremely fragile ecosystems - are now under threat. The tendency is to view deserts as hostile and useless areas of wilderness, except where they lie over rich oilfields. More deadly is their use for underground nuclear testing, which still takes place in desert areas.

The lack of water in deserts is seen as a great advantage when it comes to the dumping of industrial waste since - as long as the climate stays as it is - waste can be deposited without fear of corrosion. China, for example, has offered to take West Germany's nuclear waste from its reactors, and in return for a payment, to dump it in the Gobi Desert. Some West African countries have also offered to take hazardous chemical waste from Europe and dump it in their desert regions. This takes the Third World a step further into becoming a rubbish tip for the affluent industrialized countries.

Certain deserts, such as the Negev, have now become the playground of the affluent society, which sees their pristine emptiness as ideal for zigzagging around in four-wheel-drive vehicles. These destroy the fragile plants on which the desert ecosystem depends, leaving tyre marks that remain in evidence for years. Without bacteria, without rain to corrode, the litter and detritus of modern society remains an unsightly heap on the surface, a stark reminder that no wilderness is sacred.

ANTARCTICA

Antarctica has the reputation of being the last untouched wilderness on Earth. Yet already the hand of man is scarring the landscape. Leningradskaya is one of more than 50 scientific research stations that have now been established on this most isolated of all the continents. Manned and operated by the Soviet Union, the base is perched on top of a 300-m (1,000-foot) cliff. Strewn on the sea-ice below are hundreds of black dots which a casual observer might think were seals. On closer inspection, they turn out to be discarded oil drums. Over the years, thousands of others have been carried away with the sea ice.

The story is the same elsewhere. The seabed next to the huge US Base at McMurdo, adjacent to Winterquarters Bay, has been transformed into a graveyard for thousands of tonnes of scrap metal, beer cans and redundant vehicles. The bay is now biologically dead, poisoned by chemicals and raw sewage from the base. The destruction at the French base, at Point Géologie in Terre Adélie, is even more disturbing, if only because it is deliberate. The French came here to study the 75,000 penguins and seabirds that inhabit the archipelago. They are now systematically blowing up five islands - and with them the birds' nesting sites - in order to build an airstrip.

◀ Despite its forbidding interior, Antarctica is surrounded by seas that abound with life. Here adult and young Weddell seals rest on the ice, while a flock of cape pigeons - which breed on subantarctic islands - feed offshore.

► Airstrip construction underway at the French base of Dumont d'Urville, near Point Géologie. The scene of clashes between workers and Greenpeace protesters, Dumont d'Urville base highlights the controversy between those who believe that Antarctica should be left as a wilderness and those who wish to study the continent to discover how it may serve humanity.

▼ Dumped oil drums at the Agentinian base of Vicecomodoro Marambio, in the Antarctic Peninsula. Scenes like these have now prompted many bases to send their waste home.

▼ The Wright Valley in Victoria Land, near the Ross Ice-shelf. Like the other dry valleys in this region, its floor consists of bare, shattered rock. Although there is almost no moisture here - the little snow that falls evaporates every summer - simple forms of life can be found even in this inhospitable environment.

The continent of Antarctica once formed part of a vast landmass known as Gondwanaland, which included the modern continents of Australia, South America, India and Africa. Five hundred million years ago, this super-continent enjoyed a temperate climate and was covered with lush vegetation. Fossil evidence of this exists, as does evidence of a giant penguin which stood 1.6 m (5 feet 6 inches) tall.

As the continent began to break up, the section that we know today as Antarctica drifted southwards. Its climate changed and around 25 million years ago it began accumulating its icy mantle, which today averages more than 2 km (1 1/4 miles) in thickness and covers 98 per cent of the continent.

The plants and animals of Gondwanaland, which had once been so abundant, disappeared. Indeed, with temperatures averaging -71°C (-96°F) in the winter and wind speeds sometimes exceeding 300 km/h (185 mph), Antarctica is now a singularly inhospitable environment. On land, the only animals that can survive without recourse to the sea are small arthropods, such as springtails and mites, most of them microscopic, the largest being a wingless midge, *Belgica antarctica*, which is 1.3 cm (1/2 inch) long. The vegetation is sparse, consisting mainly of lichens and mosses, and is extraordinarily vulnerable to even minor physical disruption. A human footprint in a bed of moss can remain clearly visible for several years. The lichens grow at a snail's pace, colonizing rocks and patches of nutrient-deficient soil with their delicate crusts of sulphur-yellow and orange. Some are thought to have a 2,000-year lifespan, undergoing freeze-drying in winter and rehydrating with the return of the Sun in spring.

But the lack of land animals on Antarctica is amply compensated for by the abundance of its marine life. The high oxygen content of the cold water, coupled with long periods of sunlight during the summer months, are the ideal ingredients for a rich soup of life. Come the summer, melting snow and ice deposit up to 500 million tonnes of nutrient-rich sediment onto the seabed, providing a plentiful food source for marine organisms. Sedentary feeders, such as sponges and anemones, starved of food for most of the year, blossom into life. The abundant cells of phytoplankton, which float in the surface waters, multiply rapidly in the sunlight. They are eaten by a variety of animals, including krill, the small shrimp-like creatures that are the very foundation of life in the Antarctic Ocean.

Directly or indirectly, krill support the squid, fish, penguins, birds, seals and whales which abound in the teeming waters of the Antarctic. Indeed, so important are krill to the marine ecosystem that a predation pattern has evolved, in which different species take krill of different ages at different times, at different depths and in different locations. Truly gigantic swarms of krill have been found in Antarctic waters, frequently turning the sea red with their numbers. Their great abundance means that even baleen whales can feed on them directly, taking great gulps of sea water into their cavernous mouths then squeezing out the water to leave the krill trapped by long, flat, sieve-like structures, known as baleen plates.

Today, whale and krill alike are threatened by a relatively new and relentless predator in Antarctic waters - man.

"I watched the sky a long time, concluding that such beauty was reserved for distant, dangerous places, and that nature has good reason for exacting her own special sacrifices from those determined to witness them."

Admiral Richard E. Byrd, first man to fly over the South Pole, 1926

▲ A jungle in miniature - Antarctic lichens and mosses seen in the height of summer. The far interior of Antarctica is too cold for life, but the Antarctic Pensinsula and its outlying islands are warmed by the sea. Here, the rock near the shore is often free of ice and snow during the summer months, allowing simple land plants to survive. These plants form the base of a short food chain which includes springtails and mites.

THE GREAT WHALES

Whales are at the top of the Antarctic food chain. Early voyagers to Antarctica report veritable carpets of whales stretching from horizon to horizon. Remorselessly hunted in every ocean of the world for centuries, many species are now close to extinction. By the 1930s, 40,000 whales were being taken in Antarctic waters every year. As their numbers decreased, protection was afforded, but it was always too little too late. The International Whaling Commission's commercial whaling ban of 1985 is still being flouted by the Japanese, now the only nation that continues to send whaling ships to Antarctica.

The ban allows whales to be taken for 'scientific purposes' with prior notification. The quota is set by the country itself, which must justify its programme from a scientific perspective. To the anger and dismay of many of the other members of the IWC, Japan proposed, at the Commission's 1987 annual meeting, that it take an annual catch of 825 minke and 50 sperm whales a year for 12 years - all in the pursuit of 'science'. Legal action and scientific criticism by conservationists, however, forced Japan to revise its proposal and it has reduced its annual take to 300 minke whales.

Whatever 'scientific' procedures are carried out on the whales, they are clearly perfunctory. Within an hour of the carcase being hauled up the slipway of the factory ship, the whale is flensed and the meat is being frozen in readiness for the fashionable restaurants of Japan. Although Japan still insists that minke whales are numerous in Antarctic waters, in 1988 they were

► A Japanese whaling ship hauls in its catch. There have been worldwide protests that Japan's 'scientific whaling' is little more than a disguise for commercial hunting, but still the practice continues.

▼ Adult and young king penguins on the South Georgia islands, east of Cape Horn. Penguins breed in thousands-strong colonies, with birds returning to the same spot year after year. Penguins were once killed in large numbers for their meat, skin and oil.

unable to catch enough to meet their self-imposed quota.

Whales are not the only Antarctic species to have suffered at the hand of man. Stocks of several fish species are now at 10 per cent of their 1969 levels due to overfishing, while the Antarctic cod has already been fished to commercial extinction. Despite the 1982 Convention on the Conservation of Antarctic Marine Living Resources (CCAMLR), the fishing bonanza which began in the 1960s is still effectively unregulated. The words of the Convention promise much - "the maintenance of the ecological relationships between harvested, dependent and related populations of Antarctic marine living resources" - but in practice offer little in the way of protection.

Now the fishing fleets are turning their attention to krill. Rich in vitamins and protein, and containing calcium, copper and iron, krill have been promoted as a solution to world food shortages. But krill rapidly decompose on being caught, quickly leading to a build-up of fluoride, which can reach levels dangerous to humans within a few hours. In addition, the reproductive cycle of krill, like most creatures in the Antarctic, is slow. This means that fishing can all too easily reduce the population, and any major decrease in stocks could spell disaster for the Antarctic's wildlife.

Notwithstanding these problems, interest in krill as a possible source of food for human consumption remains high. At present, krill catches have peaked at around 500,000 tonnes a year, in part due to technical difficulties. The small mesh-size of the nets required for krill necessitates powerful ships to overcome the drag, making the operation very energy-expensive. But catches could increase dramatically - perhaps by as much as 600 per cent - unless catch limits are set and observed. New freezing and processing technologies have been developed and a new class of 'super trawlers' has now made its appearance in Antarctic waters. The ships are capable of multi-purpose fishing and of processing and canning krill at the rate of over 135 kg (300 lbs) per hour.

At present, krill is used predominantly as an animal feed, but Japan, Chile, the Soviet Union and Norway have found small markets for human consumption. Technological advances and the mistaken belief that krill swarms represent vast, untapped and sustainable yields of protein could spell disaster for Antarctic wildlife. We do not know what levels of catch represent sustainable yields or what impact such yields would have on the other animal populations of the Antarctic.

On land, the greatest threat to Antarctica comes from pressure to develop the continent's huge mineral wealth. A wide variety of metals, from copper to uranium, exist in concentrations that could become commercially exploitable, and the world's largest known deposit of coal lies in the Transantarctic Mountain range. The Dufek Massif in the Pensacola Mountains could prove to be one of the most mineral-rich areas of the world, with large deposits of platinum a high probability.

If Antarctica's resources were opened for commercial exploitation, the impact on the environment would be inestimable. The presence of people in such a vulnerable environment inevitably leads to disruption. The waste and detritus dumped by scientific research stations has caused damage enough: what could we expect from the bases and land stations, let alone the field installations and drilling rig complexes, required to exploit minerals?

But it is Antarctica's offshore oil and gas deposits that have attracted most attention. In 1973, the US research vessel *Glomar Challenger* drilled four bore holes in the continental shelf of the Ross Sea: three of the four revealed potential oil deposits. Although they later withdrew the calculations, the US Geological Survey estimated at the time that over 3,000 billion m^3 (100,000 billion cubic feet) of gas and some 50 billion barrels of oil lay awaiting extraction - twice as much oil as is recoverable in the British sector of the North Sea, but still only enough to

► Tourists get a close look at an iceberg in the Weddell Sea. Tourism is a relatively new phenomenon in Antarctica - regular cruises did not begin until the early 1960s. Exactly what impact increasing numbers of tourists will have remains to be seen. Trips around icebergs may do little harm to the environment, but by discarding waste or disturbing wildlife, tourists may have more serious consequences.

"Can he who has discovered only some of the values of whale-bone and whale oil be said to have discovered the true use of the whale? Can he who slays the elephant for his ivory be said to have 'seen the elephant'? These are petty and accidental uses; just as if a stronger race were to kill us in order to make buttons and flageolets of our bones..."
Henry David Thoreau

"'A whale! A whale! Run, Mr Audubon, there's a Whale close alongside.' I ran up, and lo! there rolled most majestically the wonder of the oceans. It was of immense magnitude. Its dark auburn body fully overgrew the vessel in size. One might have though it was the God of the Seas..."
John James Audubon

satisfy world demand for a few years. Already the Japanese and the Brazilians have financed seismic operations in the Ross and Weddell Seas.

Exploration of offshore oil deposits would bring even greater environmental risks than mineral exploration. Accidents have already happened, presenting a foretaste of the havoc that could lie in wait. During the 1988-89 season, two vessels, the *Humboldt* and the *Bahia Paraiso*, went aground, both spilling oil in environmentally sensitive areas in the peninsular region. Years of scientific work were ruined. Meanwhile, in Alaska, the *Exxon Valdez* catastrophe (*see p.230*) graphically illustrated the risks of transporting oil in polar regions.

More dramatic still is the risk of a blow-out. In 1986, an enormous iceberg - now known as B9 - broke away from the Ross Ice-shelf. It measures 90 by 20 km (56 by 12½ miles), and has been roaming the Ross Sea ever since, unstoppable and certainly no respecter of any installation put in its path. Measuring 275 m (900 feet) deep, B9 regularly scours the seabed and runs aground, as do other large bergs. The prospect of a sheared well-head gushing millions of tonnes of oil into the pure waters of the Ross Sea should be sufficient to make even the most hard-bitten oilman lose sleep. Yet, it is in the Ross Sea that some nations wish to drill for oil.

Concern over the ecological risks of mining and oil exploration has led to a growing rift between the signatories of the 1961 Antarctic Treaty, the convention which places Antarctica under international management. At issue is a proposed treaty to regulate mineral exploitation in Antarctica. Both Australia and France fear that the safeguards proposed in the treaty are insufficient to protect Antarctica's unique environment and have refused to ratify the agreement, thus angering pro-mining nations such as the UK, US, Japan and West Germany. France has now called for Antarctica to be given the status of a Wilderness Park, a position in line with conservationists who for years have campaigned for a ban on mining and the total protection of the continent.

Other nations have now joined France and Australia in questioning the wisdom of turning this near-pristine continent over to the multinationals and the issue has been taken up by the United Nations. Perhaps the tide is turning in favour of conservation.

Antarctica is not just a beautiful wilderness area worth protecting for its own sake. For 50 years, it has represented an international laboratory in which scientists from all over the world have worked co-operatively to unlock the secrets of the continent. More recently they have begun to grapple with global pollution issues, notably ozone depletion which now poses a major threat to Antarctica's marine ecosystem (*see pp.43-45*). Antarctica dictates weather patterns and its influence is felt all over the world. Cold nutrient-rich currents sweeping up from Antarctica support rich fishing areas off South America and South Africa. Apart from the Falklands (Malvinas) war, Antarctica has remained an area over which conflict has been avoided. Its current status is that of a peaceful, non-militarized area of scientific study. Mining of any description would bring with it conflict, pollution and possibly ecological catastrophe in this most vulnerable environment. Although the recent stand by Australia and France is cause for hope, the battle for Antarctica is by no means over - and will not be until the continent has been declared 'off limits' to development.

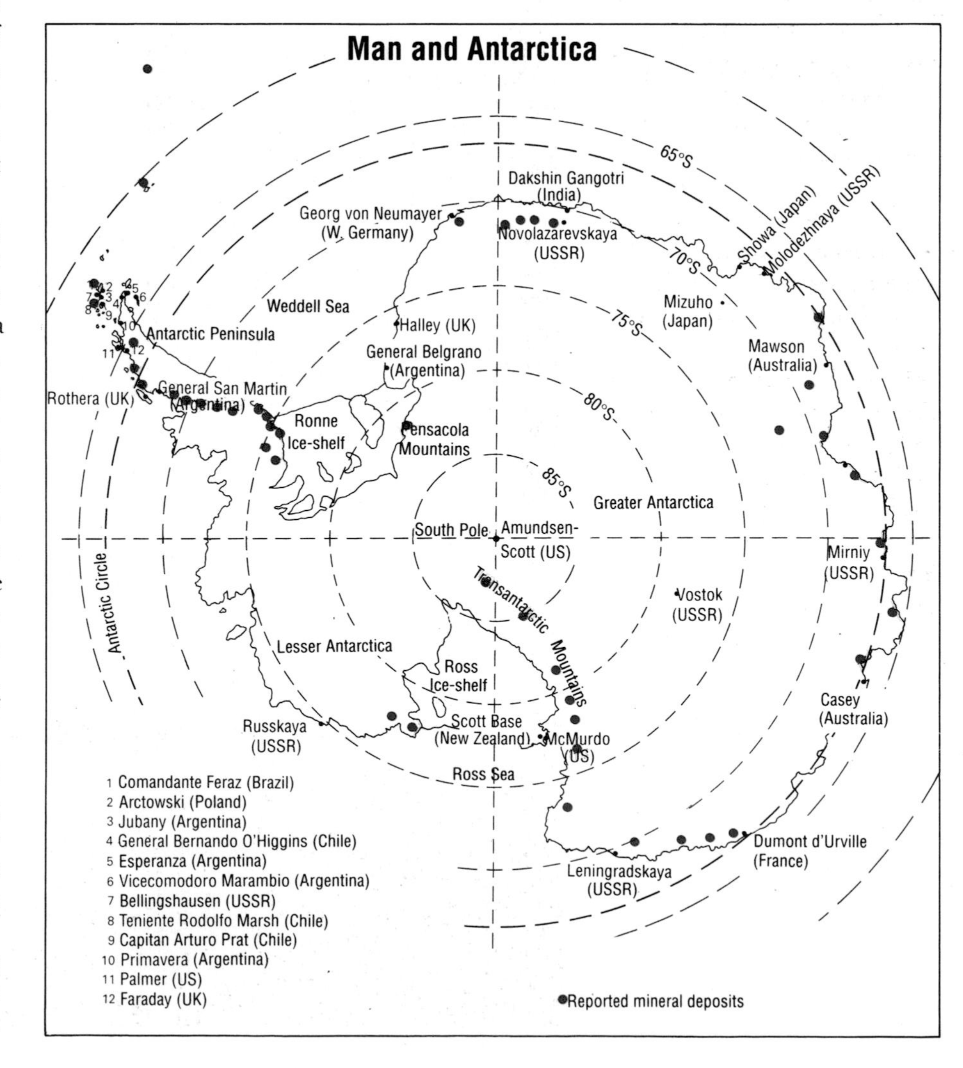

▼ Antarctica is the only continent in the world without an indigenous human population. It has only been during this century that year-round bases have given man a permanent presence in the far south. Many bases were originally set up by nations eager to make territorial claims, although at present all such claims are suspended by treaty.

The ARCTIC

The Arctic is a place of often breathtaking beauty, with magnificent mountains, plains, fjords and glaciers that produce spectacular icebergs. It is also remarkably rich in wildlife. Home of the polar bear and animals such as the walrus, caribou, musk-ox, wolf and a variety of seals, it is also a refuge for many species, all that is left of their shrinking domains. The wolf, for example, was once common throughout much of the northern hemisphere and, until the 1920s, walruses were still 'hauling out' on Scottish islands. Today, the Arctic is the remaining stronghold of both animals.

This vast region spans two continents, and covers the north of the Soviet Union, Alaska, Canada and Greenland. It is a land that is visited by a few outsiders and for many who live in lower latitudes, it has the image of a wild and forbidding icy wilderness. The Arctic is indeed cold, but the low winter temperatures are no more severe than those experienced by some central areas of North America like Minnesota or Manitoba. What makes the Arctic different is that there is little relief: the temperature is below freezing for much of the year.

In the high Arctic, the Sun sets in late October and does not return until mid-February. The land is cold, dark and devoid of birdsong, although at midday the sky to the south glows with the warm reflected light of the Sun shining in lower latitudes. By mid-April, the same region is enjoying the 'Midnight Sun' which rises and falls as it circles in the sky, but does not set until mid-August. During the brief ice-free months of summer, much of the land is covered in a blanket of brightly coloured flowers and millions of birds arrive to raise their young, exploiting the prolific food chain of the tundra and the polar seas.

◀ Dog-drawn sleds on the sea-ice at Melville Bay, in Greenland's far northeast. Traditional sleds like these are becoming an increasingly rare sight as twentieth-century technology makes its mark in the high Arctic.

▲ **Tundra bog, Spitzbergen, and one of the major threats to it - a track-laying vehicle. Tundra vegetation is slow-growing, and may take decades to recover after being driven over.**

The precise extent of the Arctic is difficult to define. The Arctic Circle, at 60°30'N, marks the point where there is one day a year on which the Sun does not rise. A more useful boundary is the treeline, the northern limit of the boreal forest. At this northern edge the trees have only 90 days or less per annum in which growth is possible. The conifers on the borderline are sparse and ragged, and often 'flagged' in the opposite direction to the prevailing wind. It may take them anything from two to three hundred years to grow just 2 m (6½ feet).

Much of the Arctic is frozen ocean, landlocked by continents. During the coldest months of February and March, there may be as much as 12 million km^2 (4.6 million square miles) of floating ice in and around the Arctic Ocean. On land, a permanent layer of frozen earth, the permafrost, lies just 2 m (6½ feet) below the carpet of summer flowers. The Arctic is often described as 'fragile', and individual plants and organisms will certainly take an inordinately long time to repair any damage done to them. But their ability to survive major seasonal stresses, as well as a succession of Ice Ages, during which they retreated southwards as the ice cap advanced, would tend to suggest their resilience. And 'resilient' is a word that could well be used about the indigenous human inhabitants of the area. The Arctic might not be everyone's idea of paradise, but for the Inuit or 'Eskimo' it is home; a beautiful land and one which, until comparatively recently, could provide them with everything they could possibly need.

THE THREAT FROM OIL

The first commercial exploiters of the Arctic came in search of animals and their produce, and the impact made on the Arctic by these dealers in furs and animal oils was considerable. But their effect was insignificant compared to the impact now being made by multinational companies, exploiting the mineral wealth that has remained locked up for so long beneath the ice and permafrost. Oil-company geologists first surveyed the area around Alaska's Beaufort Sea in 1960. In 1968, 10,000 million barrels of oil and 736,000 million m^3 (26,000 billion cubic feet) of gas - 25 per cent of proven US crude oil reserves and 10 per cent of gas reserves - were found under Prudoe Bay on Alaska's North Shore. Other substantial finds have been made elsewhere in Alaska. The rising world price of oil finally made it commercially viable to sink expensive wells in the Arctic.

Prudoe is the world's eighteenth largest oilfield and is situated at the head of the Trans-Alaska Pipeline, which runs 1,300 km (800 miles) across Alaska from the Arctic coast to Valdez in Prince William Sound in the south, where oil is pumped into storage tanks. From these tanks it is transferred to the tankers which transport the oil to American ports in the 'lower 48' states.

Many of the fears expressed by environmentalists when the pipeline was first proposed have not been realized. The pipeline, raised on stilts, has not melted the permafrost and the caribou continue to follow ancient migration routes that pass directly under the pipeline. But, despite the assurances of the oil companies, the pipeline remains an accident waiting to happen. And when that happens, the impact of the artificially warmed oil spilling out into the environment could well prove disastrous.

“As far as I can tell, we're standing in the world's greatest Arctic ecosystem. We say, if they can drill here, where can't they drill? If we're going to develop this, we might as well go ahead and dam the Grand Canyon. You can make the same arguments for national energy needs.”
Tim Mahoney, Sierra Club, on plans to drill for oil in the Arctic National Wildlife Refuge

◀ The Trans-Alaska Pipeline snakes its way across a hillside on its journey from Prudhoe Bay on Alaska's north coast, to the southern port of Valdez. The pipeline was completed in 1977 after one of the largest engineering projects ever mounted in the Arctic. The pipeline is raised above the ground to minimize warming of the permafrost, and the pillar-like objects that flank it cool the ground if the soil temperature rises. So far, the pipeline has operated without any major leaks. Ensuring that it continues to do so requires constant vigilance for any signs of corrosion.

◀ The pipeline terminal at the port of Valdez. Every day over 1.5 million barrels of crude oil arrive at the terminal, to be stored before being loaded into tankers. Like many oil installations in the far north, the terminal has been built on a shore which is, in biological terms, exceptionally productive. Fish and other marine life abound in the inshore waters, and for these animals, even minor oil spills can have fatal results.

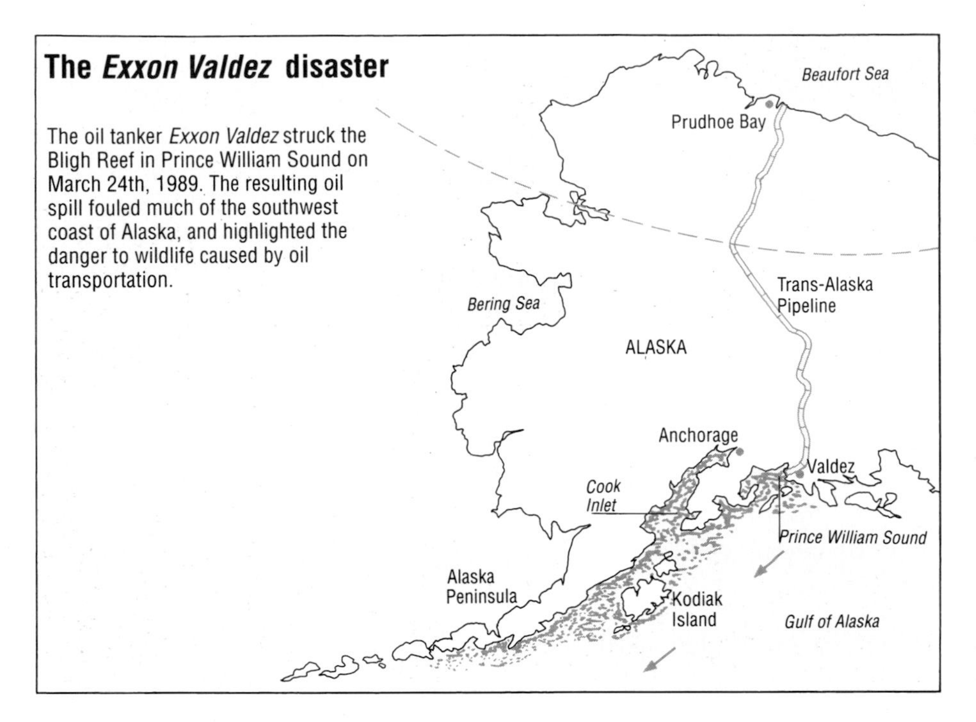

The *Exxon Valdez* disaster

The oil tanker *Exxon Valdez* struck the Bligh Reef in Prince William Sound on March 24th, 1989. The resulting oil spill fouled much of the southwest coast of Alaska, and highlighted the danger to wildlife caused by oil transportation.

▲ Cleaning up after the *Exxon Valdez* disaster. Water containing detergent is sprayed onto the oil-coated rocks, while floating booms contain patches of floating slick. The oil spill was particularly damaging because it happened so close to the shore. There was little time for the oil to break up before it was driven into countless small bays and creeks, where it wrapped coastal wildlife in a tenacious and deadly blanket.

A foretaste of the devastation that could result came on 24th March 1989 when the oil tanker *Exxon Valdez* ran aground in Prince William Sound, barely out of the port of Valdez, spilling 11 million barrels of her cargo into Arctic waters. The ensuing delays, mistakes and misunderstandings are common knowledge, but in spite of a worldwide response to the disaster, 100,000 seabirds and 1,000 sea otters died encased in oil. The coastal areas of southern Alaska will take several years to recover fully and the whole world realized that, in spite of their assurances, the oil companies never had the capacity to deal with a spill of that size.

THE IMPACT OF MINING

Mining has a longer history in the Arctic than oil exploration, the goldrush of 1896 heralding the first extensive mineral exploitation of the north. But the mining companies have not learned to live with the environment. In 1987, $80,000,000 was spent on gold prospecting in Canada's Northwest Territories alone, and today's mining activities have considerably more impact than a single prospector with a mule and a goldpan. Now they use the sort of equipment that was designed for building motorways.

Employment for the native Inuit has often been given as justification for mining operations sited close to remote Arctic settlements, but in reality few locals are employed. They have neither the skills for operating high-tech equipment nor the motivation to stay at such work on days when the caribou herds are passing nearby or when the salmon are running in local rivers.

ARCTIC PARKS - A FLAWED PROTECTION?

Some argue that the way to save the Arctic for future generations is to set up parks. The world's largest wildlife park has been established down the east coast of Greenland, and another ambitious scheme, nicknamed 'Glasnost and Glaciers', has been proposed for the wilderness areas on both the Soviet and Alaskan sides of the Bering Strait. But some fear that the parks will be used to justify fewer restrictions for oilmen and miners operating outside the protected areas. Worse still, the record shows that when areas have been designated as 'parks', it does not guarantee that they will be safe.

In 1980, for example, the US Congress extended the Arctic National Wildlife Refuge (ANWR) and declared it a 'national treasure.' Its 429,000 km^2 (165,000 square miles) are home to the 170,000 caribou of the 'Porcupine Herd'. But early in 1987, the then US Secretary for the Interior, Donald Hodel, recommended to Congress that oil development be permitted on 6,000 km^2 (2,300 square miles) of the ANWR. Tim Mahoney of the Sierra Club is quoted as saying "If we are going to develop this, we might as well go ahead and dam the Grand Canyon."

OVERFISHING IN THE ARCTIC

In the waters to the west of Greenland, fishing for salmon is undertaken on a huge scale, so much so that the stocks of salmon in Icelandic and British rivers are being threatened. Together with the Icelandic Fisheries, a Scottish conservation body called the Atlantic Salmon Trust is negotiating with the fishing fleets of Greenland and the Faroes to buy their 'salmon quota' so that the fish can safely reach maturity in Arctic waters and then return to their native rivers.

Overfishing on a massive scale is disrupting wildlife all along the Arctic fringe. The Soviet Union, for example, has a large fishing fleet based at Murmansk, a port in the Barents Sea that remains ice-free all year. Huge factory ships have decimated the sand eel population to provide fish meal for animal feed and fertilizers. By removing this link in the food chain, the fishing fleets are putting several important seabird colonies on the edge of the Arctic at risk. Every year, chicks starve to death in their nests, the parent birds having been unable to find enough food for them.

In 1988 and 1989, there were reports of starving seals tearing at fishermen's nets in the seas around northern Norway. In 1988, 56,000 seals were caught in fishing nets off Norway alone. The reasons for this were complex and not just attributable to the overfishing or the halt in the harp seal cull in Newfoundland. John Arst, spokesman for a Norwegian fishing cooperative, Norges Rafiskelag, says his members wish to reduce all quotas and obtain scientific figures on fish, seal and whale populations. His anxiety is that unless something is done soon to stabilize the marine ecology, the tiny fishing villages along Norway's coast will lose the only way of life they have known.

POLLUTANTS AND RADIOACTIVITY

Northern Arctic waters have not been as badly affected as many seas by overfishing, and for this we can thank the permanent cover of ice which protects a large area. But those creatures that live under and around this ice, apparently so far removed from the industrialized world, are subject to a far more sinister threat - that of pollutants carried by the winds and ocean currents.

In the north of Scandinavia, radioactive fallout from Chernobyl has come close to destroying the way of life practised by Lapps for for hundreds of years. Unfettered by national borders, the Lapps have followed herds of reindeer on their migrations from winter to summer pastures in the spring, and then back again in the autumn. The reindeer are their life, their food source and their culture. In 1986, however, rain contaminated with fallout from Chernobyl fell

▲ A Lapp woman with a reindeer herd in northern Norway. The Lapps' traditional nomadic way of life, based on following semi-domesticated reindeer, is gradually disappearing as many take up farming or fishing instead.

◀ The 'muskeg' of southeastern Alaska at the approach of winter. Fallen leaves carpet the blueberry bushes as another growing season comes to a close.

on their pastures. The ground, the lichen on which the reindeer feed at their winter pastures, even the lakes and the fish in them, were contaminated. In many areas, the level of radioactivity remains dangerously high.

Animals at the top of the food chain are particularly at risk. The polar bears' diet of seal has led to a four-fold increase in the levels of polychlorinated biphenyls (PCBs) in their fatty tissues between the years 1969 and 1984. If current PCB inputs continue, the polar bears will exceed the 50 parts per million limit legally designating them as 'toxic waste' by about the year 2005. Long before that, they will probably have become infertile.

Other pollutants also threaten the Arctic ecosystem. In the James Bay region of Hudson Bay, mercury levels are now so high that the Cree Indians have been warned not to eat any of the local fish. Mercury pollution has a long history in the area - due largely to the activities of pulp and paper mills and mines - but the problem has been greatly exacerbated by the flooding of vast areas under the James Bay hydroelectric development project. Bacterial activity in the newly flooded areas has released mercury from the disturbed sediment beneath the reservoirs, making it available so that it is passed up the food chain. Mercury levels in some areas are now ten times higher than those deemed safe for human consumption.

THE JAMES BAY PROJECT

Described as the "project of the century" by Quebec's prime minister, Robert Bourassa, the James Bay Hydroelectric Scheme was launched in 1971. It was to be a vast scheme flooding 176,000 km^2 (68,000 square miles) of northern Quebec's boreal forests and tundra, an area approximately twice the size of Ireland.

The bulk of the work on phase 1 of the project - consisting of three enormous dams along the La Grande River - is finished. The dams supply 10,000 megawatts of power which is carried along five transmission lines, supported by some 11,650 pylons, on its way to southern Quebec and onwards into the US, a journey of 5,562 km (3,456 miles). Phase 2, which involves damming the Great Whale River, is scheduled to begin soon, but not without opposition from the local

◀ A polar bear sorts through rubbish at Churchill, on the western shore of Hudson Bay in Canada. Intelligent and inquisitive, polar bears are all too easily attracted by the debris of twentieth-century life. As they search for food, they run the risk of injury, and they also present a very real hazard to people.

▲ An Inuit hunter hauls a seal onto the ice, after trapping it in a net. Seal hunting is an emotive issue in the far north. The Inuit have traditionally hunted seals, but only in a limited and sustainable way. They have become unwitting targets of the campaign against another kind of seal hunting, in which the animals are killed solely to supply markets outside the Arctic.

◀ A snowy owl with its chicks on a tundra nest. Living in a treeless habitat, this is one of the few owls in the world that nests in the open. During the summer, it hunts by daylight, preying on lemmings and other small mammals. Like many Arctic animals, the snowy owl is found across all the landmasses that border the Arctic.

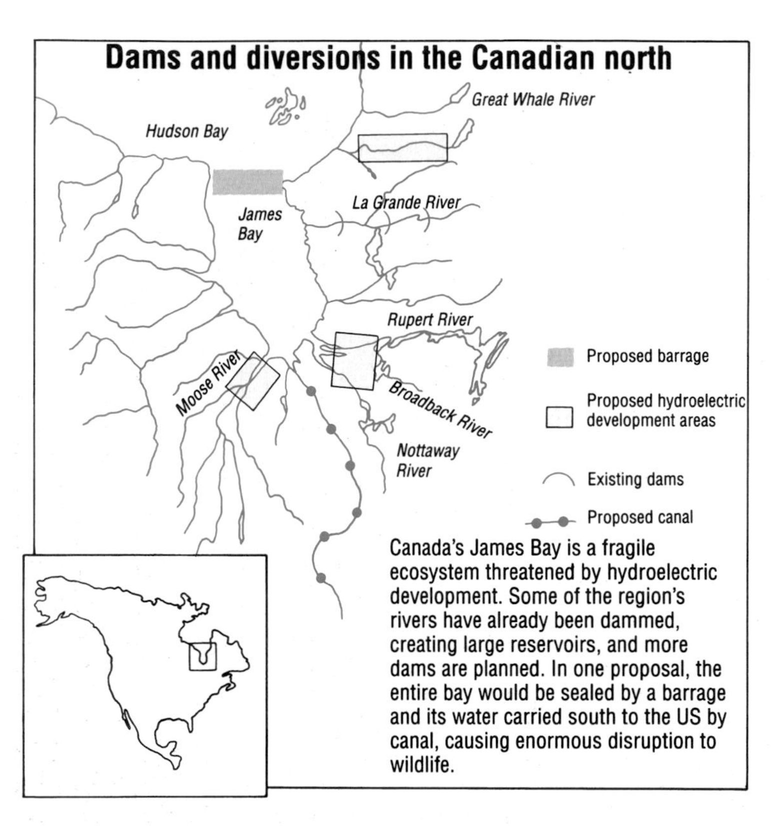

Canada's James Bay is a fragile ecosystem threatened by hydroelectric development. Some of the region's rivers have already been dammed, creating large reservoirs, and more dams are planned. In one proposal, the entire bay would be sealed by a barrage and its water carried south to the US by canal, causing enormous disruption to wildlife.

▲ Part of the James Bay hydroelectric project in northern Canada. Although water-generated power does not directly produce atmospheric pollution, the environmental cost on the ground can be high. This is especially true of the Arctic, where access roads have to be built over previously inaccessible territory.

► The tundra in late summer. In the far north, the growing season is brief but brisk. Plants bloom and set seed sometimes within days of the snow melting, and the animals that feed on them are active around the clock. By August, the Sun is already starting to slip southwards once again. The vegetation takes on a golden tint, and migrating animals depart for warmer latitudes as the Arctic once more prepares for winter.

Cree Indians, who are determined not to have more of their ancestral hunting grounds flooded by the scheme.

The second phase would be followed by further dams and diversion schemes on the Nottaway, Broadback and Rupert rivers, bringing the total area flooded to the size of Lake Ontario. In addition, there are proposals to build a series of dams along the Moose River, which feeds into the southwest of the James Bay.

Already, the environmental impact of the existing dams on the La Grande has been severe. Apart from the problem of mercury pollution, the new reservoirs have disrupted caribou migration routes (in one incident in 1985, ten thousand caribou were drowned by the sudden release of water from one of the dams) and have disturbed the timing of seasonal water flows into James Bay, destroying fish spawning grounds and affecting salinity throughout the river basin.

The coastal marshes and mudflats of the James Bay region are an important stopover point for migratory birds, many of which could be threatened with extinction if the staging area is flooded or its ecology further disrupted by the hydro project. Changes in the salinity of the bay are already having an effect on the algae, marsh grasses and other aquatic organisms on which the migratory birds feed. The habitat of a wide range of species living in the bay itself is also under threat. Affected animals include ringed seals, beluga whales (which winter in the ice-free waters around the many islands in the bay) and polar bears, which use James Bay during the warm summer months.

The commercial pressure to build phase 2 of the project is tremendous. The power generated would be sold to the US, much of it to New York. Yet, as environmentalists and the local Cree point out, it is doubtful if the power is really needed. Energy conservation measures could 'provide' an equal amount of saved energy. The cost of power supplied by the dam is artificially low, in any case, because the full social and environmental costs have not been calculated.

THE GREENHOUSE EFFECT IN THE ARCTIC

With the concentration of greenhouse gases increasing in the atmosphere, global warming *(see pp.40-53)* threatens the entire Arctic ecosystem, possibly changing it beyond recognition. Already the sea-ice north of Greenland has been reduced in thickness from 6-7 m (20-23 feet) in 1976 to 4-5 m (13-17 feet) in 1987. If drastic measures are not taken to reduce greenhouse gas emissions over the next couple of decades, it is likely that many Arctic species and the indigenous peoples which rely upon them will not survive the 21st century.

3

The Human DIMENSION

The Diminishing Quality of Life *240*

The Future in Prospect *252*

The Dynamics of Destruction *264*

Solutions for Survival *272*

◀▶ Villagers harvesting crops on the Indonesian island of Bali participate in a system of working the land that depends on cooperation and adaptability. By contrast, agriculture in industrialized countries depends not on people but on machines, degrading the land and undermining rural communities.

The DIMINISHING QUALITY of LIFE

"Why should we tolerate a diet of weak poisons, a home in insipid surroundings, a circle of acquaintances who are not quite our enemies, the noise of motors with just enough relief to prevent insanity? Who would want to live in a world which is just not quite fatal?"
Paul Shepard

▲ Homes in the sky - high-rise apartment blocks in crowded Hong Kong.

"Smog over Los Angeles used to be a disaster, but now is normal, whereas a clear day on which you can see the San Gabriel mountains is now a stroke of good fortune. Normality in Alaska is a series of toxic dumps, and soon Prince William Sound will become an 'ordinary' commercial channel like the Houston Ship Canal which periodically bursts into flame..."
Alexander Cockburn

The HUMAN HABITAT

The average city dweller in the US consumes about 568 litres (125 gallons) of water, 1.5 kg (3.3 lbs) of food, and 7.1 kg (15.6 lbs) of fossil fuels per day, and generates about 454 litres (100 gallons) of sewage, 1.5 kg (3.3 lbs) of refuse, and 0.6 kg (1.3 lbs) of air pollutants. New Yorkers produce enough garbage every year to cover Central Park to a depth of 4 m (13 feet).

City life is marked by consumption and waste. Cities, however small, have always been parasitic on the countryside around them, not least because the majority of their inhabitants must rely on farmers in the countryside to provide them with food. Where cities are small, and the demands of their citizens limited, the degradation caused need not undermine their viability. But the demands being made by city-dwellers today, particularly in the industrialized countries, are global in their reach, and global in their implications. Forty years ago, 600 million people throughout the world lived in cities. Today the figure is 2 billion and growing fast. Nor is the rate of urban expansion slowing down. At the current rate, by the year 2000, cities will have to accommodate over 75 per cent of the population of South America. In Africa, 175 million live in cities, 30 per cent of the total population; yet 368 million are expected to live in cities by the end of the century.

It seems inevitable, if current trends continue, that within the next 30 years, more than three-quarters of the people on Earth will be crammed into cities. If population levels rise as predicted, that would mean an urban population of some 7.5 billion. Many will have been born in cities, millions more will have migrated there. A high proportion are likely to be environmental refugees, people thrown off their land through debt or to make way for development projects, or those simply unable to make ends meet in an increasingly degraded environment.

The task of providing a decent standard of living for today's city dwellers, let alone those of future generations, poses seemingly insurmountable problems. Meeting even the demands of present-day cities is placing an intolerable burden on the environment and society, as more and more resources are sucked into urban conglomerations. To expect to meet them when cities have swollen to three times their present population is simply wishful thinking.

A LIFE OF SQUALOR

Urbanization in the Third World increasingly means no more than a drift to the slums. Millions now live in wretched, inhumane conditions that almost defy description: a dozen or more people, including the old, the sick and the newborn, all crammed together in a shack made of corrugated sheets, plastic bags, cardboard; the sewage from a thousand such homes running openly along the potholed tracks that separate one dwelling from the next; the children, naked, with stomachs swollen from malnutrition and never-ending bouts of dysentery, sitting listlessly on the ground; the smell, noise, smoke and vermin. This is now the lot of a considerable proportion of the world's population, and yet the numbers grow.

In the developed world, there is little to compare with the slums of the Third World. But amid the affluence of our cities, pockets of severe deprivation and poverty stand out like sores. In the US, the richest nation on Earth, an estimated 3.75 million families live on the poverty line, almost double the figure in the 1970s. Some 600,000 are homeless, forced to sleep rough, sheltering in burnt-out buildings or bedding down in cardboard boxes. Just a few blocks from Capitol Hill, Washington, are neighbourhoods where the infant mortality rate is higher than in some Third World countries.

URBAN WASTELANDS

Urban sprawl has damaged the natural environment and created a depressing habitat for human beings. It has swallowed up farmland, villages and towns. In the Midlands of England, the city of Birmingham slithers into Walsall, into Wolverhampton and into Coventry, to form a concrete slab over 30 km (20 miles) square, a megalopolis of factories, motorways, car cemeteries, box-like houses and derelict land. Meanwhile, slum clearance programmes have brought a new poverty. Stark tower blocks have replaced diverse if tumbledown buildings, creating vertical slums instead of horizontal ones. Very little account, if any, was taken of the psychological effect of destroying existing communities, nor of what it means to live high up, in a rectangular, ferroconcrete space, divorced almost entirely from nature.

“To restore a proper balance between city and rural life is perhaps the greatest task in front of modern man.”
E.F. Schumacher

In the Third World, many housing developments have been equally inappropriate. Western-style housing - seen as a symbol of development - has been imposed on communities with no regard for their culture or climate. Traditional houses in the tropics are built to maximize ventilation and reduce humidity, the building materials absorbing little heat. By contrast, the brick and concrete houses and glass-enclosed high-rises increasingly being built in Third World countries trap heat. Their interiors become oven-like, uncomfortable to live in, and, where air conditioning is installed to keep temperatures down, expensive in energy.

Increasingly, it is the motor car, rather than the needs of local people, that shapes our cities. The number of cars in the world has risen from 50 million in 1946 to over 386 million in 1986, and production is increasing by 3 million every year. In many North American cities, cars have become a seemingly indispensable part of life. In Denver and Los Angeles, 90 per cent of people go to work by car, and although the figure is lower in Europe and lower still in Japan, the car has increasingly come to dominate our lives. Even if we do not own a car ourselves, we must suffer the air pollution they cause, run the risk of being injured by them and put up with their noise and traffic jams. As more and more cars clog the roads, so our cities become increasingly congested. Despite the power of modern engines, traffic in London moves no faster than in the age of the horse and cart, at a mere 13 km (8 miles) an hour. The time has almost arrived when it would be quicker to walk.

Stark and sterile, our cities have become ugly to live in and ugly to look at. It may be impossible to put a value on the need for beauty in our surroundings, but one cannot doubt that it plays a tremendous part in our lives. Drably uniform and intimidating in their size and structure, the buildings that increasingly dominate our cities have been built with little or no consideration of those who must live and work in them.

◀▲ On a San Francisco bridge, a woman, with all her possessions piled onto a supermarket trolley, stops to rest. She is just one of many thousands who are homeless in this affluent city.

◀ Philippine slum-dwellers search for food, clothing or other useful items on Manila's main tip.

SOCIAL ALIENATION

It is traditional, in the West, to look back over human history as a steady but inevitable march of social improvement. The line from hunter-gatherer to man-the-cultivator and builder-of-cities and now to technological man has seemed a logical sequence of events that will ultimately end in universal prosperity and 'progress'.

Yet somehow the dream has turned sour. It is not simply that we are destroying our environment in the quest for higher and higher standards of living: it goes deeper than that. For amid increasing material wealth, we see increasing evidence of human misery. On our television screens, we can watch the social fallout of progress: psychiatric patients, heroin addicts, teenage alcoholics and others unable to cope with the stresses of modern life. Almost a century ago, the poet A. E. Houseman summed up a similar feeling of isolation with the line, "Alone and afraid in a world I never made".

LIVING IN THE MATERIAL WORLD

In today's world we measure success by such criteria as how much we earn, where we live, the size and speed of our car, and where we go on holiday. Life is increasingly governed by money and how to acquire it, as we battle against inflation and the ever-escalating cost of living.

The sense of alienation felt by modern city-dwellers, deprived of any sense of community, is also shared by many who live off the land. Two generations ago, a medium-sized farm of 80 ha (200 acres) employed half-a-dozen labourers, and a farmer expected to run the farm without needing to borrow from the banks. A generation ago, a farmer on the same land would have halved his workforce while increasing the number of animals he kept, and he would be lucky to get through without borrowing. Now, with one worker beside himself, the farmer is forced to borrow heavily to pay for all the equipment required. And, to generate the profits necessary to pay the interest, he will probably have to double again the number of livestock.

Such production, sustained by large feed and veterinary bills, turns farming into an industrial process and much less a way of life. It is basically anti-community as each farm becomes an isolated unit, removed from the tight connections with the rest of the neighbourhood. Moreover, it fuels the exodus of people from the land; people who no longer have a role must seek work in the offices and industries in cities. As for the farmer, he is left struggling with mounting debts and the need to increase his cash flow. At the end of the day what has he gained? A family car, a more modern house. And at what price? The loss of the community feeling which marked traditional farming, and the loss of solidarity between people in the countryside.

INCREASING CENTRALIZATION

Modern industrial society thrives on the fragmentation of community. It needs workers, operators, managers and salesmen from wherever it can get them. It is not interested in such things as traditional allegiances between people, or in long-standing communities. On the contrary, when unemployment became rife during the 1980s, the unemployed were exhorted to pack up their bags and move to where they could find employment, irrespective of having to find new homes and disrupt their children's schooling - or leave their families behind.

As families and communities break down, so the state takes on the roles which the community traditionally performed: care of the elderly, care of children when their parents are ill and care of the sick. Increasingly, the decisions that affect our lives are outside our control. The community becomes redundant and instead of people helping each other when they have problems, they are now handled by some distant, centralized bureaucracy, with a thousand different demands for its limited resources of time and money. We become faces in the crowd.

The policeman on the beat gives way to the policeman in the squad car; the local hospice closes and is amalgamated into a massive regional hospital; the local school shuts its doors and the children are moved into a bigger school, often far away from where they live. Our lives become dominated by more and more impersonalized institutions.

In our work, too, we are at the mercy of economic forces that take little account of social needs. The decision to open a factory or close it is not made by those whose lives depend on the work. We may not even know the name of the parent company that takes the decision.

And, perhaps most alienating of all, we must rely on experts to tell us whether our food is safe

"The trouble is that the large-scale mechanized agri-businessman, because of his scale and high capitalization, has been forced above all to be a specialist. If you have to pay twelve thousand pounds for one combine harvester you can't afford to keep chickens. The one thing you must do above all others is to save labour. The question the agri-businessman is for ever asking himself is: 'How can I dispense with another man?"
John Seymour

"...the wider human habitat, far from being humanized and ennobled by man's activities, becomes standardized to dreariness or even degraded to ugliness. All this is being done because man-as-producer cannot afford 'the luxury of not acting economically' and therefore cannot produce the very necessary 'luxuries' - like health, beauty and permanence - which man-as-consumer desires more than anything else. It would cost too much; and the richer we become, the less we can 'afford'."
E.F. Schumacher

▶ On the island of New Britain, in Papua New Guinea, women coming to market spend some of their precious money on soft drinks. Advertising has a powerful effect, especially in developing countries, undermining traditional values and creating a demand for unnecessary goods. Soft drinks are especially pernicious, and sometimes lead to malnutrition if they are given to babies instead of milk.

"Our questing eyes are programmed to seek rewarding habitats, blending fertile open spaces, rich in living creatures, interspersed with sheltering woodland or rocks, and with streams, pools or expanses of refreshing water. Instead, our civic designers present us with the precise opposite of our profound human need - an environment that is hard, sterile and lacking in greenery, or offered only in stiff mathematical unnatural strips, and which threateningly overhangs us with structures too tall, too geometrical, too uniform and too coldly surfaced to give any human satisfaction."
Max Nicholson

to eat or whether the air is safe to breathe. Our instincts are increasingly redundant. We cannot tell what chemicals have been sprayed on the food we give to our children. We have only the word of some anonymous committee of experts to assure us that the pesticides are safe. Then we read that the expert advice has changed, or that the chemical has been banned in some other country, or that important information about the dangers of the chemical have been deliberately suppressed. Where does that leave us? Alone and afraid in a world we have not made?

THE BREAKDOWN OF COMMUNITY

In most industrialized countries, the social and economic changes that have led to increasing centralization are history. But, in the Third World, the process is happening now. In the name of development, Third World governments have set about the systematic transformation of their societies. It is a transformation that has not only brought dramatic change to agricultural practices *(see p.109)* but also to land tenure systems, local political systems and to the fabric of local communities.

Money, rather than traditional social bonds, becomes the currency that determines social relations. Life is no longer seen as a pattern of recip-

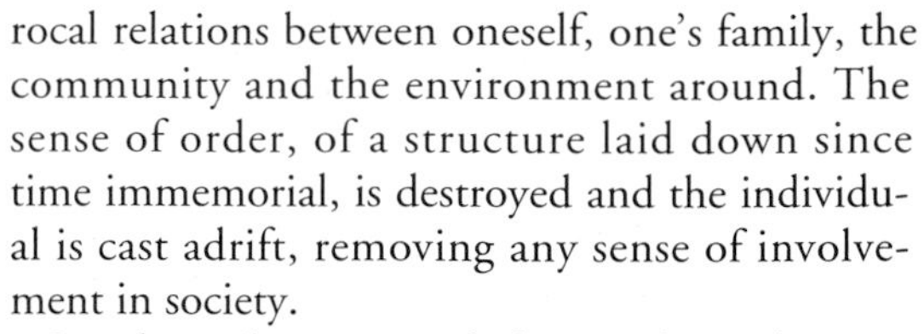

rocal relations between oneself, one's family, the community and the environment around. The sense of order, of a structure laid down since time immemorial, is destroyed and the individual is cast adrift, removing any sense of involvement in society.

People are being cut adrift, too, from the past, which is denigrated as primitive and backward. In Indonesia, the government has embarked on a project aimed at the total 'development' of the Indonesian people. Tribal groups in West Papua, whom the government describes as "still living in the Stone-Age era", are being forced into mainstream society. Tribesmen have been forcibly shorn of their hair to conform to national standards. Their animist beliefs have been denounced as irreligious, and their traditional medical practices banned.

A SENSE OF ISOLATION

During the Industrial Revolution, the idea emerged that nature was there to be manipulated for mankind's benefit. The traditional way of life, as it then was in Europe, with its rigid hierarchy of nobles, clerics and the common man, was seen as a poverty trap for the masses and a block to progress. New ideas gained ground about the individual and his or her role in society as a free agent, able to move physically and socially. Those ideas are still inherent in all development programmes. Thus, industrialization is seen to offer choice and opportunity as well as access to the markets with their enticing consumer goods.

Ironically industrial society, with its promise of individualism and the freedom to choose an occupation and a place to live, has created a fragmented mass culture from which traditional cultural diversity has largely been eradicated. Far from generating job satisfaction and contentment, in the main it has generated boredom. People now live for weekends and holidays, compartmentalizing their lives rather than feeling their life to be part of a continuous process in which work is as satisfying as play.

It is little wonder that the divorce rate is high, that children suffer from psychological problems unimaginable among well-structured communities, that the suicide rate has soared, crime, delinquency and murders are widespread, drug abuse has already reached epidemic proportions. Our attempts to build a world free of suffering and want through increased material wealth has certainly made life easier for some, but it is a life that is increasingly empty for all.

AIR, WATER and FOOD

Londoners awoke on the cold, still morning of Friday, 5 December, 1952, to one of the worst smogs that the city had ever seen. With over a million households and factories burning coal, the smoke from their chimneys had combined with fog to form a thick and deadly mixture that hung in the stagnant air. By Saturday afternoon, the visibility on central London's roads had fallen to 5 m (16 feet) and the transport system came to a standstill. The health effects of the three-day 'Killer Smog' were appalling. An estimated 4,000 people died from heart and lung ailments and thousands more fell ill.

Legislation was adopted to clean up Britain's cities. The emission of dark smoke was prohibited, and smoke-control zones were gradually introduced, within which only smokeless fuels could be burned. Tall chimneys were built to carry away smoke emissions from coal-burning power plants and factories from urban areas.

Although these measures greatly improved urban air quality, like most attempts to halt air pollution, they did not solve the basic problem. They merely transferred it elsewhere. The valleys of South Wales, where the smokeless fuel is produced by partial burning of coal, now choke from the pollution which the cities have been spared. Emissions from tall chimneys in cities are carried away by the wind to pollute areas hundreds of kilometres away.

Clean air and clean water should be basic human rights the world over, but they are becoming increasingly scarce - and expensive - commodities. In the slum districts of many Third World cities, one-fifth or more of a family's income can be spent on buying water from vendors who come around with buckets or donkey carts. For these impoverished people, stranded in an unplanned city with no access to a clean water supply, buying water in containers is a symbol of their helplessness and lack of control over their environment.

In the West, by contrast, buying drinking water in bottles has become a status symbol for the affluent, or a way of partially escaping pollution for those concerned about their health. Drawn from springs that are supplied by glaciers or by sources deep underground, 'mineral water' is at once a mark of wealth and a powerful symbol of the environmental damage which generating that wealth has caused.

Packaging water in plastic or glass, then transporting it thousands of kilometres to be sold in supermarkets, creates pollution in itself. It is no more a solution to the problem of water pollution than Britain's Clean Air Act was an answer to the problem of air pollution. Rather than tackling the cause of the problem, it simply shifts it somewhere else, postpones it for a while, or makes it bearable for the wealthy minority. The price of this ostrich-like attitude to our poisoned environment will be paid by the next generation in mounting ill-health.

The question of how far pollution is already affecting health is a controversial one. Sometimes the impact on health is immediate, obvious and undeniable, as with the Bhopal accident *(see p.270)*. But in most parts of the world, the average person is unlikely to feel that their own health is affected by pollution, and governments are always anxious to deny any such possibility. Yet there is a good case for believing that pollution does have insidious effects even at 'low' levels.

One well-defined health problem that has shown a steady increase due to pollution of our air, water and food is cancer. Pesticide residues in food are one factor. In 1987, the US National Research Council warned that 90 per cent of the fungicides in use in the US could be carcinogenic, along with 30 per cent of insecticides and

▼ Photochemical smog hangs over the city of Sydney, Australia. Even in this thinly populated continent, man is exerting a powerful impact on the environment.

"...I personally have vomited on a number of occasions from the chemical smells from Leigh. Instant headaches are another thing, instant sore throats and gritty eyes. When the east wind blows it sets in for a couple of weeks at a time. The chemical tanker lorries smell so foul you can't describe how horrible they are. I have walked along Mill Road and a lorry has passed me and by the time I have got indoors I have been vomiting."

Ken Wilkes, campaigner against local waste dump

"Yes, the children are often ill and sometimes can barely breathe. We want to live in another place, but we cannot afford to."

Resident of Brazil's industrialized Cubatao district, where nine out of ten babies have to be given daily doses of oxygen

60 per cent of herbicides. The yearly increases in the incidence of asthma are likely to be linked to steadily increasing air pollution, particularly car exhaust fumes. In other diseases, such as lung infections, pollution is a contributory factor. In the US, deaths from chronic lung disease have increased by more than a third since 1970.

Other signs of increasing ill-health may not even be seen by doctors nor form any part of official statistics: headaches, mild diarrhoea and other digestive disturbances, fatigue, coughs, runny noses and skin rashes are trivial everyday ills that some doctors suspect are increasing in frequency. Some of these symptoms may be due to increasing exposure to pollutants. Those who dismiss any such link take the view that everyday pollutants such as exhaust fumes and pesticides cannot be harmful, because if they were, everyone would be ill. Yet research into basic metabolism shows that people are enormously variable in the way they respond to alien chemicals. What is a toxin or irritant for one person may be harmless, at the same dose, for another.

KING COAL

The major air pollutants, in cities around the world, stem from the burning of fossil fuels. In Eastern Europe and much of the Third World, the use of coal for heating and industry is the main culprit. In western cities like London, coal-burning is restricted, but motor vehicles produce millions of tonnes of pollutants as they burn up fossil fuels in the form of petrol.

The burning of coal releases large quantities of sulphur dioxide and particulates (soot and other small solid and liquid particles in smoke). These, either alone or in combination, increase the number of people who suffer from breathing complaints such as coughs, asthma, bronchitis and emphysema.

Airborne particulates can make breathing difficult and can carry toxic heavy metals and other materials deep into the lungs. In the US, this combination may cause as many as 50,000 deaths every year. In Eastern Europe and the Soviet Union, the problem is far worse, as much of the coal in these countries is high-sulphur 'brown coal', which is burned with outdated and notoriously inefficient technology.

Sulphur dioxide is also one of the gases which causes acid rain. By building tall stacks to clean up urban areas, industry has merely succeeded in dispersing acid rain to rural areas and other countries. Acid rain affects human health because it leads to metals in soil - such as lead, cadmium and mercury, which are bound to particles of soil - being washed out into water supplies. Aluminium, which could be a factor in Alzheimer's Disease (premature senile dementia), is released by acid rain from its normally harmless state in bedrock. High concentrations of aluminium are now found in the drinking water of southern Norway, which receives most of its acid pollution on the winds blowing in from Britain over the North Sea. The incidence of premature senile dementia is now increasing in the region.

POISONS FROM TRAFFIC

Another gas which leads to the formation of acid rain, and is also a direct health hazard, is nitrogen dioxide. This is released both by burning coal and by motor vehicles. Nitrogen dioxide irritates the lungs and can cause bronchitis and pneumonia. It can also make people more vulnerable to viral infections such as influenza.

Nitrogen oxides also contribute to the formation ozone in the air around us, which is a major health hazard as well as being poisonous to trees and crops. At low altitude, ozone is formed by sunlight acting on hydrocarbons (which are released with industrial and vehicle emissions) mixed with nitrogen oxides in the air. Ozone is the major constituent of photochemical smogs which are notorious in cities like Los Angeles and Mexico City, where there is heavy traffic and long periods of windless, sunny weather. Although in the stratosphere, ozone protects life, at ground level it interferes with breathing, causing coughing and choking. It aggravates asthma and other chest complaints, and reduces resistance to colds and pneumonia. The US, where the automobile runs unchecked, suffers particularly badly from ozone pollution: more than half of all Americans live in counties where the federal ozone standard is exceeded.

Other dangerous pollutants which contaminate city air include carbon monoxide, lead and toxic hydrocarbons, such as benzene, toluene and xylene. Some of these can cause cancer, reproductive problems and birth defects. Motor vehicles are responsible for producing 85 per cent of carbon monoxide pollution which, at high concentrations, deprives the body of oxygen, impairs perception and thinking, slows reflexes and causes drowsiness. Around a half of all city dwellers in Europe and North America are exposed to unacceptably high levels of carbon monoxide according to the WHO.

The concentration of lead in the air, which affects the nervous system and lowers children's learning ability, is declining in many parts of the developed world, owing to the phasing out of lead in petrol, which was originally added to make engines run more smoothly. In the US, the average lead level in Americans' blood fell by over one-third between 1976 and 1980. The use of unleaded petrol is also being encouraged in many European countries. In the Third World, however, lead levels are rising fast.

To make up for the loss of performance in cars using unleaded petrol, the hydrocarbon composition of petrol has been changed, and this leads to an increase in the levels of toxic hydrocarbons released in exhaust fumes. The growing use of unleaded petrol has thus increased emissions of such potent carcinogens as benzene and toluene.

A CHEMICAL SOUP

Emissions of hydrocarbons and other toxic chemicals from industry are also a cause of rising concern. A 1989 report estimated that about 1 million tonnes of toxic chemicals are pumped into the air every year in the US, a figure which may be three times too low since the study did not include emissions from vehicles, waste dumps, dry cleaners, and chemical releases into soil or water that end up in the air through evaporation. Of the 320 chemicals which were surveyed for this report, only seven were subject to emission standards.

Data which has recently emerged from Eastern Europe indicates that the situation there is even worse. Each year, belching smoke stacks in Silesia, the industrial heartland of Poland, cover the region's cities with up to 1 kg per m^2 (about 2 lb per square yard) of chemical 'fallout'. The Polish government is considering a ban on vegetable growing in the area, because the levels of cadmium and lead in the soil are some of the highest in the world. Among children, deformities and disabilities are alarmingly high.

As industrialization in the Third World continues apace, there is little doubt that the amount of poison in the air is increasing, especially as these countries have even more permissive anti-pollution laws than in the developed countries. As dirty industries in the industrialized world become slowly squeezed by increasing environmental standards, they are gradually shifting operations to the Third World.

CLEARING THE AIR

In 1970, the US Congress passed the much-heralded Clean Air Act, claiming that it would restore the air quality of America's cities. Two decades later, however, 487 counties had failed to comply with the Act and, under heavy pressure from the polluting industries, Congress is attempting to draft a revised version.

Air-quality legislation has failed because it has simply shifted pollution somewhere else or has been too limited in its ambitions, aiming to reduce specific pollutants only. Such laws have often worsened the impact of other pollutants. A case in point is the introduction of catalytic converters in North America, Japan, South Korea and several European countries. These are fitted to vehicle exhausts and greatly reduce emissions of hydrocarbons, nitrogen oxides and carbon monoxide. However the massive growth in numbers of cars around the world has completely wiped out the benefits to air quality.

The US, recognizing that catalytic convertors cannot solve their air pollution problems, is now considering forcing vehicle manufacturers to switch to alternative fuels such as methanol and ethanol. Although this would help the ozone problem, it could result in increased emissions of carcinogens and carbon dioxide, the main greenhouse gas *(see p.17)*.

It is now clear that any serious attempt to deal with air pollution must tackle the root of the problem, namely the constant drive for increased industrial growth and the parallel rise in road use that this brings in its wake.

"Levels of cancer, skin disease, respiratory problems, hypertension and premature births are all higher than average here. A man will come in with bronchitis but we have real difficulties diagnosing it because his problems are compounded by so many other diseases. One thing is certain - when these people retire at 50 they have been totally destroyed."
Dr Ion Luca, doctor, Copsa Mica industrial centre, Romania

▼ Congestion in Bangkok, where a pall of black traffic fumes often covers the city streets. Traffic pollution is now a problem shared by industrialized nations and countries in the Third World.

WATER FIT TO DRINK?

A supply of clean drinking water is possibly the most precious resource on the planet. Luckily, freshwater is available in prodigious quantities, and is continually renewed by the global water cycle of rainfall and evaporation, powered by the energy of the Sun. Only around a third of the 9,000 trillion litres (about 2,000 trillion gallons) of freshwater available in rivers, lakes and groundwaters is currently used, mostly for irrigation and industry.

The global totals, however, mask wide variations in the local availability and use of freshwater. Much of the water is in areas like Siberia and Amazonia where few people live, while some regions with very little freshwater, such as the southwest of the US, have large and growing populations which demand vast amounts of water *(see p.132)*. Many areas of North Africa and the Middle East are reaching, or have already reached, the point where demand is outstripping the natural supply.

In other parts of the Third World, the problem is not the total amount of water available for use, but the filthiness of the water and the lack of access to it. At least one in five city dwellers and three-quarters of the villagers of the Third World are without reasonably safe supplies of water.

In most of the developed world, piped water and adequate sewage facilities exist, but noxious chemicals in the drinking water of Europe and North America now reach dangerous levels. Similarly, with the spread of industry and chemical farming methods to the Third World, freshwater resources in these areas are also becoming contaminated with chemicals.

▲ Dogon women fetching water from the local well in Mali. Clean, fresh water, available in abundance, is becoming increasingly scarce in every part of the world.

WATERBORNE DISEASES

In the developed world, over a century of public health legislation and improved water treatment has drastically reduced the incidence of waterborne diseases. Now, however, over 120 years since the last major cholera epidemic in London, the sewage system built by the Victorians to solve Britain's urban sewage disposal problems is overloaded and decaying, and pollution incidents from sewage plants are increasing.

Outbreaks of waterborne disease spread by *Cryptosporidium*, a parasitic microorganism, have occurred several times in Britain since 1985. *Cryptosporidium*, which causes severe diarrhoea and vomiting, and which is potentially lethal to young children, is spread mostly in farm slurry. The risk of *Cryptosporidium* entering the water supply cannot be entirely eliminated with current chemical methods of water treatment. These have replaced the slow sand filtration method which was much more effective in dealing with the parasite.

In the Third World, particularly in urban areas, the discharge of raw sewage to waterways is the rule rather than the exception. The bacteria, viruses and parasites in untreated human waste make it the world's most dangerous environmental pollutant. Typhoid, cholera, amoebic dysentery, polio and hepatitis are all transmitted in human excreta. Nearly two billion people in the Third World are exposed to these diseases by drinking contaminated water, and every day an average of 25,000 people die from them. Four out of five deaths of children in developing countries are due to water-related diseases.

In 1980, in a characteristically unrealistic gesture, the UN General Assembly proclaimed that the 'International Drinking Water Supply and Sanitation Decade' had begun, with the aim of "Clean Water and Adequate Sanitation for All by 1990". Needless to say, the UN's grandiose plans have largely come to nought. Only a tiny fraction of the $300 billion which the project was to have cost was spent on water schemes around the world. At the end of the decade, it is likely that drought and water-related diseases affect as large a proportion of the world's population as they did at the beginning.

Even if more of this money had been spent it is debatable what good this would have done the world's poor. Most foreign aid for water projects is spent on buying unsuitable and expensive equipment made by the country which has given the funds. In 1970, a $3.5 million sewerage system was built in Accra, Ghana. Designed by foreign consultants and supported by foreign aid, the system was excessively elaborate and too expensive to use, and by 1977 only 171 houses had been connected.

CHEMICAL CONTAMINATION OF DRINKING WATER

The main threat to water quality in industrialized countries undoubtedly stems from chemical pollution. Industrial discharges, the run-off from land treated with agricultural chemicals, leakages from landfill sites, and accidental spillages of chemicals have all combined to render many drinking water sources increasingly unfit to drink.

Twenty million Europeans obtain their drinking water from the Rhine, which is heavily contaminated with chemical wastes *(see p.140)*. In spite of the most up-to-date treatment of water supplies, more and more chemicals are being found in drinking water from the Rhine. The head of the Dutch waterworks in the Rhine treatment area has admitted that just about every substance present in untreated water can also be found in drinking water.

Many rivers in the eastern US are so polluted by industrial chemicals that they can no longer serve as sources of drinking water. The water of other less polluted rivers has to be treated at great cost, but this has not stopped nearly 130 dangerous chemicals being detected in US drinking water. Only 14 of these contaminants are regularly monitored.

Not only do water treatment processes fail to remove hazardous chemicals, but they also pose health risks themselves. The chlorination of water supplies kills most harmful germs but the chlorine can react with various organic pollutants to form chlorinated hydrocarbons, several of which are suspected of causing cancer. In the US, increased cancer rates have been found among people supplied with chemically polluted and chlorinated water. Many German treatment plants now use ozone instead of chlorine as a disinfectant. Groundwater, which provides drinking water for many millions of people around the world, including a half of the US population, are also becoming polluted with industrial and agricultural chemicals. Whatever legislation is brought in to deal with the problem of chemical pollution, it cannot do anything to offset the pollution that has already occurred.

POISON ON OUR PLATES

Scares about agrochemicals contaminating food hit the headlines regularly. As with the growth regulator, Alar, there may be an official or voluntary ban following such publicity, and the general public feels reassured by this, assuming that their food has once more been made 'safe'. The truth is that the safety testing and approval of pesticides and other agrochemicals is a shambles. No one knows what effect most pesticides have on health. Residues are present in all foods - often at unacceptably high levels - throughout the industrialized world and also in parts of the Third World where pesticide use is common.

In the early 1980s, doctors at the Central Emek Hospital in Israel saw a number of patients with a collection of similar symptoms, including stomach pains and diarrhoea, general weakness and 'nervousness'. Detailed enquiries revealed that they all ate diets with a high proportion of fruit and vegetables. When blood samples were tested they showed clear signs of having been affected by insecticides known as anti-cholinesterases. These interfere with the body's ability to break down the chemical messenger, acetylcholine, which carries impulses from one nerve cell to another, and from nerves to muscles.

The effects of large, single doses of such insecticides are known from accidents involving agricultural workers, but this was the first time that anyone had studied the effects of eating small amounts over a long period of time. All the

“There have long been rumours in Rovigo that the water from the Adige, Italy's second largest river, was polluted. It gave off a vile smell 'like rotting rats or mice' which some said was from factory waste dumped upstream. But the local authorities insisted that the water taken from the Adige was safe to drink, and Rovigo was resigned to the noxious fumes - until a chemist sponsored by a conservationist group produced scientific proof of what many had long suspected. He found that a sample from the Adige contained potentially lethal levels of pollutants.”
Dalbert Hallenstein

“Like the constant dripping of water that in turn wears away the hardest stone, this birth-to-death contact with dangerous chemicals may in the end prove disastrous... As matters now stand, we are in little better position than the guests of the Borgias.”
Rachel Carson

“...the advancement of analytical techniques has, in fact, dispelled the myth that these new age anti-cholinesterase pesticides rapidly detoxify into harmless residues. Conversely, many of these chemicals have now been shown to fall out into an unpredictable range of by-products, some being more highly toxic than the original chemicals applied. Much of this research has been ignored.”
Mark Purdey

▲ Pesticide spraying of cropland by plane. This method of spraying is particularly hazardous, as the spray may drift in the wind to affect nearby homes.

patients were advised to eat no fruit or vegetables for a while, and this cleared up the symptoms. None of the patients concerned ate fruit or vegetables without washing them first, and there is no reason to suspect that the Israeli farmers were misusing these pesticides any more than farmers elsewhere in the world. Would cases of this sort be recognized as pesticide poisoning by most doctors? The answer is almost certainly "no".

TESTS AND SAFETY LIMITS

Apart from this study, research on the effects of eating ordinary pesticide-contaminated food is scanty. As the UK Ministry of Agriculture has admitted, there is no such thing as pesticide-free food any longer. Even organic food may acquire some pesticides through spray drift and soil contamination, although the amounts are relatively low.

The safety tests that are carried out before agrochemicals can be approved for use always involve laboratory animals, which may or may not react in the same way as humans. Studies of *known* human carcinogens, carried out by the drug company Pfizer, showed that less than half produced cancer in laboratory rats and mice. The company concluded that tossing a coin would have been just as reliable a guide to human health effects as testing on rodents. Rodent-based tests are equally unreliable as a guide to toxicity to humans.

It is ironic that pesticides that have gone through laboratory trials with animals are considered to have been rigorously tested compared with others. Some pesticides, especially those introduced many years ago, have not even had this form of screening, or have only had very short-term, inconclusive tests. In the US, 1,400 basic pesticides are permitted, with 600 of these in common use. Yet, according to the US National Academy of Sciences, only 37 have been adequately tested. Information on the health effects of the rest was insufficient to complete a full assessment of their health hazards. For over a third, there was no safety data at all.

Although the Environmental Protection Agency plans to review the safety data on pesticides, this will not be complete before the turn of the century - and in the intervening years, untested and potentially hazardous pesticides will continue to be applied.

In Britain, untested pesticides are also widely used. The Ministry of Agriculture has plans to test about 250 long-established pesticides. These include various suspect items, such as maneb, mancozeb and zineb, three of the most widely used fungicides in agriculture, which have been positively identified as carcinogens in the US. Lack of staff and funds means that the retesting will take about 40 years, and meanwhile such pesticides continue to be used.

Even when pesticides are tested, official secrecy in the UK prevents the public from knowing the results, which are protected by law and cannot be published. Of the 400 pesticides permitted, there are only 13 for which the original safety studies have been released. Fourteen permitted pesticides are banned in other countries, suggesting that the interpretation of safety data is sometimes arbitrary. But lack of access to the basic data makes it impossible to question the decision of the Ministry of Agriculture on permitting a given pesticide. As one British Member of Parliament remarked "We know more about what's in a pair of socks than what's in the food we eat."

Another worrying aspect of conventional tests for pesticide safety is that they only look at a single pesticide at a time. The 'cocktail effect' of eating two or more pesticides at once, or of eating pesticides in conjunction with food additives, are never considered in these tests, despite the fact that the average meal can contain dozens of different residues. Yet it is well known that 'cocktail effects' can occur, one toxic chemical making the effects of another worse.

A research team working in Vienna has found yet more cause for concern, since the standard tests employed may severely underestimate the amount of pesticide present on stored grain. They showed that insecticides, sprayed on to the grain after harvesting, bound tightly to the grain and could not be extracted by solvents, making them undetectable by conventional tests. The pesticide dosage from bread, cereals and other forms of grain may be far higher than scientists previously supposed.

The extent of pesticide contamination of food in Third World countries is largely unknown, but the widespread use of pesticides and the frequent disregard for recommended safety measures (such as not spraying just before harvest) make it likely that food is contaminated. Many pesticides that have been banned in developed countries, such as lindane (banned in Germany, Holland and Japan) and DDT (widely banned) are exported to the Third World, and residues are detected on imported food.

The chances of pesticide poisoning from contaminated food being noticed and reported in Third World countries is remote indeed. Even when agricultural workers are made ill by pesticides this is rarely recognized or recorded officially. A study in Nicaragua looked at 15,000 workers in contact with pesticides and found 396 poisoning incidents in the space of a year. Of these, only seven - under 2 per cent of sufferers - were correctly diagnosed and notified to the health ministry by local doctors.

THE HORMONE-FED HEN

Pesticides are just one group of items that can contaminate food. As agriculture becomes more and more dependent on drugs, these too can find their way on to our plates. Hormones and antibiotics are increasingly present in meat produced by 'factory farms'.

To maximize production and minimize labour costs, farmers have increasingly turned to intensive methods of livestock rearing. Gone are the days when chickens were free to wander around the farmyard, or when calves were reared with their mothers. Today, they are crammed into cages, fed hormones to boost their growth and antibiotics to stave off the diseases that come with overcrowding.

Several growth hormones have now been banned in the EC and North America as carcinogens but there is strong evidence that the illicit use of banned hormones still thrives. The consumption of hormone residues in meat has been linked to premature breast development among young girls and to other signs of precocious sexual development. Studies in Puerto Rico reporting cases of children as young as three years developing pubic hair.

Cooped up in cages so small that they can barely move, and eating feed made up of fish, bones, feathers, dried blood, offal and even human effluent, it is small wonder that the animals raised in today's battery units are often severely infected with bacteria such as salmonella. In Britain, the incidence has now reached epidemic proportions, cases of salmonella infection rising twelve-fold between 1982 and 1988. In the US, salmonella is rife in egg and egg-based products. To combat salmonella and other infections, farmers have dosed animal feeds with antibiotics. As a result of their widespread and indiscriminate use, more than eight antibiotics are now ineffective, bacteria having acquired resistance to them. Livestock farmers risk becoming trapped in an 'antibiotic treadmill' in the race to keep ahead of the rapidly evolving agents of disease.

Rather than attack the problem at source, however, many governments are looking towards food irradiation as an easy technological solution to the problem of bacterial contamination in food. Food irradiation involves exposing food to ionizing radiation at doses high enough to kill the bacteria responsible for food poisoning. The practice is already permitted in the US and most EC countries. Irradiation causes foods to undergo some chemical change, and may produce residues known as 'radiolytic products'. Some of these are suspected of being carcinogens, mutagens or teratogens.

Were the subsidies which intensive farming receives from government removed, the price of such food would rise dramatically (some estimate by a quarter or more) making organic produce, free of pesticides, hormones and other chemicals, competitive in price. In effect, the justification for adulterating our food with chemicals and for keeping animals in horrendous conditions is a myth. Chemically grown crops, processed foods and intensively reared livestock are not cheap. On the contrary, they are unacceptably costly in terms of energy, pollution, nutrition and cruelty to animals.

▲ Hens in battery cages. These hens have not been in the system long, and are still in relatively good condition. The same will not be true when their productive life nears it end.

“Why should the profits of chemical companies and convenience of farmers outweigh the importance of children's health?”
Pamela Stephenson, Parents for Safe Food

THE FUTURE in PROSPECT

▲ Smoke from a factory in Leningrad blows downwind, contaminating the surrounding land and housing.

“Man's conquest of Nature turns out, in the moment of its consummation, to be Nature's conquest of Man. Every victory we seemed to win has led us, step by step, to this conclusion.”
C. S. Lewis

“To what extent we humans are on the brink - a decade, fifty years, a century away - is a matter of dispute; but not that the nemesis is near in any real time scale.”
John Fowles

The ENERGY EXPLOSION

From pole to pole, our demands on the environment are causing irreversible destruction. Few now doubt that changes to our way of life are necessary if we are to avert a global catastrophe, and a variety of international and national programmes have been launched in response to the environmental crisis. But will these prove adequate to the problems they address?

A major problem with predictions of the future is that they cannot account for unknown thresholds, those points at which the effects of pollution or habitat destruction become irreversible, or at which a chain reaction comes into operation. However, governments rarely allow for such uncertainty. All too often, they tackle environmental problems in isolation from each other, waiting until overwhelming evidence has been amassed that major damage is already underway. This kind of approach is especially evident in the fields that we will examine in this chapter - energy, the effects of economic growth, and the allied problems of population growth, famine and the struggle for resources.

ENERGY SOURCES

Every day, the Sun bathes the Earth in plentiful supplies of energy, without which life would be impossible. Eventually, in some two billion years time, the Sun will die, and with it, life on Earth. But, for all practical purposes, the resources that are sustained by the Sun's energy - trees, plants and the animals that feed on them - may be considered renewable.

The same is not true of the deposits of minerals, metals, ores and fossil fuels including oil, gas and coal. Laid down millions of years ago, their stocks are strictly finite: they cannot grow, they can only decrease. Such resources are thus non-renewable. Although improvements in technology can stretch the amount of time they last, they will eventually run out. As society has industrialized, so it has become increasingly dependent on non-renewable resources, and nowhere is this more apparent than our use of energy.

Without radical changes in current economic policies, global energy demands look set to increase dramatically. Energy analysts currently talk of world energy consumption doubling within 30 years. But such predictions have to be tempered by reality. A doubling in oil consumption, for example, would mean that, over the next 30 years, we would consume as much oil as has *ever* been consumed during the entire history of mankind. Within a matter of a few decades, the total world resources of petroleum will have been exploited.

The environmental implications of a doubling in energy use are horrendous. To produce twice as much energy as we currently consume would mean doubling the infrastructure for its supply - coal mines, oil wells, refineries, power stations, pylons, filling stations and the like.

THE SEARCH FOR SUPPLIES

The world's stocks of fuels can be divided into reserves and resources. Reserves are known deposits that can be extracted with current technologies and at acceptable costs, whereas resources are somewhat more speculative, because they contain estimates of as yet undiscovered deposits, as well as allowances for improved extraction techniques in the future.

Current reserves of crude oil are approximately 100 billion tonnes, whereas total resources lie in the range of 220 to 300 billion tonnes, which at current rates of production would meet 80 years' supplies. Coal, with some 800 billion tonnes of reserves and 10 to 15 times that of resources, could last 2,000 years at current rates of production and natural gas, some 150 years.

Although new sources of oil and other fossil fuels are constantly being discovered, the rate of new discoveries is declining. During the first decades of this century, discoveries of oil deposits came thick and fast. By the 1950s, only a small proportion of the total land surface of the US had been explored and most petroleum experts assumed that many large discoveries were still to be made. But the US is now importing nearly half its crude oil requirements, despite new technologies to improve energy efficiency. Eighty per cent of the oil discovered in the US has now been consumed - and what remains would only supply domestic needs for 9 years at current rates of consumption. The US experience is not unique. Since 1973, despite intensive exploration throughout the globe, proven oil reserves have increased by a mere 5 per cent.

THE NUCLEAR OPTION

The inevitable prospect of oil running out - and the fear of future embargoes - has led many

The energy boom

World energy consumption is now far outstripping renewable supplies, and is increasing at a rate which exceeds growth in population. Coal consumption has doubled during this century, and energy analysts predict that, over the next 30 years, we will use as much oil as has ever been been consumed in the entire history of our planet.

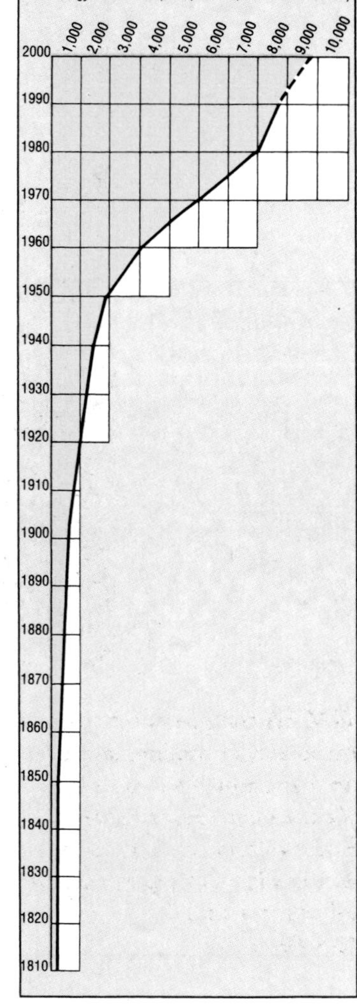

"A citizen of an advanced industrialized nation consumes in six months the energy that has to last the citizen of a developing country his entire life."
Maurice Strong

planners to look towards nuclear power as a solution to our energy problems. One country that has adopted the nuclear option with zest is France. Today, more than 70 per cent of France's electricity is generated from nuclear power plants - and the percentage is likely to increase as still more nuclear plants are commissioned over the next half-dozen years.

France's nuclear programme was based on the assumption that fossil fuel use would be reduced both by replacing oil-fired power stations with nuclear ones, and through more and more consumers turning to electricity. However, as the true cost of electrification has begun to bite, consumers have realized that they are better off burning fossil fuels, such as natural gas.

The irony is that whereas other countries in western Europe have managed to cut their consumption of fossil fuels, including oil, by 14 per cent between 1975 and 1984, France managed a mere 7 per cent, due to an 83 per cent boom in the consumption of natural gas. The savings made in fossil fuels through the massive nuclear programme amounted to some 6 billion francs per year, but the cost of constructing those nuclear plants averaged over 30 billion francs per year during a 15-year period.

The nuclear programme in France has had momentous economic repercussions. By 1987, the French electricity board, Electricité de France (EdF) had accumulated debts of more than 220 billion francs, a sum greater than the country's entire income tax receipts in 1986, and one-and-a-half times greater than the total annual investment in private industry.

Nuclear power is not only an economic disaster but also a major threat to the environment and public health. Mining and enriching the uranium which fuels reactors requires considerable quantities of energy and is highly polluting. Once a reactor is operating, there is always the chance of an accident which leaves enormous swathes of land uninhabitable and virtually unusable, for tens if not hundreds of years.

Official statements have claimed that the risk of a serious accident per operating reactor is once every 100,000 years. The head of nuclear safety in France, however, puts the chance of a serious accident as several per cent over the next 20 years. That view is shared by officials at the US Nuclear Regulatory Commission. They acknowledge that the odds on a major reactor accident in America before the end of the century are evens. Meanwhile, minor accidents and

▼ Drums of low-level radioactive waste being disposed of in a landfill site in the US.

► An Indian woman shapes cow dung into compact cakes for use as a domestic fuel. Lack of timber forces Indian villagers to use the dung in this way, depriving the soil of valuable natural fertilizers which might otherwise have gone onto the land.

everyday discharges continue to pollute the environment. As for the low-level and high-level wastes that are produced, including the reactors themselves, no-one yet has a satisfactory solution for their long-term storage or disposal.

Concern over the costs of nuclear power and over safety saw a growing disenchantment with nuclear energy in many countries during the 1980s. In the US, no new plants have been ordered since 1978, while 108 have been cancelled since 1974. The greenhouse effect and atmospheric pollution from fossil fuels, however, have revived the nuclear industry's fortunes, because unlike coal or oil, nuclear fission does not generate carbon dioxide.

However, this is only half the story. Nuclear power stations, with their massive structures for containing radioactive materials, demand a heavy investment in energy - and the energy required for construction has to come from fossil fuels. As more uranium is mined and the quality of the ore falls, so the energy cost of extraction goes up. A mass nuclear power programme would exhaust high-quality ores so quickly that, within a generation, the uranium being mined would provide no more energy for each tonne of rock than would mined coal.

THE EFFECTS OF ENERGY ADDICTION

Renewable energy resources - wind power, wave power and solar power - combined with improved energy efficiency, are certainly well-developed enough to ensure that no-one need freeze in the dark. But despite the cost-effectiveness of energy efficiency, industrialized countries have shown little enthusiasm for energy conservation, with notable exceptions such as Denmark and Sweden. Furthermore, between 1972 and 1980, more than 90 per cent of the aid supplies to the Third World for energy programmes went to large-scale electricity generation projects. Less than one per cent was devoted to energy efficiency.

Fears over the greenhouse effect may alter that. But it is not just in industry and the home that changes in energy use are needed. Oil is essential to the manufacture of pesticides and fertilizers and as farmers have adopted modern chemical-intensive agriculture, so they have become increasingly hooked on oil. In the US, the amount of oil used to produce one tonne of grain has risen two-and-a-half times between 1950 and 1985. Farmers worldwide now consume sixty times more oil than they did in the 1920s, and 100 times more electricity. Much of the increased consumption comes from the massive expansion in the use of tractors, but the switch from organic fertilizers to artificial fertilizers has also been key to agriculture's increasing dependency on oil.

But, most critically of all, the energy inputs used by modern farming have helped to mask the growing ecological damage being done to agricultural lands, and in particular, soils. When the oil runs out, farmers will be faced with land that is often too degraded to farm. And here we come to the nub of the problem. It is not just how energy is produced that matters, but also what we do with it.

The ecological damage being caused by modern consumer society is placing an intolerable burden on the natural cycles that we rely on for our health and well-being. We may be able to stretch the supply of non-renewable resources - be it oil or iron - for a little longer, and we may be able to find substitutes for those resources that are running out. But if the only outcome is to prolong our assault on the environment, we shall have gained nothing. Ecological collapse poses a far more immediate threat to our way of life than depleting resources ever will.

"In the extreme, rather than build a new power station, one California utility just gave low energy light bulbs away"
Dr Bob Everett, Open University Energy and Environment Research Unit

The energy cost of farming

Throughout the developed world, millions of tonnes of non-renewable fossil fuels are consumed every year to grow food. Some of this is used to power the machinery needed to plough fields, sow seed and harvest crops, while additional 'hidden' energy is used to manufacture fertilizers and agricultural chemicals. This diagram shows just how much energy is used in agriculture per head of the population in a range of developed countries.

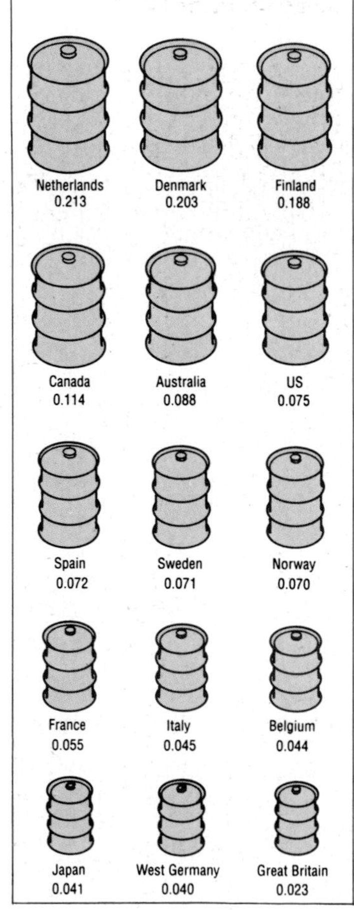

The INDUSTRIALIZED ENVIRONMENT

▲ **Microchip components being assembled in Singapore. Spiralling industrial output, shown by many countries in the Far East, poses as many problems as it answers.**

The world's leading policymakers foresee an unprecedented worldwide expansion of industry and western-style consumerism in the coming decades. The liberalization of the Soviet and Eastern European economies, the impressive economic performance of the 'Asian Tigers', such as Taiwan and South Korea - often held up as models for other developing countries - and the forthcoming economic union of the European Community, offer the prospect of a massive increase in markets and a surge in the performance of the world economy.

If this surge in economic activity does occur, it could only be short-lived, as the environmental devastation which it would cause would almost certainly be enough to push the world into an ecological catastrophe. The main cause would almost certainly be global warming.

The state of the climate today results from human activities in the recent past, and our activities now will affect the climate of the future. We are already committed to a global warming of 1°C (about 2°F) or more, owing to the delay between the emission of greenhouse gases and their effect upon our climate. The UN expert committee on climate change calculates that just to stop the atmospheric concentration of the main greenhouse gases rising any further would require emissions to be cut immediately by over 60 per cent. They also warn that if emissions continue to increase at present-day rates, then even more drastic reductions will be needed in the future. Because the emissions of greenhouse gases from the non-industrialized countries are very low and cannot realistically be reduced much further, industrialized countries will have to bear the brunt of these changes.

But huge reductions in global emissions of greenhouse gases are simply not on the international agenda. Although international leaders pay lip service to the importance of dealing with the greenhouse threat, the measures which have so far been proposed are hopelessly inadequate. At a 72-nation governmental conference held in Holland in 1989, a proposal for the industrial countries to cut their carbon dioxide emissions by 20 per cent by 2005 was blocked by the US, Japan, the Soviet Union and Britain. These four major polluters, which between them emit around 50 per cent of the world's carbon, refused to agree to more than a loose commitment to stabilize emissions "as soon as possible".

On a global scale, carbon dioxide emissions could double by 2010, largely due to the expansion of industry in the Third World. If India's economic goals are met, increasing coal and oil use will lead to a 150 per cent increase in carbon dioxide emissions by 2010, although per capita these will still be less than a tenth of the emissions from the First World.

A report commissioned by the Dutch government in 1989 concluded that, with our present economic structure, only a dictatorship could implement the kind of long-term plan needed to bring about the required cuts in levels of carbon dioxide.

THE DIMINISHING OZONE LAYER

As if global warming was not a terrifying enough prospect , the next few decades are also likely to see a rapid erosion of the stratospheric ozone layer due to the continued release of CFCs and related gases into the atmosphere. This will put ecosystems and human societies under increasing stress from growing exposure to harmful solar ultraviolet-B radiation.

So what is happening to tackle this problem? In 1987, an international agreement, the Montreal Protocol on Substances that Deplete the Ozone Layer, was reached, under which it was agreed to cut the use of CFCs by half by the end of the century. Within months, however, the agreement was declared hopelessly inadequate, and in 1990 its measures were tightened at a meeting in London. CFCs are now to be banned altogether by the year 2000, as are several other ozone-depleting chemicals that were not previously covered by the Protocol - including carbon tetrachloride and halons. Contrary to expectations, it was also agreed to phase out methyl chloroform, an ozone-depleting solvent currently produced in large quantities, by 2005. Moves to bring in an earlier ban on CFCs were blocked by Japan and the US, under intense pressure from industry. The USSR also opposed an earlier ban, although some countries stated that they will cease CFC production and use well before the end of the century.

Joe Farman, the British scientist who discovered the ozone hole, has warned that even these revised measures are still "too little, too late". Under the agreement, chlorine levels in the atmosphere will be allowed to increase by 1.2 parts per billion over the next decade. This, Farman calculates, will cause the ozone layer over the northern hemisphere to thin by 20 per cent.

Millions of kilogrammes of CFCs are currently in use in equipment such as refrigerators and air conditioners, and ultimately, much of this will leak or will be deliberately discharged into the atmosphere. Once CFCs are released, they can take as long as a decade to reach the stratosphere and they may take a century to break down. The CFC substitutes that the chemical industry is keenest to promote, although less damaging, still deplete ozone, yet their use is not controlled under the Protocol. Also, like CFCs, they are greenhouse gases. Moreover, although India and China have agreed to the phasing out of CFCs, they will only do so if supplied with the technology to manufacture alternatives.

EXTINCTION IS FOREVER

If present trends continue, a quarter of the world's species of animals and plants could vanish within 50 years. The impact of global atmospheric pollution upon the environment will exacerbate the already high and rapidly accelerating rate of the destruction of 'biodiversity' - the variety of living things on the planet.

The local impact of such a devastating loss of natural wealth can only be guessed at, but a telling analogy has been made by the biologists Paul and Anne Ehrlich. They compare individual species - whether they be bacteria, herbaceous plants, worms, mites, insects, frogs, lizards or small mammals - to the rivets that hold together an aeroplane. Although we know that every species has its position in a food chain and has evolved to play its own unique ecological role, ecologists can no more predict the consequences of the extinction of any one species than an airline passenger can tell the effect of losing any one rivet. But, both can see the eventual consequences of continually losing species or rivets is disaster.

Between 50 and 90 per cent of all species live in tropical forests which, as we have seen, make up one of the world's most endangered ecosystems. If virtually all of the remaining tropical forests do indeed disappear before the middle of the next century, this would not only mean the extinction of their biological wealth and the human cultures which depend upon this resource, but would also provide yet another shock to the world's climate system.

The response of governments and international agencies to the destruction of the tropical forests and the biodiversity crisis is, as with their responses to global warming and ozone depletion, hopelessly inadequate. In fact, the proposed solutions, far from solving the problem, or even slowing the devastation, are likely to accelerate it. The Tropical Forestry Action Plan *(see pp.84-85)* is a case in point.

Undeterred by criticisms of the Tropical Forestry Action Plan - especially from environmental and social activists working in the Third World - the World Bank, the UN Environment Programme and other institutions are preparing 'A Global Strategy for the Conservation of Biodiversity'. This, too, is likely to exacerbate rather than solve the problem, largely because it wrongly interprets the causes of the crisis.

Instead of making sure that the principles of conservation are included in all agricultural and industrial practices, as they have traditionally been in village-based societies in the Third World, the Global Biodiversity Strategy proposes that species should be protected in national park reserves. But the record of national parks around the world in protecting species is dubious to say the least. While the pressures on land and resources continue to build up outside the national parks, it is inevitable that they will eventually be destroyed.

"What is the best environment? The simple-minded answer is: probably the one we have evolved with over the last few thousand years. If you change it, it may in some cases get better, but do you want to risk it? The probability is much higher that it will get worse."

Dr Robert Watson, NASA scientist

"Acid rain has been a training ground for us. It has been our kindergarten, for scientists, for policy makers, for engineers, for economists, for legal experts... But nature has intervened, and we are not allowed the normal years to develop our skills. We're confronted with a curriculum which includes primary and secondary school, university courses and graduate studies, simultaneously."

Dr Hans Martin, on the challenge posed by climatic change

The greatest extinction?

Studies of the fossilized remains of marine animals provide evidence that on at least five occasions in the Earth's history, 'mass extinctions' have wiped out whole families of plant and animal species. In the greatest of these mass extinctions, half the animal species in the seas disappeared, and it was millions of years before numbers recovered. Species are now disappearing faster than ever before, and if mankind continues on its present course, a devastating mass extinction could be about to take place

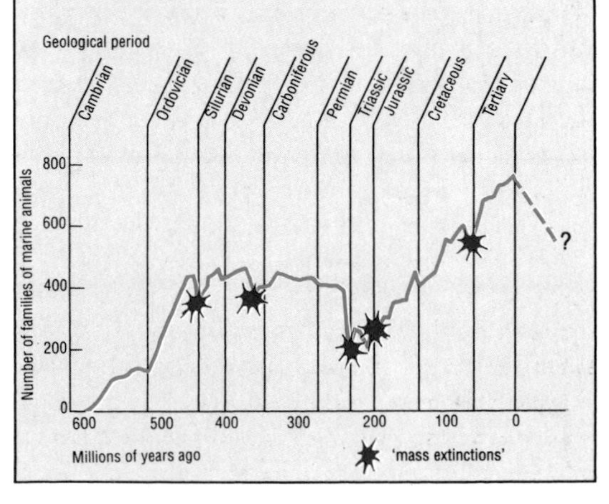

Adapted from "Threats to Biodiversity", by Edward O. Wilson.

OVERLOADED EARTH

▲ **Children of the Muthare Valley near Nairobi in Kenya. In most of the Third World, a large segment of the population is under 35, and the population problem is set to worsen as these children reach child-bearing age.**

Somewhere in the world, two babies are born within the space of a single heartbeat - three every second, 180 every minute, 259,200 every day, 94,600,000 every year - a figure equivalent to the combined populations of Scandinavia, Belgium and Britain. Having grown from 2 billion in 1930 to over 5 billion today, the world's population is set to double again within 40 years. By then, Kenya's population will have grown by 13 million, Brazil's by 160 million and that of the US by 53 million. The population of Nigeria alone will be equal to the present population of the whole of Africa.

Population growth, like many problems that beset our planet, cannot be dealt with in the short term. Even if birth rates dropped dramatically, the world's population would continue to grow for some time into the future. The relative youth of the majority of the population in Third World countries - in Africa, 45 per cent of people are under 15 - provides a momentum that will not be stopped easily as increasing numbers of young women reach child-bearing age. Ninety per cent of the predicted population growth will occur in the Third World, Africa alone passing the billion mark by 2010.

Many of the countries which are expected to see the greatest increases in population are already unable to feed themselves, and are locked into a cycle of spiralling environmental degradation and grinding poverty. Although more people could be fed if land, currently used to grow cash crops for export, was released for growing food for local consumption, the land's capacity to support people has already been exceeded in many areas. Indeed, some countries are already on the brink of disaster. In Egypt, for example, the population increases by a million every ten months. In the Nile Delta, where most of the country's people live, much of the best farmland is already severely degraded and water supplies are stretched to the limit. By the end of the century, land will be in short supply, and there will be scarcely enough water to supply the average Egyptian with more than half a bucketful a day.

It is small wonder, therefore, that the rapid growth in population is often viewed as the single most important problem facing humanity. Whether the population bomb can be defused in time will depend to a large extent on how far we can bring about a decline in birth rates.

POPULATION AND INCOME

Because population growth is most rapid in those countries with low per capita incomes, rapid industrialization has been promoted as the lynch-pin of any strategy to slow population growth. The assumption is that the shift from a rural society to an urban, industrial one will bring higher incomes, greater financial security and thus a decline in birth rates.

In support of this view, population experts point to the experience of Europe. Here, industrialization initially brought a massive surge in population which, after decades of increase, is now stabilizing and, in the case of West Germany, even declining. Using this example, it is expected that population growth in the Third World (now running at 2.4 per cent a year) will undergo a similar demographic transition as countries industrialize, with birth rates declining as material standards of living increase. On that basis, it is confidently predicted by many demographers that population levels will eventually plateau at 10 billion.

But although several Third World countries have seen steady declines in fertility, the link with higher per capita income levels - let alone the level of industrialization - is often tenuous. Mexico, for example, has achieved a 37 per cent decline in fertility since the 1960s and India a 32 per cent decline, but, in terms of income, the mass of people live in abject poverty despite industrial growth. China, Costa Rica and Sri Lanka, on the other hand, have all seen a reduction in fertility but little industrial growth. Conversely, in many parts of Africa, large families are seen as a sign of wealth. Rising affluence could thus boost population growth.

By western standards, for example, the !Kung bushmen of the Kalahari traditionally lived in absolute poverty. Until recently, they had little or no income, few material goods, no permanent housing, no running water, and no savings. Yet, until western values began to take hold, their population levels remained stable. Although infant mortality was undoubtedly a factor, more important were a wide range of practices, which kept births widely spaced. Breast-feeding, for example, reduces a woman's fertility and the !Kung, in common with many other tribal groups in Africa and elsewhere, breast-fed their children for three years or more. Taboos on sex during this period decreased still further the likelihood of breast-feeding women becoming pregnant. Natural contraceptives, too, played their part.

But perhaps most important of all, the tight-knit web of relationships within traditional societies such as the !Kung provide parents with an extraordinary degree of security. Aunts, uncles, grandparents and cousins make up an extended family that not only shares the daily chores but also gives comfort and help in times of need and, in particular, in old age.

The changes wrought by modern development policies have shattered much of this social and economic security. The imposition of new land tenure systems - especially the introduction of private ownership - has led to the widespread dispossession of peasant farmers. Denied access to land, the chances of going hungry are greatly increased. As the extended family breaks down, children become not only an important source of labour, but also the only guarantee of security in old age.

Many demographers now argue that social factors play a far more important role in determining population levels than income alone. Indeed, the most successful family planning programmes have not been in countries with high growth rates, but in those where a concerted effort has been made to improve health care, education and the status of women. The Indian state of Kerala, for example, has some of the highest rates of unemployment in the country. Economic growth rates are low, poverty widespread, and population densities extremely high. Yet, fertility levels are declining. Kerala's infant mortality rate is a third that of the rest of India, while its birth rate is only a little higher than that in the US. Behind such statistics lies a radical programme of land redistribution, unemployment welfare, free education, subsidized food, and food programmes for nursing mothers and schoolchildren. Health services are also widely spread throughout the state, rather than being concentrated in the capital. In effect, the state government, through a thorough-going welfare programme, has ensured that families have social and economic security.

Yet despite their success, such programmes are the first to be cut as Third World countries become increasingly unable to service their massive burden of debt. Austerity measures imposed by the International Monetary Fund have forced African countries to slash their health and education budgets. Meanwhile, in an attempt to boost economic growth, development programmes have been increasingly orientated to the production of goods for export. The result is a massive loss of land to plantations, further aggravating the problems of landlessness and promoting precisely those conditions that most favour population growth.

THE OVERPOPULATED NORTH

Even in those countries where the demographic transition has been achieved (and significantly this does not include the US, the richest country on Earth, which is expected to add 50 million to its population before reaching zero growth levels) it would be wrong to conclude that a *safe* level of population has been reached. The Netherlands has a population of 398 people per km^2 (1,031 per square mile). Such a high population density is only possible because the Netherlands, in common with many other northern countries, is able to import much of its food and resources - including 4 million tonnes of cereal, 130,000 tonnes of oil and 480,000 tonnes of pulses a year. Without such imports, it would unable to feed itself.

The Netherlands may thus be said to be even *more* overpopulated than many Third World countries. Most of the rich nations of the industrialized world are in a similar position. More than 1 million km^2 (400,000 square miles) of arable land in the Third World are exploited in this way. The UK farms two hectares abroad for every hectare of land under the plough in Britain itself. The US also exploits vast tracts of such 'ghost land'.

And so we arrive at the flaw in the 'development' solution to the population crisis. The levels of affluence apparently responsible for the demographic transition in the North have only been achieved at tremendous cost to the environment - not only in the industrialized world but also in the South. If every citizen in the world were to reach the same standard of living as in the North, the stress placed on the environment would be intolerable. Attempting to *solve* the population explosion through increased per capita incomes - and hence increased consumption - can only increase the impact of people on their environment. It is therefore self-defeating. If the present elusive goal of achieving the demographic transition is continued, the fate of billions will be left to the four horsemen of the Apocalypse - foremost among them being the agent of famine.

"The spectacle of the tropical forests being invaded by the landless has led many to believe that sheer population pressure is the principal cause of tropical deforestation. But the most rapid clearance of tropical forests today is taking place in Brazil, a country twice the size of Europe, immensely rich in natural resources, and with a population only one quarter that of Europe. Thus to attribute tropical deforestation to population pressure alone is to argue that spots cause measles... the policy of transferring hundreds of thousands from the Brazilian north-east to the Amazon was conceived not to relieve the situation of the poverty-stricken landless, but to avoid the redistribution of resources through land reform..."

Jack Westoby, former Senior Director of Forestry, FAO

"While overpopulation in the poor nations tends to keep them poverty-stricken, overpopulation in rich nations tends to undermine the life-support capacity of the entire planet."

Paul R. Ehrlich, Professor of Population Studies, Stanford University

FEEDING the WORLD

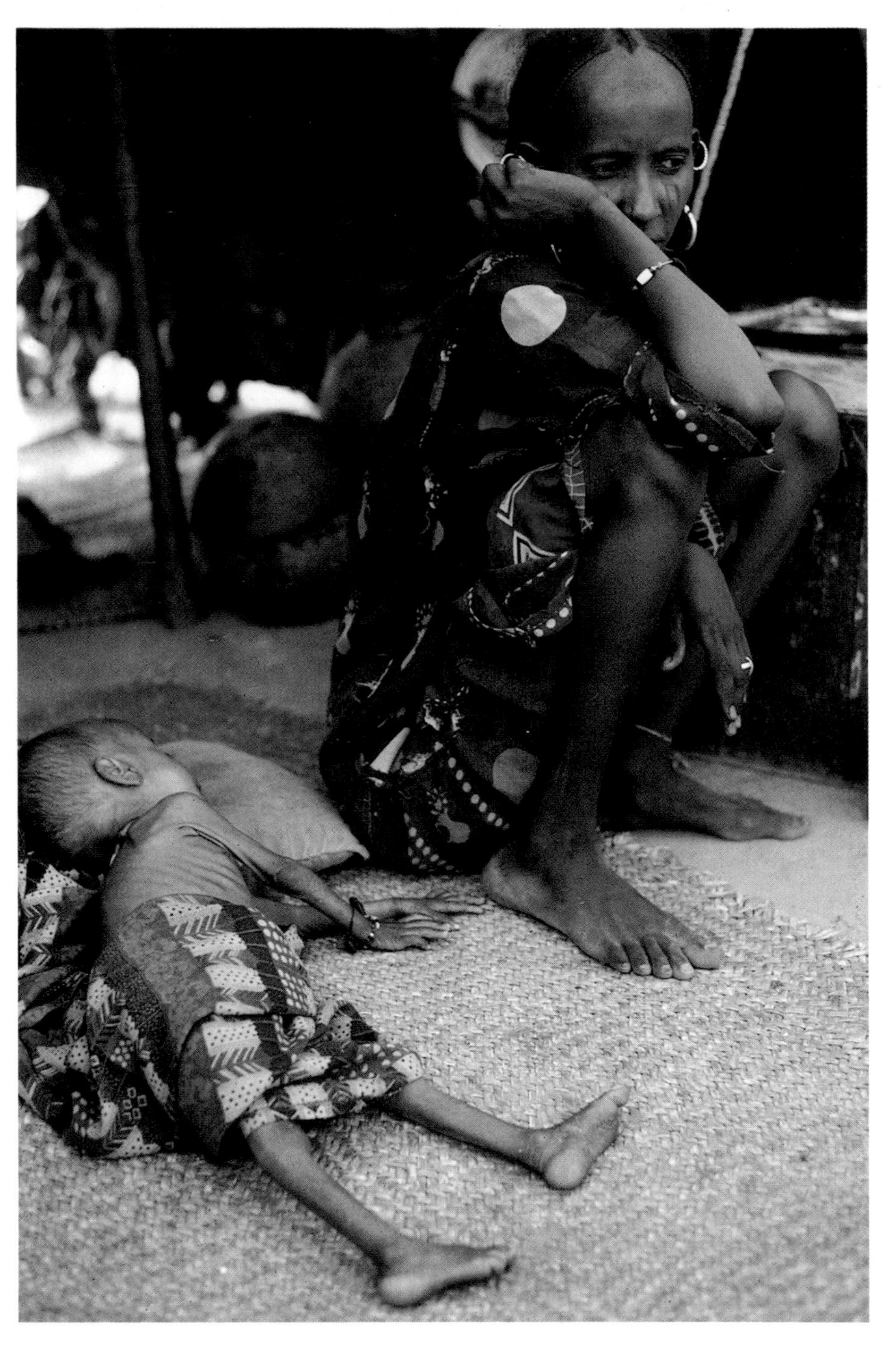

▲ **A mother and her sick child, displaced by the drought in northern Chad.**

In 1974, just a year after Ethiopia was ravaged by one of the worst famines in its history, the UN Food and Agriculture Organization (FAO) hosted a major international conference in Rome. Its agenda was ambitious: to draw up a strategy for solving the world food crisis once and for all. Attending the conference were agricultural experts, representatives of the major development agencies and government leaders, including the then US Secretary of State, Dr. Henry Kissinger. Despite the growing incidence of famine around the world, the mood was optimistic. Provided the new high-yielding varieties developed under the Green Revolution *(see p.109)* were rapidly introduced by developing countries, there was no reason why famine should not become a plague of the past, as remote from the lives of future generations in the Third World as it now is in the West. In his keynote speech at the conference, Dr Kissinger went so far as to vow, "Within a decade, no man, woman or child will go to bed hungry".

Over fifteen years later, Ethiopia is still wracked by famine, now an annual fact of life. Refugees streaming into the Sudan find the situation little better there. Indeed Africa as a whole teeters on the edge of continent-wide famine, two-thirds of countries suffering from chronic food shortages and the bulk of their populations grossly malnourished. Although the international spotlight has focused on Africa, many countries in Southeast Asia, and South and Central America are in a similarly desperate straits. In 1987, more children died from malnutrition in India and Pakistan alone than in all the 46 nations of Africa put together. In the meantime, the world population continues to grow. Against that background, the promise of a world free of hunger begins to look remote indeed.

The world food crisis has never been more desperate than it is today. In 1988, world grain stocks declined dramatically. At their peak, they would have fed the world for 101 days. In 1988, they would have lasted half that period. Rebuilding the stocks will take time, if indeed they can be rebuilt.

Between 1950 and 1985, world grain output increased two-and-a-half times, growing at a steady 3 per cent a year. With population growth running at less than 2 per cent, grain production more than kept pace with increasing human numbers, the amount of grain available per head rising by a third. But since the mid-1980s, there has been a negligible increase in output. Indeed, production actually fell in 1987 and again in 1988. The 1989 harvest was only 1 per cent higher than in 1988, while the world's population grew at 1.7 per cent. In effect, per capita output fell by 7 per cent. In some areas, the decline has been even greater. In Africa, for example, the output of grain per person has fallen by 20 per cent since the late 1960s, while in India, China, Indonesia and Mexico - some of the world's most populous areas - overall output has now levelled off.

NO MORE LAND

Faced with the imminent threat of mass starvation, the FAO is pressing to increase the land under cereals from 650 million ha (1.5 billion acres) to 730 million ha (1.8 billion acres), an expansion equivalent to the total agricultural area of Western Europe. But almost all the arable land suitable for permanent and sustainable agriculture is already farmed or has been lost to urbanization. In the US, some 20 million ha (50 million acres), set aside under government programmes aimed at maintaining prices, could be brought into production, but even so it would only increase the world's cropland by 2 per cent.

Any attempt to bring more land into production would mean ploughing up land unsuited to permanent agriculture - rangelands and forest in particular. At best, it would bring only short-term relief. Although much of the increased grain production since the 1950s has been achieved by opening up marginal land, large areas have now had to be abandoned.

In the USSR, vast tracts of forest in Kazakhstan were cut down and converted to agriculture under the 'Virgin Lands' programme launched during the 1950s. The land proved to be quite unsuitable for agriculture, eroding very quickly. Yields fell off and the land was abandoned. But the lesson was not learned. In the early seventies, more land was brought under the plough, with similar results. Since 1977, 1 million ha (nearly 2.5 million acres) a year have been lost to production. As a result, the area of land in the USSR under cereals has *fallen* by an estimated 13 per cent.

THE LIMITS TO INTENSIFICATION

Since the 1960s and the beginnings of the Green Revolution, the intensification of agriculture has formed the major plank of successive international programmes to feed the world. FAO actually blames the increasing inability of farmers to keep pace with population growth on a failure to apply the Green Revolution both widely enough and vigorously enough. To remedy that, FAO proposes to double the amount of fertilizer and numbers of tractors used by farmers, to double the use of improved varieties of seeds, to increase pesticide use by about 50 per cent, and to extend the area under perennial irrigation by 20 per cent. Africa, in particular, has been targeted for a massive expansion of Green Revolution technologies.

Although the Green Revolution has brought increased yields of rice and wheat, it has only done so at the expense of the long-term fertility of the land. Sooner or later, yields begin to level out and decline, no matter how much fertilizer is applied to buoy them up.

Intensification will bring other problems too. The loss of crops to pests is likely to increase, both because high-yielding varieties are particularly susceptible to pests and because of the need to grow them in monocultures. More tractors mean more soil compaction, and putting more land under perennial irrigation will result in more being lost to salinization and waterlogging. Yields are also likely to be affected by increasing water shortages.

Future food supplies will also be affected by pollution. Increasing levels of ozone at ground level have already led to a decline in yields in many parts of the world. Global warming could have an equally dramatic effect on agriculture, undermining the productivity of many areas. The main cereal growing areas of the US and of the USSR look set to become drier and warmer. Some areas will become so arid that rain-fed agriculture is no longer possible: one-third of the rain-fed farmland in the western US could be lost to production.

Agricultural experts are now increasingly looking towards biotechnology as a means of increasing the world's food supplies. Through genetic engineering, for example, they hope to alter the genetic make-up of crops such as rice and wheat to enable them to 'fix' nitrogen directly from the atmosphere, just as beans and other legume crops can do. But even if this proves possible (and scientists estimate that it will be at least 20 years before commercially viable plants are produced), the principle gain would be in reducing the need for fertilizers. It is unlikely that output would be increased, because plants can only convert so much solar energy into living matter.

"We now know, alas, that nature is finite in her beneficence, and that we have come close to her limits... Most parts of the world capable of sustaining a high level of agricultural productivity are already developed - those areas not already developed (such as the Amazon basin) are probably not suitable for it. So many apparently valuable development projects... have had appalling and unexpected results."
Dr Bernard Campbell, Professor of Anthropology, University of California

"When the next agri-businessman tells you 'only we scientific agriculturalists can feed the millions in modern populations', *do not believe him.* He and his like will, if not prevented, end up by starving the millions. Their land is running out under them. Their production per acre is on the downward curve. They are bankrupt of ideas of good husbandry."
John Seymour

Unless the rate of photosynthesis can be artificially increased, there is a limit to a plant's output. Unfortunately, it is possible that the limits to photosynthesis have already been reached in many agricultural regions.

WHO WILL EAT?

Even if the constraints on increasing food production could be overcome - and the likelihood of this happening is extremely slim - the world food crisis would be far from solved. As history repeatedly teaches us, hunger is as much a result of unjust social and economic systems as it is of crop failures or declining yields.

History books are full of tales of feudal courtiers holding massive banquets while their peasants went hungry. Today, their place has been taken - albeit unwittingly - by consumers in the industrialized North. The peasants are still with us, but they are out of sight in the Third World, and our distance from them cocoons us from the devastating impact our way of life has on theirs.

"In the 1990s and in the 21st century, the greatest food security challenge will be economic and/or ecological access to food, arising from inadequate purchasing power on the one hand and environmental degradation on the other. Global agriculture is at the crossroads. We need to meet the challenge."
M.S. Swaminathan, President of the International Union for Conservation of Nature and Natural Resources

The Third World exports more food to the industrialized countries than it either imports or receives in food aid. In 1973, 36 of the nations most seriously affected by hunger and malnutrition nonetheless exported food to the US - a pattern that still continues. Moreover, the food exported from the Third World is of a higher protein value than that imported. Within the Third World itself, the same pattern is repeated. India proudly proclaims itself self-sufficient in wheat, yet thousands starve every day because they are unable to buy food. The rich in the cities, however, lack for nothing.

Current agricultural development policies do little to address the primary causes of hunger, and they often make matters worse. At the village level, they have thrown many into debt, reducing once self-sufficient farmers to penury. Likewise, on a national scale, they have increased the Third World's crippling debt to the West, so adding to the pressure for developing countries to export their crops to earn the hard currency needed to pay the interest on the debt. This reduces still further the amount of food available to local people.

In the Philippines, half the country's prime agricultural land is used to grow export crops, such as pineapples and sugar cane. Even where agricultural development schemes have been specifically intended to benefit small farmers, the pressures of debt repayment frequently lead to the land being turned over to export crops. As originally conceived, Africa's massive Senegal River Valley Scheme was intended to provide thousands of hectares of irrigated land to small farmers. For all the talk of rural development, however, the local population has seen few benefits: the irrigation water from the first of the dams to be built under the project is going almost exclusively to large mechanized farms producing crops for export.

▶ Sacks of grain in a warehouse in Mali, waiting to be distributed to those worst affected by the drought. Overall, more food flows from developing countries to developed ones than is returned as food aid.

The CONTEST for RESOURCES

The polluted, unhealthy, overcrowded and highly technological world that we are creating is one that is bound to breed tension, and ultimately conflict. Although enough weaponry already exists to blast all life to oblivion, more and more lethal conventional, chemical, biological and nuclear weapons are constantly being built. And while the killing power of our weaponry increases, so the depletion of resources, population growth and climate change increase the risks of it being used.

Among the superpowers, confrontation has given way to mutual acceptance. But there is little indication that the nuclear powers are ready to completely relinquish their weapons of mass destruction, and indeed more and more countries, such as Israel, South Africa, India, Pakistan and Iraq have either joined or are on the brink of joining the 'nuclear club'. Even more countries are in possession of chemical weapons, which because of their relative cheapness have been described as the 'poor man's nuclear bombs'. The spread of nuclear and chemical weapons, and the instability of the Third World, are seen by First World governments as a reason for retaining a nuclear capacity and powerful armies, while the arms industries in the same countries see the Third World as a growing market for their deadly products.

If the East-West rapprochement continues, the missiles of the West are likely to turn 90 degrees, from pointing East to pointing South. The First and Third Worlds are becoming increasingly interdependent, but although the First World needs the resources of the Third, the real power lies in the North. When Third World countries attempt to challenge northern governments or multinationals, then force is often used to make them toe the line.

As we reach the end of the twentieth century, two major causes of tension between the First and Third Worlds will be the supply of resources and the emission of pollutants. By the year 2000, the US and the UK will be minor oil producers, and, the Gulf States will effectively control the world's oil supply. The willingness of the industrialized world to use force to maintain the flow of oil through the Gulf and the rapid build-up of arms in the Middle East, means that the region is likely to become ever more of a threat to world peace.

Control over the Third World's soil and water resources is also likely to be another cause of tension. If global warming and rising populations cause worldwide food shortages, the First World will continue to use its greater wealth to buy the fish and crops of the Third World, even at the cost of mass starvation in the countries where the food is produced. This will produce unrest among the hungry, and political instability which could result in the North using force to ensure that international agreements are obeyed.

Regional resource conflicts, especially within the Third World, are likely to be intensified by environmental degradation. Struggles over land and water resources could spark off serious disputes. For example, Turkey is building a vast complex of dams on the Euphrates which will drastically reduce the flow of water to Syria and Iraq which lie downstream - a development likely to be greeted with hostility by those countries. By increasing soil erosion and silt accumulation in rivers, and changing local weather patterns, deforestation could similarly create tensions between neighbouring countries.

Degradation of agricultural land, pressure on water resources and flooding due to rising sea levels, could lead to tens of millions being forced to flee their homelands over the coming decades. Two of the countries most vulnerable to sea level rise - Bangladesh and Egypt - both have rapidly growing populations, severe land and water shortages, and neighbouring countries that will be unable to support hordes of hungry refugees fleeing over their borders.

If we keep on our current course of environmental exploitation, the question is not will modern society survive the next century, but will it disappear with a bang or a whimper? If the former occurs, it will most likely be due to a 'nuclear winter' following a nuclear exchange in some regional conflict. If industrial society goes out with a whimper, it will be due to a combination of environmental degradation, food shortages and disease.

Although the Third World may initially suffer most from environmental shocks, ultimately it is better equipped for survival. With the supermarket shelves empty, the inhabitants of New York, London or Paris would starve. But those who knew how to live off the land, however poor they might be, would inherit the Earth.

The DYNAMICS of DESTRUCTION

“Most of sub-Saharan Africa, vast expanses of South America and central China are stark in their black fastness. North America, Western Europe and Japan, where a quarter of the world's population uses three-quarters of the world's 10,000 million kilowatts of electricity, shine out as if we are hell-bent on advertising our profligacy.”
Malcolm Smith on pictures of the Earth, seen at night from space

▲ Times Square at night, with most of the scene dominated by advertising.

“We are living in a world locked into a suicide pact, with deeply entrenched ways of thinking about energy and security which are designed, unwittingly, to keep us on course for disaster... The ways out of the suicide pact are obvious, yet are deemed visionary by many people. We can only pray that the visionary becomes seen as the imperative before it is too late.”
Dr Jeremy Leggett, Director of Science, Greenpeace

In the summer of 1972, the United Nations held the first-ever international conference on the environment. Thousands of people flocked to Stockholm, where government ministers, scientists and activists had gathered to discuss growing public concern over the state of the global environment. Media interest at Stockholm was intense, and the conference was followed by a flurry of political activity. A new international agency, the United Nations Environment Programme (UNEP), was set up and national governments introduced a spate of legislation to tackle the problem of environmental degradation. Within the decade, UNEP had launched a variety of action programmes that were designed to clean up the world's regional seas and to combat desertification.

But the rhetoric and legislation has brought few concrete results. UNEP's desertification programme foundered on political indifference, with governments in both the Third World and the industrialized world refusing to provide more than a fraction of the funds required. Most of the projects launched under the programme have involved monitoring the extent of desertification rather than combating its spread. Similarly, plans to clean up the Mediterranean have achieved little. Despite several international conferences and endless declarations, the Mediterranean is no cleaner. Indeed, many warn that it is on the brink of ecological ruin.

Equally depressing is the fate of national legislation on the environment. In Britain, the Labour government brought in the 1974 Control of Pollution Act. Labour ministers promised that the Act would bring fish back to even the dirtiest rivers and estuaries within a decade. Initially, the government promised to implement the Act by 1976. Then, in 1975, it was announced that implementation would be delayed owing to spending cuts. In 1978, a new timetable was drawn up, and the deadline for implementation was deferred until the end of 1979. A change in government in May 1979 brought further delays. For the next six years, the most important sections of the Act simply gathered dust.

The delay in implementation was exploited to the full by Britain's most polluting industries and sewage works. By the time the Act came into force, its teeth had been effectively pulled. Those companies that knew they would have difficulty in meeting the terms of their discharge permits simply negotiated with the water authorities to have the permits relaxed. Factories which would have been in breach of the new legislation were thus let of the hook. The Confederation of British Industry admitted: "The water authorities are basically fudging the books. It's a very sensible thing to do".

Within the major international development agencies, too, promises of action have often come to nothing. In 1970, well before the Stockholm Conference, the World Bank announced that it would take steps to ensure that its projects did not have serious adverse ecological consequences. Yet 14 years later, an internal report admitted: "As a matter of routine, environmental issues are not considered... The bank does not have the capacity to conduct sector work on environmental issues on a routine basis". Until 1987, the bank, with more than 6,000 employees, had an environmental unit of just six people, of whom only three were full-time. This team was charged with reviewing over new 300 projects a year, and monitoring hundreds already in operation.

The World Bank was one of the first development agencies to have its own environmental department - a decision which, at the time, was welcomed by environmentalists as a breakthrough. But as Catherine Watson, who worked in the department, recalls, it was never more than a token office which could never have any real effect on the bank's policies. "We were treated like scourges. When our proposals threatened the future of a project, or had major implications for Bank practice, they were dismissed as unrealistic and impractical. Reform was possible, but only insofar as it left the Bank's basic structure unchanged."

Watson left the bank, thoroughly disillusioned. Following renewed public criticism in the mid-1980s, the environment department was expanded to 70 people. Most of the new staff were recruited from the Public Relations department. Although projects now require an accompanying environmental impact assessment, the requirement is often little more than a formality. In many cases, the impact assessments are not completed until after the project has been approved by which time it is usually too late to halt or change it. In terms of destructiveness, the projects now being funded by the World Bank differ little from those in the past. The emphasis is still on large dams, massive plantations, industrial complexes, commercial forestry operations and the like. All these projects produce profits, but bring little benefit to the poor and often cause environmental damage.

"In a week's time all cleaning up work will halt on the massive *Exxon Valdez* oil spill in Alaska... While green issues promise to be increasingly important, President Bush, who declared his passion for the environment during his election campaign, has failed to visit Valdez or declare a national emergency. Now, a planned presidential visit after Exxon's departure has been cancelled as a political 'downer'."
Christopher Reed

"We are not going to do away with the great car economy."
Margaret Thatcher, 1990

"Eighty per cent of our pollution is caused by flowers and trees."
Ronald Reagan

"I have never pondered such matters... nor do I ever intend to."
Enoch Powell, senior British politician, on environmental issues

"Let every man make known what kind of government would command his respect, and that will be one step toward obtaining it."
Henry David Thoreau

▲ In the weekly market at Antananarivo, Madagascar, thousands of parts from broken-up cars are offered for sale. In a more affluent society, most of these parts would go to waste.

Looking back on the Stockholm Conference, the issues then on the agenda appear depressingly familiar: pollution, the depletion of the ozone layer, acid rain, the greenhouse effect, the depletion of resources, the destruction of habitats and the loss of biodiversity. Two decades later, att another global conference on the environment, the issues were the same, and the rhetoric was identical. But whereas in the 1970s, ozone depletion was little more than a hypothesis, today the ozone hole is with us. The same goes for the greenhouse effect. And the assault on ecosystems continues unabated.

The question is, why? Why, given all the knowledge at our disposal, have we been unable to translate the rhetoric of environmental action into reality? What are the forces that are driving us towards nemesis?

POVERTY AND POPULATION

"Poverty is the worst form of pollution", Mrs Indira Gandhi, then Prime Minister of India, told the Stockholm Conference. That view lies at the heart of conventional thinking on the environmental crisis, shaping policies and determining official solutions. In 1987, the UN World Commission on Environment and Development (WCED), chaired by the Norwegian Prime Minister, Gro Harlem Brundtland, published the results of a four-year study on environment and development under the title *Our Common Future.* Like Mrs Gandhi, the Brundtland Commissioners blamed poverty as a major cause and effect of global environmental problems.

Warning that the very survival of the planet was at stake if current environmental trends continued, the Brundtland Report argued that the roots of the crisis lay "in the lack of development and... the unintended consequences of some forms of economic growth". Our future well-being, the report argued, could only be ensured by adopting sustainable development to meet the needs of the present without compromising the ability of future generations to meet their own needs. To achieve that goal, "a new era of economic growth would be required, one based on policies that sustain and expand the environmental resource base." The Commission claimed that a "five- to ten-fold increase in world industrial output can be anticipated by the time world population stabilizes some time in the next century." This increase in output, it was suggested, would also help to solve the other main factor in environmental degradation - population growth: "Economic development, through its indirect impact on social and cultural factors, lowers fertility rates."

The argument sounds humanitarian and plausible. In its concern for the world's poor, it puts the emphasis firmly on their right to 'development'. Anyone visiting a Third World country soon becomes aware of the vicious circle of poverty and environmental degradation to which the Brundtland Commission refers. Arriving in Bombay by air, you see the filth and squalor of the cardboard and corrugated-iron shanty towns, which stretch right up to the perimeter fence of the airport. Without running water or sanitation, the people who live in these slums have little choice but to use any nearby rivers and open spaces as toilets.

Again, a trip by boat along the River Niger south of the Sahara shows how rural Africans are forced to use up the sparse local vegetation for fuelwood and browsing for their goats, accelerating the desertification of their land. A bus through the countryside in Honduras reveals poor farmers burning the forested mountainsides to grow their beans and corn, and the red gashes which soon appear on the sides of the slopes as the precious topsoil is washed away in the tropical storms.

When people are miserably poor, it is obvious that they will put the search for fuelwood and their families' next meal before the long-term sustainability of the forest or savanna they live in. But claiming that poverty is the 'biggest polluter', and that only development can solve poverty, implies both that the people of the Third World have always lived in miserable poverty - as they have never before had development - and also that they have continually been destroying their environment over the millennia. This is both insulting and untrue.

WORLDS APART

The destruction of the rainforests only began on a wide scale during the colonial period, when the European powers cut down the forests of their tropical dominions to provide timber for their fleets and railway lines, and to grow plantation crops such as opium, tea and sugar. The population of the Amazon before the Spanish and Portuguese arrived is estimated to have been around 10 million - far larger than today, yet the forests were very largely intact. Similarly, the early European explorers in West Africa did not record desertification and drought, but instead, rich cultures with cities surrounded by fertile agricultural land. The literary, artistic and architectural achievements of the Buddhist and Hindu cultures of Asia were not the product of impoverished societies. Likewise, the terraces sculpted from the rocky slopes of the Philippines, which have provided the indigenous people with centuries of rice harvests, were not produced by poverty-stricken peasants who cared only for their next meal.

Conversely, higher material standards of living have not prevented environmental destruction in the First World, even though the bulk of the population are wealthy beyond the wildest dreams of an Indian peasant. The US and Canada are logging their old-growth forests *(see p.88)* almost as fast as Brazil is destroying its rainforest. In Europe, the North Sea is severely polluted, yet it is surrounded by some of the richest nations on Earth. The acid pollution which is killing vast areas of Europe's forests is not occurring because the power-station workers or car-owners of England or Germany are poor.

Blaming poverty in the Third World for our environmental problems is a gross oversimplification. The truth is that the material wealth which the majority of people in the industrial world enjoy is possible only because of the impoverishment of the majority in the Third World and our plundering of the global environment. In seeing poverty as the cause of environmental destruction, the Brundtland Commission, together with mainstream economists and policymakers, have been looking at the symptom, not the cause of the problem.

In India, the bulk of the shanty-town dwellers are 'development refugees', made destitute by economic growth in the countryside. Since Indian independence, six million peasants have been thrown off their land by the construction of large dams alone. Many more small farmers have been ousted by the advent of the Green Revolution *(see p.109)*. In Africa, the pressure on vegetation is in large part due to the introduction of cash crops and modern farming methods. In Senegal, African farmers were alarmed at the promotion of the 'western' type of plough, as they knew that it was unsuited to the fragile local soil and harsh and unreliable climate. But the European 'experts' were assumed to know best, and 20 years later, the Africans were paying the price - their best farming land was being lost through wind and water erosion. Central American farmers do not cultivate precarious hillsides because they are unaware of the destruction they are wreaking on their environment, but because there is nowhere else to farm. In most of Central America, a tiny elite owns the best land. Instead of growing staples for their poor compatriots, they grow crops such as pineapples and bananas for export to the US.

Likewise, singling out population growth as the cause of environmental destruction glosses over several important issues. To its credit, the Brundtland Report recognizes the vital connection between population and levels of consumption. "An additional person in an industrial country consumes far more and places far greater pressure on natural resources than an additional person in the Third World." The US, for example, with just 4 per cent of the world's population, is responsible for some 24 per cent of global carbon dioxide emissions.

Population growth has not caused these problems. In fact, the cure which the Brundtland Report recommends for population growth - namely economic growth - has caused them. The report claims that economic growth lowers fertility rates, but as we have seen, there is much evidence to the contrary *(see pp.258-259)*. If one accepts that the Third World has to reach the level of material affluence of the First World before its population growth will slow down to

"If a high-growth economy is needed to fight the battle against pollution, which itself appears to be the result of high growth, what hope is there of ever breaking out of this extraordinary circle?"
E.F. Schumacher

▼ Old furniture and other unwanted items dumped beneath Manhattan Bridge in New York. Poorly made goods with short lifespans contribute to the wastefulness of our society.

“The underdeveloped countries cry out to the West to help them achieve further ‘development’... Today ‘development’ is in most instances a euphemism for environmental exploitation. The building of the great Aswan Dam across the Nile in Upper Egypt is probably the most recent and striking example of a supposedly wonderful development project which is proving to have devastating side effects (only some of which were foreseen by ecologists).”
Professor Bernard Campbell

“The Third World is littered with the results of crass environmental blunders often perpetrated by engineers, economists and others who ought to have known better... large sums were expended on unwise investment and many innocent men, women and children starved.”
Max Nicholson

First World rates, it follows that population growth will never stabilize. Even the World Bank, the body whose chief purpose is to help the Third World develop, admits that most of Africa, Asia and Latin America will never reach North American, European or Japanese levels of economic output.

THE IDEAL OF DEVELOPMENT

The word ‘development’, like ‘freedom’ or ‘democracy’, expresses a lofty ideal, yet it is has been used so often to justify so many different goals and strategies, that it has become largely meaningless.

In colonial times, the western powers did not seek to ‘develop’ their colonies which were essentially sources of natural resources. During the Second World War, however, the seeds were sown for a new world order. At an international conference of the allied powers held in 1944, the World Bank and the International Monetary Fund were established. The aim of these institutions was to ‘develop’ the world economy so as to prevent a recurrence of the Great Depression, and to ensure the post-war expansion of American capital in an increasingly integrated global economy based on free trade and free enterprise.

To the experts from the World Bank and the newly established development agencies, the rural peoples of the Third World were considered uneducated because they had no schools. Because their houses were made out of mud and thatch, they were considered to live in hovels. Because they produced food without money, chemicals or modern machinery, their sustainable farming systems were considered unproductive. Because their decision-makers did not wear suits and ties and sit in offices, they were considered primitive. The leaders of most independence movements endorsed the analysis and in doing so effectively accepted that their indigenous cultures were inferior to those of the industrialized world.

But since the 1960s it has been obvious that few of the runners in the great race to develop will ever finish the course. Most have limped off to the side, their economies burdened with unpayable debts and their environment seriously degraded. The development industry has recognized failures. First it redirected aid towards rural farmers who were becoming impoverished, because it was thought too much aid was going on urban industries. But rural development was seen to be impoverishing women, so development strategies were aimed at women. Next it was realized that the environment was being devastated by development, so the aid agencies then lent money to improve the environment.

But only the way in which development is implemented, not the process of development itself, has been questioned.

LIFE ON THE MARGINS

Despite the high ideals of many of those most involved in economic development programmes, the current process of economic development is one that inevitably removes control over resources and decision-making from local communities and transfers it into the hands of government and corporations.

In the rural areas of the Third World, the loss of land rights is particularly devastating. Many land tenure systems do not recognize private ownership of land but instead rest on rights of use during an individual’s lifetime. By imposing private ownership, state authorities have dispossessed millions of peasants. This process has been further encouraged by land speculation, indebtedness and the confiscation of land where farmers cannot prove outright ownership.

Modern economics defines people as poor if they do not take part in the cash economy. They are poor if they provide for themselves, instead of buying commercially produced and distributed processed foods. They are poor if they wear handmade clothes instead of factory-made garments. They are poor if they concentrate on guaranteeing the survival of the whole community rather than maximizing production.

In trying to make these people ‘better off’, development destroys the rural economy which has provided villagers with a secure livelihood for hundreds and in some cases thousands of years. And, like environmental destruction, this is not a by-product of development but is an integral feature of the process. The farmer who barters his grain for a cloth from a local tailor adds nothing to ‘development’ as measured by the gross national product. His transaction cannot be taxed and will not help earn foreign exchange to pay the foreign debt and buy technology from the West.

THE CASH ECONOMY

Throughout the Third World, the cash economy has had a profound impact on village life. Cooperation in clearing fields or helping in the harvest, for example, often becomes less and less necessary: labourers can be employed instead.

Without cooperation, the communal activities that help ensure it become redundant. The community loses its cohesion and becomes unable to resist yet further encroachments by market forces and bureaucracy. Those who benefit are businessmen, government officials and large landowners, living in what the Indian ecologist Vandana Shiva terms "the visible enclaves of economic development". Meanwhile, "the invisible hinterlands of economic underdevelopment, the homes of the silent majority, are left with shrinking access to a shrinking resource base."

In the industrialized world, the majority of people have long since lost their self-sufficiency, but the same process is still at work. Small companies are bought up by multinationals. If their profits start to fall, they prune back their operations, putting thousands out of work and destroying communities over which they have total economic power but no long-term interest. Small shopkeepers who fall into debt are bought out by chain stores. Local land is bought up for development by anonymous offshore companies. In a democracy, the only influence that individuals have over the political decisions which decide their future, and the future of the community in which they live, is a vote, perhaps every four or five years. It is a long way from the day-to-day participation in decision-making which is a feature of the traditional village community.

Within the First World too, economic growth also provides benefits for fewer and fewer people. In the US, the most economically developed country in the world, it is conservatively estimated that 100,000 children are homeless and 500,000 malnourished. Sociologists in both the US and Europe are increasingly worried about the emergence of an 'underclass' of illiterate, unemployable young people who are totally alienated from mainstream society.

OUR COMMON FUTURE?

In effect, development involves a shift in power from community-based institutions to special-purpose organizations, such as government departments, corporations and international agencies. Their decisions affect every aspect of daily life. But do their interests broadly coincide with the public's, or with the requirements of maintaining a healthy environment? The chairman of a multinational corporation may indeed share the same planet as a Sri Lankan fisherman, but no amount of talk of sharing a common future can disguise their differences or the fact that they have conflicting interests.

▲ Australian Aborigines demonstrating in support of their claim to land rights. Like many minorities, they have been dispossessed as large companies seek to exploit natural resources.

Because they are set up to manufacture or sell specific goods, to provide specific expertise, or to perform specific services, special-purpose organizations are primarily concerned with promoting their own interests, perpetuating themselves and increasing their power and influence. Their employees and even their chief executives are often naively oblivious of the wider implications and damage created by their daily work. This explains the hurt response of a World Bank employee when confronted with protesters demonstrating against the impact of the Bank's policies on tropical rainforests: "Why are you attacking the Bank? We are not timber merchants. We are bankers. We have nothing to do with the environment. We simply make loans".

Large institutions generate a tunnel vision that has a dynamic of its own. Decisions are not made because they are desirable on social or ecological grounds, but because they serve particular vested interests. Having developed a new product, for example, a company's overriding priority is to market it. In that context, the public has importance principally because of its purchasing power: the environment only because it supplies raw materials and offers a convenient repository for wastes. Those who seek to question the need for the product or raise doubts about its safety are seen as trouble-makers or scaremongers. Indeed, time and again, we find that special-purpose organizations have manipulated research, distorted cost-benefit analyses and suppressed information in order to sell harmful products or to continue activities which are detrimental to the environment.

So great is the power of many special-purpose organizations, from the military to multinational corporations, that they have effectively been able to capture the agencies that supposedly regulate them. The extent to which the livestock industry in the US controls the Food and Drug Administration has often been severely criticized. A Congressional Committee recently noted: "The FDA has consistently disregarded its responsibility... repeatedly putting the interests of veterinarians and the livestock industry ahead of its legal obligation to protect consumers... jeopardizing the health and safety of consumers of meat, milk and poultry".

International agencies are also controlled by powerful special-purpose organizations. In the 1960s, the agrochemical industry formed a lobbying organization called GIFAP (Groupement International des Associations Nationales de Pesticides). GIFAP soon persuaded the UN Food and Agriculture Organization to form a joint bureau called the Industry Cooperative Programme (ICP), in which GIFAP representatives could work hand-in-hand with technicians from FAO. By the early 1970s, joint FAO-ICP seminars had been organized in various parts of the Third World to promote new and better ways of distributing agricultural pesticides. At the same time, GIFAP was asked by FAO and World Health Organization to play an active role in the two agencies' consultations and other international meetings.

Industry lobbyists openly dominated several of the sub-committees which were responsible for formulating UN policy on agriculture. Eventually, the hundred or so agrochemical corporations that made up the ICP came to enjoy a semi-official status in FAO. The chemical giant Hoechst, for example, was brought in as adviser on a Tanzanian agricultural development project. The UN representative on one Bangladesh anti-malaria project was also a consultant to a large European company supplying the insecticide, malathion. In situations like this, a clash of interests is inevitable, and the large sums of money at stake are bound to influence decision-making.

▲ The Union Carbide plant at Bhopal, India, a year after the catastrophic accident which claimed so many lives, and left countless people blinded or disabled. The survivors and victims' families later had to seek compensation from an employer based on the opposite side of the globe.

GREEN GROWTH?

Given the power enjoyed by organizations such as multinational companies, the belief that the development process can be 'greened' in any meaningful sense of the word ignores political and economic realities. Foresters who call for sustainable logging in the rainforests, for example, are undoubtedly able to work out sustainable systems in their test plantations. But they disregard the fact that the logging industry is not only corrupt, but would have little incentive to adopt such systems, even in theory.

In Sarawak, Malaysia, logging concessions are often handed out by the government to their own friends and families. The politicians know that they may only be in office for several years and so it makes sense for them to log the forests while they control them. It makes economic sense for a company to devastate a forest in Sarawak and then invest its profits in real estate in Hong Kong. When the local people who need the forests for food, fuel and shelter controlled them, the forests survived. But when development puts the control of the forests into the hands of corporations and governments whose survival is not tied up with that of the forest, the trees soon disappear.

The same criticism can be made of proposals to use market forces to phase out environmentally damaging products. Incorporating the environmental costs of products into their price can only be a good thing, but the nature of vested interests makes it inconceivable that the true costs will ever be reflected.

Take the environmental cost of the car, for example. It is built with finite resources. Iron, aluminium, copper, lead, platinum and zinc must all be mined and processed, using up large amounts of energy and causing environmental degradation near the mines. Plastics, rubber and glass are also needed, using up more fossil fuels. The materials must then be transported to the factory. The factory then needs more energy, and also land and building materials. Roads must be built with aggregates, gravel and tarmac involving more mining, transport and energy. The roads destroy agricultural land, wildlife habitats and houses. Producing petrol to run the cars involves exploring for oil and then drilling, refining and transporting it, causing environmental damage from oil spills, the building of roads and pipelines into wilderness areas, and the emission of fumes from the refineries. A network of petrol stations must then be built.

When a car is running, the burning petrol produces a cocktail of pollutants. Catalytic convertors fitted to car exhausts can reduce noxious emissions from cars but themselves require platinum, causing more mining. When tyres are worn out they are dumped in huge tips which can catch fire, often emitting toxic fumes and polluting groundwaters. Waste engine oil is dumped down drains, eventually reaching rivers and the sea.

If all these costs had to be accounted for - especially the cost to our climate of emitting carbon dioxide - then cars would probably never be built. But that is an option which neither car manufacturers, nor the road industry, nor government agencies, nor indeed the general public, whose lives increasingly depend on private transport, are prepared to contemplate.

The nature of political and economic power in modern industrial society dictates that measures to combat environmental destruction are only acceptable if they do not interfere with the workings of the economy. In that context, 'sustainable development' or 'green growth' in fact amount to little more than having one's cake and eating someone else's.

THE REAL AGENDA

Despite the rhetoric from politicians and the leaders of the business community, the real agenda that they intend to follow in the next decade can be seen from the negotiations currently taking place under GATT, the General Agreement on Tariffs and Trade. This lays down the rules for approximately 90 per cent of world trade among nearly 100 countries. GATT, originally drafted in 1947, is periodically amended by complex negotiations that may take several years to complete.

The main goal of the US, EC and Japanese negotiators who dominate the current discussions is to remove all duties and quotas that restrict 'free trade'. Agricultural produce and textiles, which were previously outside the scope of GATT, will now be brought within it.

Under GATT, Third World governments will no longer be able to restrict in any way the import of cheap agricultural produce in an attempt to protect their own farming communities. It is already 'GATT-illegal' for a government to prevent the export of food, except in times of famine, and the US has proposed that even this condition should be scrapped. It will also be illegal for any government to take vulnerable and eroding land out of agricultural production or to take any measures to preserve scarce resources if these are judged to be in restraint of trade. Already, on the basis of the Free Trade Agreement between Canada and the US, Canada has been forced to abandon environmentally sound measures to protect the threatened Pacific salmon, and is being prevented from restricting the sale of its water resources to the US, even in times of local water scarcity.

It will no longer be GATT-legal to prevent foreign investments in environmentally destructive operations nor to prevent the import of goods that are damaging to health or the environment. It would even be illegal to restrict the present practice of exporting hazardous waste to the Third World.

Environmental and food safety standards will be 'harmonized' - in effect, reduced to their lowest common denominator. If these GATT proposals are forced through, then the world will be transformed into a vast 'free trade zone'. Within it, individuals, communities and even whole nations will no longer be able to decide how to use their resources and protect their environment. Power will be given over to multinational corporations and the western governments which represent their interests.

In an economic system which, as one commentator puts it, "values American cats over West African people because the former can pay for the food while the latter cannot", giant organizations now dominate our lives. They are the force propelling us towards ecological catastrophe, and the challenge is to bring them back under social control.

"After 300 years of virile growth, and a long boom in which doubt was inconceivable, the growth and greed society has suddenly burst into an area of problems, chaos and uncertainty, giving us an unprecedented opportunity to move to more sensible arrangements. Whether or not we manage to take that opportunity will depend on whether enough of us devote ourselves to the crucial educational task. Unless more of us do adopt this as our chief long-term priority, there is little chance of us making it to a just, peaceful and ecologically sustainable world order."
Ted Trainer

▼ A new superstore with trolleys parked outside. The food in stores like this comes from every continent in the world, except Antarctica, some of it being imported from countries that cannot feed themselves.

SOLUTIONS for SURVIVAL

“Everywhere people ask: ‘What can I actually *do*?’ The answer is as simple as it is disconcerting: we can, each of us, work to put our own inner house in order. The guidance we need for this work... can still be found in the traditional wisdom of mankind.”
E. F. Schumacher

▲ How we live in the last decade of the twentieth century has consequences that stretch far into the future. It is today's children - and their children after them - who will suffer if we cannot take immediate steps halt the destruction of environment.

“Our movement grew out of the needs of the rubber-tappers. We made a lot of mistakes but we learned from them. You know, people have to look after themselves, they have to fight and be creative. That's how we built this movement.”
Chico Mendes

It is midday and in the village of Nagami, high in the foothills of the Himalayas, a group of villagers sit cross-legged in the heat of the Sun listening as Sudesha, a woman in her late thirties, recounts how she was jailed for attempting to stop timber merchants from felling the trees near her village. Her story is told in song and she sings of her willingness to go to jail again to prevent further deforestation. Sudesha is prominent in Chipko, a village-based movement that takes its name from the Hindi word for "to hug", villagers literally hugging trees to prevent them being felled. She is speaking at the end of a two-day meeting, organized by the villagers to discuss how best to prevent the destruction of their forests and to encourage the regeneration of areas which have been logged. Already many villages have set up tree nurseries, growing fruit and other trees of the villagers' own choice, and hundreds of hectares of denuded hillside have been replanted. In one village, a single villager called Saklani has planted more than 20,000 trees, all native species, creating a little oasis of ecological sanity in an area which has been devastated by the timber industry. Streams have begun to flow again, birds and other animals have returned in large numbers and the hillside no longer suffers from erosion.

On the southern edge of Brazilian Amazonia, where the savanna of Mato Grosso meets the rainforest, the Xavante Indians are slowly bringing their ancestral lands back to life. Years of illegal ranching and logging by large landowners, whom the authorities refused to expel from the Xavante reservation, had devastated the area. In 1982, the Xavante took matters into their own hands. Having ousted the loggers and ranchers, they burned the sawmills and let nature take course, planting trees and leaving the degraded land to regenerate. The savanna ecosystem is naturally rich in a wide variety of nutritious plants and fruits, which the Xavante have begun to grow and market. They have also set up a research centre in the small village of Pimental Barbosa to process forest products and to develop new techniques for encouraging forest regeneration. Efforts are also being made to restore the local wildlife, which has been seriously depleted through illegal hunting by settlers. Once the population of wild animals has been restored, the Xavante hope they will be able to give up keeping cattle and obtain their protein through hunting, as they always have in the past. "We must bring the forest and its animals back to life," says Syboopah, a renowned local shaman, "we must help the different forest people to regenerate their forest."

In East Africa, villages along the Tanzanian coast have allied together to protect their fishing grounds from commercial boats, most of which are owned by entrepreneurs in the city looking for a quick and easy source of income. The commercial fishermen use dynamite to stun or kill fish, increasing their catches but destroying the coral reefs where the fish live and breed.

In the US, a powerful movement of citizens' groups has emerged to combat the threats from toxic waste dumps and to fight for safer methods of waste disposal. In 1980, there were just a few hundred groups: today there are almost 5,000, a network that stretches from coast to coast and covers every state in the Union. The movement grew out of the Love Canal disaster when 900 families were evacuated from a housing estate built over a leaking toxic waste site near Niagara Falls in upstate New York. The evacuation came about only after a prolonged campaign by local residents; few of them had previously been involved in politics and many had family members working for the companies which had dumped the polluting waste. Stonewalled by officialdom, the local residents resorted to demonstrations and even civil disobedience to apply pressure on their elected representatives to take action. Only then were they evacuated.

One of those at the forefront of the campaign was a local housewife, Lois Marie Gibbs. Galvanized into action and deluged by telephone calls from other groups living near toxic waste dumps, Gibbs went on to form the Citizens Clearing House for Hazardous Wastes (CCHW). The CCHW set as its goal an end to the landfilling of chemical wastes in the US by the end of the 1980s - a goal dismissed at the time as a pipedream. Yet, against all the odds, it has succeeded: a ban on the land disposal of liquid hazardous waste came into force in the mid-1980s and 1990 saw an end to the landfilling of untreated hazardous solid waste. The alternatives put forward by the CCHW - the four 'Rs' of Recycling, Reduction, Re-use and Reclamation - were also dismissed as impracticable a decade ago. Now they are increasingly accepted as the way of the future.

For Lois Gibbs, the rapid growth of the movement and its success is evidence of a "yearning that exists in society, both for change and a new way of winning that change." The victories have not come about "through slick lobbying techniques, clever research or 'magic facts' but

"They say the trees are unproductive until they are cut; that the land is needed for development. What is development? For most of us development means affluence. You have to produce more and more things for your luxury and comfort. Ultimately this tells on your resources because everything has to come from nature. You have to decide whether development means affluence or whether development means peace, prosperity, and happiness. I think peace and happiness have gone away..."

Sunderlal Bahuguna, Chipko movement

"...if the remedies the fishermen themselves have promoted are put into operation then it might save the fishery. What is so commendable is that the people who actually earn a crust of bread there said they'd stop. There are three water bailiffs, including me, working out on the fishery and checking that the [self-imposed] regulations are carried out, and the oysters are the right size... Please God we have caught the problem of overfishing here in time."
Captain John White, Truro harbourmaster

through trusting in people's common sense and willingness to act once they are aware of the issues." Along the way, she believes, important lessons have been learned: that law does not equal justice; that real change only comes about when local people become involved in the issues and begin to build up a grassroots political base; and, perhaps most important of all, that there are only two sources of real power - people or money. Environmental groups, and concerned citizens, can never match the financial power of those vested interests - from chemical corporations to government agencies - against which they are invariably pitted. But they have one resource whose strength, once tapped, should never be underestimated. People.

A TIME OF OPPORTUNITY

If there is one lesson that Chipko, the Xavante, the fishermen of Tanzania and the Citizens' Clearing House on Hazardous Wastes have in common, it is this: that individuals, joined together with others, can bring about change. "Grassroots activism wins," says Gibbs. "And by winning, the entire environmental cause is strengthened. And that's why I feel my children and their children will have a chance for a better environment in the 21st century."

In Chinese, the word for 'crisis' is not only the word for 'danger' but also for 'opportunity'. And so it is with the ecological crisis that now confronts us. The dangers are clear enough: if left unchecked, global warming alone could render the planet uninhabitable for higher forms of life. Add to global warming the threats of ecosystem destruction, the loss of species, pollution, land degradation and the growing marginalization of the poor and dispossessed, and one is presented with a picture of a society deeply at odds, not only with its environment but with itself.

Such a society cannot survive indefinitely. Twenty years ago, in the opening sentence to its "Blueprint for Survival", *The Ecologist* magazine warned: "The principal defect of the industrial way of life with its ethos of expansion is that it is not sustainable. Its termination within the lifetime of someone born today is inevitable - unless it continues to be sustained for a while longer by an entrenched minority at the cost of imposing great suffering on the rest of mankind. We can be certain, however, that sooner or later it will end (only the precise time and circumstances are in doubt) and that it will do so in one of two ways: either against our will, in a succession of famines, epidemics, social crises and wars; or because we want it to - because we wish to create a society which will not impose hardship upon our children - in a succession of thoughtful, humane and measured changes."

Therein lies the opportunity presented by the current crisis. For there is nothing ordained about the way of life we live today. There are alternatives open to us - provided we have the will to initiate them. Our use of energy is profligate, yet there are numerous technologies that we could employ to use that energy more efficiently, reducing emissions of carbon dioxide. Better systems of public transport could wean us away from our reliance on private cars, further reducing greenhouse gases but also encouraging a society that is less aimlessly mobile. Our cities do not have to be the anonymous urban conglomerates they are today, nor their suburbs simply dormitories for commuters: by encouraging more people to work near where they live and by giving more power to local neighbourhoods, our cities could become mosaics of vibrant communities. We do not have to produce throwaway goods that last a minimal amount of time, squandering resources and creating problems of waste: we could, if we were so minded, produce goods of quality, built to last, encouraging craftsmanship, creativity and a sense of pride in our work.

Technologies also exist that would enable us to reduce dramatically the wastes we produce. In agriculture, there are tried-and-tested methods of farming that do not rely on chemicals and which are as productive, in the long term, as modern intensive methods, producing healthier food into the bargain. We do not have to accept the unequal systems of land tenure that lie behind so much of the landlessness in the Third World and which are indirectly the cause of much ecological destruction. Similarly, our political system is not immutable. We do not have to live under a system of government where decisions are made by distant bureaucrats and which is dominated by single-purpose organizations. We could decentralize power, giving local communities ultimate control over decisions that affect their future and over their own resources. And, in doing so, we would make it possible to live by that guiding precept of a sustainable society: "Think globally, act locally".

AN AGENDA FOR ACTION

If we approach the problems facing us piecemeal, without any overall strategy for change, and above all without tackling the underlying forces that have brought us to the brink of disaster, we shall surely muddle our way to nemesis. Action is urgently needed on five broad fronts: to reduce our impact on the environment; to regenerate and restore damaged ecosystems; to phase out activities which are inherently destructive of the environment and to phase in sustainable methods of production; to move away from a throwaway economy that relies on maximizing the through-put of goods towards a conserver society that minimizes wastes and reduces its consumption of resources; and to increase the security of those who have been marginalized by our activities.

Such changes cannot be achieved overnight. But the task will be greatly eased by the interlocking nature of the problems confronting us, for in solving one problem we will frequently find that we are solving numerous others. Recycling aluminium, for example, not only saves huge amounts of energy - the production of aluminium from bauxite requiring 20 times more energy than recycling used aluminium - but would also cut down on the environmental damage caused by strip-mining bauxite. Since many bauxite deposits are in tropical forests and many aluminium smelters are powered by large hydroelectric dams, the wider benefits of reducing aluminium use are obvious.

A similar 'solution multiplier' effect can be seen in the measures needed to combat global warming. To curb carbon dioxide emission, we need to reduce dramatically our use of energy, primarily through improving energy efficiency. In doing so, however, we will also reduce emissions of sulphur dioxide, the major cause of acid rain, and at far less cost than installing pollution control scrubbers. By combating acid rain, we will remove a major threat to forests and an obstacle to reforestation. Through reforestation, we will be able to reduce still further carbon dioxide levels. Each solution reinforces the other, creating a dynamic of change, in which each step makes the next easier.

TECHNOLOGIES FOR CHANGE

Much could be done using existing technologies to reduce our impact on the environment. Such 'engineering solutions' represent an immediate tool for change.

The scope for improving energy efficiency, for example, is enormous. About one-third of the energy consumed in modern industrial societies is used in the home, yet much of that energy is squandered. In Britain, about three-quarters of the energy used by domestic householders is devoted to water and space heating, and about a third of this energy is wasted due to poor insulation. If British homes were insulated to Danish standards, carbon dioxide emissions would be reduced by over a half a million tonnes a year. Simply lagging a hot water tank with a thick jacket can reduce heat loss by three-quarters, covering the initial cost in just a few weeks. About half the houses in the United Kingdom have cavity walls. Insulating these walls can halve heat loss. With better design, energy consumption in the home could be reduced still more dramatically. Houses in Sweden have already been designed with heating systems that require a third of the energy used in the average house in the US and, with super-insulation, they can be designed to use 90 per cent less energy. The use of heat pumps alone can reduce the energy required for water heating by as much as 50 per cent: if they were installed in all homes throughout the US, this would eliminate the need for 15 large thermal power plants - reducing carbon dioxide emissions by as much as 112 million tonnes.

More efficient lighting could eliminate the need for yet another clutch of power plants - perhaps as many as 120 in the US. Fluorescent-tube light bulbs are already on the market that use up to 80 per cent less electricity than traditional light bulbs - and last five times longer. Simply replacing one traditional filament light bulb by an energy-efficient one could reduce carbon dioxide emissions by one tonne per year. Energy use in freezers and refrigerators could also be cut dramatically. New efficient models, designed in Denmark, consume between a fifth and a seventh of the energy consumed by conventional fridges and, according to US and Danish engineers, their efficiency could be improved still further, reducing current energy consumption by 95 per cent. In a country the size of the UK, the use of such fridges could eliminate the need for two large power plants.

Even without buying new equipment, householders could improve energy efficiency around the home. Regular defrosting and stuffing empty spaces within freezers with newspapers cuts electricity use - and hence reduces carbon dioxide emissions. Similarly, by turning off lights in rooms that are not in use and in reducing heating levels by a small amount, energy will be saved. Turning down the central heating system by just 1°C (about 2°F) reduces the amount of energy used by almost 10 per cent.

“For the past 10 months, the location of the proposed dam at Serre de la Fare has been a campsite, a campaign headquarters, a spiritual home for a tiny local group called Loire Vivante. A few dozen people from the local communities have decided that they wish to preserve their river valley without an enormous reservoir. They are passionately devoted to stopping the dam... the strength and sincerity of Loire Vivante has come from a special relationship, a love that can grow between people and the little corner of the world in which they live.”
Brian Leith

"What I hope will happen is that in between the coming great storms and droughts and atmospheric phenomena never before seen, there will be time to think and the will to react. As Dr. Johnson said 'nothing so concentrates the mind as the prospect of hanging'."
Professor James Lovelock

"...the alternative to progress is not stagnation but *renewal*."
Richard Mabey

Using energy more efficiently not only saves money for the individual householder, it also saves money for the companies supplying energy. Building new power plants is expensive and many energy supply companies, particularly in the US, now recognize that it is cheaper to improve the energy efficiency of their customers than to invest in new plant. According to Pacific Gas and Electric, a major Californian power company, it costs seven times as much to produce one kilowatt-hour of electricity from a new source as it does to save a kilowatt-hour through conservation. The company is now actively encouraging its customers to conserve energy, offering low-interest loans, rebates and grants to those who invest in energy-saving measures. The company estimates that its conservation programme has saved $5 to $7 billion, by not having to build the new power plants.

Massive savings in energy could also be made in transport, which at present accounts for one-third of energy used in industrialized societies. In Britain, the use of cars and other vehicles releases 100 million tonnes of carbon dioxide into the atmosphere a year - a figure that could rise by 40 per cent by the turn of the century if present trends continue. Cuts could be made through improving the efficiency of cars - at present, the average new car can travel between 9 and 12 km per litre of fuel (25 and 33 miles per gallon), but Toyota has a prototype that can travel three times as far on the same amount of fuel. However, energy use in transport could be reduced most effectively by improving public transport, a measure that would also improve the quality of life in cities. Given that three-quarters of the road journeys in the UK, for example, are under 8 km (5 miles) long, the shift from private to public transport need not be onerous, provided that public transport was improved.

The final third of the energy used in a modern society is by industry - in the developing world, the figure is closer to 60 or 70 per cent. In the US, energy efficiency in the chemical industry - the biggest consumer - has been improved by as much as 34 per cent since 1973. The scope for further improvement in this and other industries is still enormous, in particular through the recycling of basic industrial materials. Recycling paper uses a third less energy than producing paper from trees, while recycling iron and steel brings an energy saving of 60-70 per cent.

Improving the efficiency of power generation could also save large amount of energy. Sixty per cent of the energy generated by conventional power plants (both nuclear and fossil fuel) is lost directly into the atmosphere through cooling towers. Using cogeneration or Combined Heat and Power (CHP) systems allows that waste heat to be captured and used to provide hot water and space heating for the local community. The overall efficiency of energy use can thus be increased to 75 per cent or more. Where CHP systems use coal, they can be made still more efficient if they are equipped with technologies which involve using a flow of gases to create a combustion bed. By putting limestone in the combustion bed, 90 per cent of the sulphur in the fuel can be captured, thereby curbing acid rain. CHP systems, which already operate in several European cities, have a further advantage of being small, so that they can be sited within settlements that can absorb all the waste heat they make available. They are also easy to operate, so they can be managed by local councils or even by neighbourhood committees.

The prospects for expanding CHP schemes are bright. In the US, it is estimated that, even under current economic conditions, the market for cogeneration could reach some 100,000 megawatts by the turn of the century, supplying some 15 per cent of US energy requirements.

RENEWABLE ENERGY SOURCES

The technologies also exist to enable us to meet a large proportion of our energy needs through renewable, non-polluting sources of energy. Moreover, the opportunities for introducing such technologies will grow with increasing energy efficiency. A small hydroelectric turbine or a single windmill may not be able to meet the needs of an energy-intensive community but it can meet the needs of an energy-efficient one.

For countries which border the sea, wavepower offers the greatest potential source of renewable energy. Studies by the UK Wavepower Consortium estimate that the wavepower resource of the Irish coast during summer months is treble the current installed generating capacity of the Irish Electricity Supply Board - and during winter months, it could be six times higher. Off the Atlantic coast of Scotland and Cornwall, wavepower has the potential to generate enough electricity to meet 40 per cent of the peak electricity demand of England and Wales during the summertime and 60 per cent during the winter. Were windmills to be built in conjunction with wave machines - a feasible technology - offshore power generation could satisfy an even higher percentage of demand.

For countries that are landlocked, there are numerous other renewable technologies - powered by water, sun or the wind - which offer alternative sources of energy. Small hydropower too has a considerable future. China has already installed some 90,000 turbines in rural areas, while in the US, about 3,000 megawatts of generation capacity has been installed since 1985. In several other countries, small dams that had been abandoned are now being rehabilitated. Tidal power, on the other hand, usually involves building large and ecologically disruptive installations, and is of doubtful value.

In Israel, solar power provides heating for some 700,000 homes, satisfying 65 per cent of all domestic water heating requirements. In Japan, some 4 million are in use. In the US, the sale of solar collectors increased by 44 per cent a year between 1980 and 1985 but the market collapsed when oil and gas prices fell and subsidies on renewable energies were withdrawn by the Reagan Administration. However, developments in photovoltaic cell technologies, first used to power spacecraft, could bring a vast expansion in the use of solar power as the efficiency of collectors increases and their cost falls.

Even in colder, northern countries, solar power can be used seasonally to great effect. In the Third World, solar energy could have a particularly important role to play. Although initially expensive, solar power units need little or no maintenance, since they have no moving parts, and can operate unattended for 10 to 20 years. They thus offer an ideal energy source for remote villages, ruling out the need for trips to the nearest town for diesel or bottled gas, and providing a power source that can be operated without mechanical expertise.

Wind farms producing energy from giant windmills have been established in California and the costs of wind-generated electricity are now just about competitive with the costs of conventional power. In 1987, wind-turbines produced more than 5 per cent of the electricity generated by Pacific Gas and Power, one of the largest power companies in the US. The company sees that contribution expanding, less than half of the potential sites for windmills in its territory having been developed. In the US as a whole, the potential for wind power is enormous: according to the Washington-based Worldwatch Institute, wind-power could provide a trillion kilowatt hours of electricity a year - enough to satisfy a quarter of the projected energy demands at the end of the century.

▲ Wind turbines on a 'wind-farm', one of the promising developments in renewable energy. Enthusiasm for nuclear power, as an alternative to fossil-fuel power stations, has led governments to neglect such renewable sources.

In Europe, too, wind-power offers a major alternative to conventional sources of power. Denmark already has an installed wind-power capacity of 100 megawatts, enough to power a city of 30,000 people. By 1995, the government aims to produce 10 per cent of the country's electricity by wind-power. In Britain, it is estimated that wind-power could satisfy a fifth of energy needs, reducing carbon dioxide emissions by 50 million tonnes. For the Third World, too,

▲ Bales of waste paper destined to be recycled. Recycling paper at home and work is something everyone can do. Buying recycled paper is equally important - prices are artificially high at present simply because so few people buy recycled products. Virgin paper, although more expensive to produce in real terms, sells more cheaply because it is produced in bulk. For the same amount of paper produced, recycling uses up only half as much energy, and a tenth of the water. It also reduces the need for intensified wood production, which damages wildlife.

"Either we change now, by choice, or we change later through necessity and suffer the consequences."
David McTaggart, Greenpeace International

wind-power is now seen as an increasingly viable energy source, particularly for arid regions where water supplies are scarce. India has embarked on a major wind energy campaign and now has an installed capacity equal to that of California. In temperate regions, one advantage of wind-power is that the energy produced is potentially greater in the winter, as the atmosphere is more turbulent at this time of year.

Alternative energy sources (with the exception of wave - and tidal power) offer power systems that can be tailored to specific locations and needs. They are a decentralized power source and indeed operate best when small-scale: the record of 'mega wind machines', for example, has been dismal. Small windmills can be built and operated without major investment, and they are also powered by an energy source which it is hard to monopolize. As Daniel Deudney and Christopher Flavin put it, "There are no Saudi Arabias of renewable energy. Heat for buildings in North America will come from the rooftops, not from the Middle East. Villagers in India will light their houses with electricity generated using the resources of the local environment instead of with kerosene from Indonesia's outer continental shelf. Energy production will thus reinforce, rather than undermine, local economies and local autonomy."

REDUCING POLLUTION

By far the most effective means of pollution control is to avoid generating the pollution in the first place. The potential for change is enormous. Indeed, the technologies exist to treat, recycle, re-use or eliminate 70 per cent of the waste we currently generate. Yet, we have barely begun to exploit the range of possibilities for reducing waste in industry and in the home - and hence of reducing pollution. INFORM, a New York-based research organization, has identified more than 40 different ways in which industry could reduce its waste; 11 of them cut individual pollutants by 80 per cent or more. With notable exceptions, however, most industries have yet taken up the challenge. Of 29 major US companies surveyed by INFORM in 1986, only 12 had waste-reduction programmes - and in most cases, these only covered a fraction of the wastes generated.

Relatively simple changes in production methods could substantially reduce pollution, in some cases eliminating it altogether. Already there are some major success stories. In the US, the 3M Corporation, long recognized as a pioneer in the field of pollution control, initiated a "Pollution Prevention Pays" campaign in 1975. By 1984, the company had eliminated 10,000 tonnes of water pollutants, 90,000 tonnes of air pollutants and 15,000 tonnes of sludge from its manufacturing processes. The financial rewards were considerable, saving the company some $192 million in less than 10 years. In Sweden, one leading metal-working company found that by switching from petroleum-based lubricants to vegetable-based ones, it not only eliminated the need for highly toxic degreasing agents but also the problem of disposing of contaminated machine oil. In the electronics industry, several companies have now begun to replace CFCs, used in cleaning microchip circuit boards, with a water-based degreasing agent, extracted from the rinds of citrus fruit. This saves money and causes no atmospheric pollution.

Many waste products can be recovered, recycled and re-used. Electrolysis has been used successfully in the electroplating industry to recover gold, silver, tin, copper and other metals from waste. Similarly, trichloroethylene, widely used for degreasing machinery, can be recovered from contaminated waste oil. Even where industries cannot re-use their own wastes, they can sometimes recycle them for others to use: indeed, solvent wastes from the electronics industry are often purer than the technical grade solvents used in heavy industry. In Holland, a waste exchange network has been operating successfully since the early 1970s, with over 150 wastes listed for exchange.

On the home front too, the potential for recycling domestic waste is enormous. In Britain, the average family of four throws away six trees' worth of paper every year, while in the US a city the size of San Francisco disposes of more aluminium than is produced by a small bauxite mine. Yet, in the EC as much as 80 per cent of the 2,000 millions tonnes of household waste generated every year could be re-used or recycled, saving both resources and money. In Britain, the value of the materials dumped every year that could be reclaimed has been estimated at £750 million. Several countries now operate successful domestic recycling programmes. Denmark, where throwaway containers have been banned, recycles 90 per cent of cardboard products and 99 per cent of glass products, the average bottle being re-used 30 times. In Norway, 40 per cent of the country's municipal waste is recycled, with sophisticated equipment separating out plastics, scrap metal and paper for re-use. Japan, for all its wasteful use of tropical timber, recycles twice as much paper as the US, including 90 per cent of newspapers and 81 per cent of cardboard products.

If the environmental impact of a product were considered at its design stage, the scope for pollution prevention would be even wider. Raw materials and manufacturing processes can be selected that minimize the risk of pollution and waste throughout the entire life-cycle of the product. Indeed, Mercedes-Benz has now announced plans for a car that can be entirely recycled, the components either being used in new cars or in other industries. Mercedes is already collecting accident-damaged plastic bumpers from dealers and, where not repairable, grinding them down to be processed into other components. According to the company, almost all the steel in its cars could be re-used; brake-fluid can be potentially used as a solvent; and engine oil can be recycled. The platinum and rhodium in catalytic converters can also be recovered for re-use.

THE ORGANIC ALTERNATIVE

In agriculture, the way forward is also clear. There are numerous systems of farming - broadly termed 'organic systems' - which are both productive and ecologically sustainable, because they farm with nature rather than against it. Pests and weeds are not controlled through waging chemical warfare against them, but by reducing vulnerability to pests with such techniques as polyculture *(see pp.104-105)*, where the wide range of plants grown maximizes biological diversity and minimizes the niche for pests. Maintaining the fertility of the soil relies on careful husbandry, farm wastes being recycled in the form of manures, mulches and composts. A variety of crop rotations are also used to rest the soil and to restore the necessary levels of nitrogen and minerals.

The principles behind organic agriculture are as old as agriculture itself, although farmers throughout the world have adapted them to local conditions. The ecological sustainability of organic farming is beyond doubt, and recent reports also attest to the productivity of organic systems. A 1989 study by the US National Academy of Sciences found that organic farms in the US not only produced yields comparable to those from conventional farms using artificial fertilizers but at much lower cost. Most significant of all, given the threat from global warming, organic farmers were two-and-a-half times more productive per unit of energy than conventional farms. The study concluded that organic agriculture offered a viable alternative for the US. Similar conclusions have been reached by studies in France and Britain.

By far the most productive organic systems are the smallest. The most productive units of all are gardens. Except on the very best land, even conventional farms rarely obtain yields of more than 7.5 tonnes per ha (3 tons per acre) - and such yields require massive applications of artificial fertilizer. An organic vegetable garden, by contrast, can produce 20 tonnes of varied vegetables per ha (8 tons per acre) - and this without the use of chemical inputs or the need for much fuel. Indeed, it has been calculated that a family of four, eating an entirely vegetarian diet, could derive all its protein needs and a third of its energy needs from just 0.1 ha (0.25 acre). In the USSR, 68 per cent of the country's food is grown on private vegetable gardens, which together occupy just 2 per cent of the land. In the UK, during the Second World War when the government initiated its "Dig for Victory" campaign, 10 per cent of the food produced came from private gardens.

Switching from monoculture to polyculture not only reduces the need for chemical pesticides but also boosts productivity, particularly when intercropping is practised. This does not mean planting different crops at random but planting specific combinations of crops that enhance each other's productivity. In a well-planned intercropping system, for example, early-established

“Meanwhile, forest-dwelling people in the Narmada valley go about their simple lives trusting that good sense will prevail and they will not witness the destruction of the few remaining forests upon which their survival hinges. Such peaceful people hold several answers to the world's ecological crisis in the crucible of their civilizations. It is *their* attitudes to the Earth which we need to imbibe, rather than foist our death wish on them.”

Bittu Sahgal, editor of India's Sanctuary Magazine, *on the planned Narmada dams*

"Each of us is responsible for the ongoing debt crisis, for our nations' short-sighted energy use policies, for trade policies and for all other policies affecting... the environment. The most highly educated and richest people on Earth are the most wasteful consumers of resources and the worst polluters of the atmosphere. In this our governments back us."
Lloyd Timberlake and Laura Thomas

plants will reduce soil temperature and create the appropriate microclimate for other plants. Similarly, deep-rooted plants act as 'nutrient pumps', bringing up minerals otherwise locked in the subsoil and making them available for shallower rooted plants. Minerals released by the decomposition of annual plants are taken up by perennial ones, while the heavy nutrient demands of some plants are compensated for by the inputs of others.

The resulting increases in yield can be spectacular. In Nigeria, yields of maize and sorghum can be four times greater under some conditions when grown together than when grown separately. In Mexico, interplanting maize with beans and squash enables the farmer to produce almost 70 per cent more food than when maize is grown by itself. In general, intercropping boosts food production from farms by 50 per cent. Part of the benefit comes from the role intercropping plays in reducing pests.

In the North, a shift from intensive agriculture to organic systems is likely to be a lengthy process, not only because the land must be weaned away from its reliance on chemicals but also because organic techniques will be new to many farmers. In the Third World, however, there are numerous traditional systems of agriculture, many of them dating back for thousands of years, which are still operational. Such systems have evolved to suit local conditions and have a proven record of sustainability. Much could be accomplished, and in a short space of time, just by putting a halt to the imposition of unsuitable modern agricultural methods on Third World farmers. Instead, the policy should be to build on traditional systems, relying on the proven inventiveness, knowledge and adaptability of local farmers to make whatever improvements they deem necessary.

The shift to organic agriculture would have a far-reaching 'solution multiplier' effect. Reducing the use of artificial fertilizers would not only help curb emissions of nitrous oxide - a powerful greenhouse gas - but would also help prevent the contamination of rivers and groundwaters with nitrates, improving the quality of drinking water and preventing pollution of lakes and rivers. Reducing energy consumption in farming would further help combat the greenhouse effect. The use of manure and compost, together with rotations and intercropping systems, would improve the quality of soils and do much to restore the viability of degraded agricultural land. Reducing the use of pesticides would not only curb pollution on the farm but also reduce the health risk to chemical workers, many of the most hazardous chemicals being pesticides. The quality of food would improve, organic produce being richer in vitamins and trace elements and uncontaminated by pesticides and nitrates. Organic systems reduce costs and, by enabling farmers to get off the 'pesticide and fertilizer treadmill', would do much to relieve the indebtedness of farmers. Finally, by favouring smaller, more labour-intensive farming systems, organic farming would help revive rural communities and curb rural depopulation, so relieving the strain on cities, especially in the Third World.

THE CONSERVER SOCIETY

Improving energy efficiency, moving towards renewable energy sources, reducing waste and making farming sustainable are just four ways in which we could reduce our impact on the environment - and do so without major changes in lifestyle. However, there are many features of our society today - nuclear power plants, toxic waste dumps, CFCs and battery farms, for example - that cannot be 'greened'. We have no option but to campaign for them to be phased out. Even where ecologically sound technologies can be introduced, they are unlikely to be adopted on a wide scale without public pressure - either through the market-place or through legislation. While landfill remains a cheap and legal disposal option, for example, few industries will opt for the more difficult route of waste reduction.

More generally, any gains made by adopting more ecologically sound technologies and practices will be eventually overtaken if we remain committed to an expansionist economy. If we reduce the pollution from cars by half, but double the number of cars, we shall be back to square one. And however energetic we are in recycling non-renewable resources, there is a physical limit to the number of times any product can be recycled: some of the original material is always lost in reprocessessing. Even if we achieved an across-the-board reduction in pollution of 80 per cent, we should be producing as much pollution again in 25 years if we had an annual economic growth rate of 7 per cent. It is critical, therefore, that we push for wider changes in society that will enable us to move away from an expansionist economy to one that maximizes the conservation of resources. In effect, we must move from an economy of flow to an economy of stock.

Here fiscal measures could play a major role. One possibility is an Amortization Tax, set according to the estimated life of a product - 100 per cent for products designed to last no more than a year, zero for those which last 100 years or more. The aim would be to penalize short-lived products, especially throwaway items, thereby reducing resource use and pollution. Plastics, for example, which are remarkable for their durability, should only be used in products where this quality is valued, and not for 'single trip' goods such as packaging. Such a tax would also encourage craftsmanship and employment-intensive industries where quality is valued more than quantity.

A second possibility is a Greenhouse Tax, operating on the same sliding scale as the Amortization Tax, but specifically targeted at reducing emissions of greenhouse gases. Such a tax would not only penalize power-intensive activities - generally the most polluting - but would further discourage short-lived products. It would also penalize the needless transport of goods, reducing the size of markets and encouraging local self-reliance. Because it would be broader than a simple energy tax, it would also cover goods whose production resulted in deforestation - tropical timber products, for example - or in the generation of methane. As such, the tax would have a wide impact: by discouraging the eating of beef (cattle being a major source of methane), it would release large areas of land for crop production. At present, some 40 per cent of world grain output is fed to cattle, as is 40 to 50 per cent of world fish catches.

Such taxes would start off at low levels, but it would be made clear from the start that they would increase every year until such time as the offending activities were appropriately reduced.

Changes in subsidies could also make a contribution towards encouraging the move towards a conserver society. At present, conventional agriculture is heavily subsidized to the detriment of organic agriculture, with the result that organic food is more expensive.

Equally, public transport in many countries suffers from the subsidies to private motorists: in Britain, the tax concessions supporting company cars amount to five times the subsidy to British Rail. Similarly, subsidies to the nuclear industry have been massive, involving not only direct government grants but also indirect subsidies in the form of government-sponsored research and development programmes and other favours. At the height of the French nuclear programme, it is estimated that government subsidies to the industry amounted to 5,000 million francs a year. The effect has been to stifle many alternative energy projects and to give electricity that is generated by nuclear power an unfair (and totally misleading) economic advantage over far more desirable alternatives.

DECENTRALIZING POWER

The transition to an ecologically sustainable society will not be complete, without empowering local communities, both politically and economically. This is not some anarchist whim: it is fundamental to a sustainable society. For if the dynamic behind the current ecological crisis involves the transfer of power and resources from local communities to single-purpose organizations, the solution lies in reversing the process. Indeed, in the Third World, there are few environmental problems that can be remedied unless this issue is addressed. For many Third World environmental groups, giving local communities a decisive say in matters affecting their livelihoods is seen as an essential precondition for achieving a sustainable society, second only to a moratorium on debt.

Land rights - and thus land reform - are central to the problem. Without land security, the dispossession and marginalization of people is inevitable. While half the population of Brazil remains landless, and the best land is used to grow crops for export, there is little hope of stemming the flow of settlers into Amazonia. Equally, while forest dwellers are denied land rights and are marginalized politically, there is little hope that they can resist the imposition of development projects that undermine their livelihood. Chico Mendes, the rubber-tappers' leader whose murder did so much to focus international concern on rainforest destruction, was fighting to give the rubber-tappers control over resources that they require to maintain their way of life. Mendes set out their objectives plainly: "We demand a development policy that favours forest people and not large landowners and multinationals. We rubber-tappers demand to be recognized as true defenders of the forest."

In claiming to be the "true defenders of the forest", Mendes was making more than a rhetorical point, for those who have a direct interest in an area are the best people to care for it. That interest cannot be measured in conventional economic terms alone. For forest dwellers, the value of the forests is not to be judged by the profit and loss accounts of timber companies or by the

"The components of the natural world are myriad but they constitute a single living system. There is no escape from our interdependence with nature; we are woven into the closest relationship with the Earth, the sea, the air, the seasons, the animals and all the fruits of the Earth. What affects one affects all - we are part of a greater whole - the body of the planet. We must respect, preserve, and love its manifold expression if we hope to survive."
Professor Bernard Campbell

"Never doubt that a small group of thoughtful, committed citizens can change the world. Indeed, it is the only thing that ever has."
Margaret Mead

money that could be made by exploiting the forests for their pharmaceutical potential: it is paid in the currency of running streams, wildlife, productive soil, flood control and a stable climate - the numerous ways in which the forest ecosystem supports them. The same could be said for how pastoralists view their rangelands, and peasant farmers their land.

While people can have their land taken over at the whim of a government agent or a landlord, it is inevitable that there will be landlessness - and with landlessness comes insecurity and pressure to encroach on areas that are unsuitable for farming. Combined with a "Food First" policy, in which land currently used for export crops would be released for growing food, land reform would do much to reduce our impact on the environment. It would also help resolve the problem of hunger - according to the UN Food and Agriculture Organization, a radical programme of land reform could increase food output in Pakistan by 10 per cent and in Colombia by 25 per cent - and thus further enhance the security of those who have been marginalized by the development process. That in turn would do more than any other measure to combat the population explosion.

But land reform without wider political change is likely to be doomed to disaster, as it nearly always has been in the past. In particular, the conditions need to be created in which small farms can survive as viable units, both economically and socially. If they are producing for distant markets over which they have no control, or reliant on chemical inputs which put them further and further into debt, many small farmers eventually have no option but to sell up. Land holdings become increasingly concentrated and the process of marginalization begins all over again. Similarly, if farmers cannot rely on the cooperation of their community and of their family, many small farms cease to be viable units. Irrigation farmers, for example, are utterly dependent on the cooperation of their fellow farmers to maintain their tanks and irrigation canals, so much so that, with the disintegration of rural communities, irrigation practices have always come to an end.

Re-empowering communities thus involves more than simply devolving control of resources to local communities or power to local officials. It involves creating those conditions in which communities can act as communities: caring for their members' welfare, educating their children. Indeed, as a rule of thumb, nothing should be done at the level of the state or the region that can be best done by the community, and nothing done by the community that can be done by the family.

It is in the face-to-face context of a community that decision-making is potentially most democratic and fairest. Blanket directives from above to cut carbon dioxide emissions are unlikely to be as effective as local communities themselves working out ways to address the problem. Indeed, the solutions drawn up by local communities are likely to be radically different to those drawn up by government or industry. Experience shows that local people do not want reforestation programmes that force people off their land or which involve planting monotonous rows of pine or eucalyptus to cut down every decade. They want trees that will meet their local needs, trees that are planted to last. Similarly, it is at the local level that the power of public opinion - rather than coercive legislation - can work most effectively, both as a force for change and for policing that change. It is within a community, too, that individuals have the greatest sense of identity, supported by other members and without that sense of alienation that comes from living in a rootless and mobile mass society.

TAKING ACTION

There is no magic wand that will bring the environmental crisis to an end overnight. The opportunity to move towards a less destructive society, where people have greater control over their lives, and where one person's wealth does not come about by impoverishing others, is there for us to take.

It may well be that we prove unwilling to change our lifestyles in the interest of future generations. If that is the case, then the prospects are indeed grim. But the indications are otherwise. Throughout the world, and increasingly in the developing countries, numerous local groups are now demanding a halt to the destructiveness of modern society. It is through such groups - and by taking individual responsibility for our actions - that change will come about, change that is not imposed from above but which is created from below - by ordinary people.

Daunting as the task ahead might seem, individual actions - through lobbying, changing lifestyles, political action, boycotts and the like - can make all the difference in the world. But only if we are willing to act. And act now, before it is too late.

INDEX

A

Aborigines, 118, *119*, *269*
acid rain, 21, 93-4, 206, 246
advertising, *244*, *264*
agriculture, 97-115
  battery farming, 251, *251*
  chemical pollution, 112-14
  deforestation and, 75-80
  diversity, 104-5
  effects of, 101-2
  energy consumption, *255*
  fertility, 106-7, 110-11
  fertilizers, 21, 31, *107*
  food supplies, 260-3
  green revolution, 109, 261
  intensification, 261
  interplanting, 104-5
  irrigation, 111-12, 157
  and the local community, 282
  land use, *100-1*
  modern, 243
  modernizing, 107
  mountain areas, 203-5
  nutrients, 102
  organic farming, 114-15, 279-80
  origins, 98
  pastoralism, 122-4
  pioneering ecosystems, 98-101
  pollution, *138*, 139
  rangelands, 117-27
  soil erosion, *111*
  soil salinization, 97, 111-12, 132, 157
  swidden, 85
  wetland pollution, 159
  wetland reclamation, 158
  wetlands, 156-7
  see also pesticides; hormones
agrochemicals, 112-14
  food and, 249-51
  water pollution, 145
  see also fertilizers; pesticides
Ahlen, Ingmar, 94
aid, 124
algae, water pollution, 31, 171
Alar, 249
Alaska, oil, 228, *230*
Aldabra, 195, 197
Alps, 203
  destruction, 206-7
aluminium, 275
  in the soil, 21
  in water supplies, 246
Amazon Basin:
  agriculture, 104
  dams, 84, 129
  destruction of the forest, 75, 76, *76*, 77, *79*
  fishing, *128*
  fragmentation, 36-9
  greenhouse effect and, 75
  harvesting minor products, 86, *86*, 87
  mining, 82
  natural drugs from, 70
  saving the forests, 87
  soil erosion, 73
  varzeas, 152
  wild fruits from, *86*
  Yacuna Indians, 12
Amortization Tax, 280-1
Andes, agriculture, 205, *205*
animals:
  coasts and estuaries, 166-7
  deserts, 209, 211-14
  extinction, 39, 257
  fragmented habitats, 36-9, *37*
  global warming and, 49-50
  interrelationship with plants, 28-9
  islands, 195-8
  mountains, 202, *202*
  see also wildlife
Antananarivo, *266*
Antarctica, 219-25, *225*
mineral exploitation, 223-5
  ozone layer, 41-3
  whaling and fishing, 222-3
Aral Sea, 97, 99, *178*
Arctic, 227-34
  Arctic parks, 230
  greenhouse effect, 234
  oil exploitation, 228-30
  overfishing, 230
  pollution, 231-2
  radioactive contamination, 231-2
  wildlife, 234
Arctic National Wildlife Refuge (ANWR), 230
Arst, John, 231
Aswan Dam, 135, 157, 171
Atacama Desert, *216*
atmosphere:
  balance of gases, 16-21
  burning coal and, 246
  industrial pollution, 256
  motor vehicle fumes, 246-7
  ozone layer, 41-5, 256
  planetary, *17*
  pollution, 114, 245
atolls, *189*
Australia:
  Aborigines, 118, *119*, *269*
  Antarctica, 225
  bush, 118, *119*
  deforestation, 58, 90
  ecological invasions, 30, *31*
  forests, *60*, 61, 67
  introduction of the prickly pear, 30, *30*
  wetlands, *153*
  wildlife, 197
Austria, polluted rivers, *15*
automobiles, 242, 246-7, 270-1
  recycling, 279
Avdat, 214

B

Babylon, 13
Bacon, Francis, 12
bacteria, 29, 110, 152, 163
Badlands, North Dakota, *18-19*
Bahuguna, Sunderlal, 55
balance of nature, 27-39
Bali, erosion, *191*
Baltic Sea, pollution, 171, 177
Bangkok, *247*
Bangladesh:
  flooding, *136*, 263
  mangrove swamps, 163
Bascuit Bay, 191
BASF, 140
Basle, 140
battery farming, 251, *251*
beaches:
  pollution, 171
  tourism and, 171
  see also coasts
beavers, *135*
becquerels, definition, 33
Bedouin, *214*
beechwoods, *62*
bees, *29*, *74*
Belgium, river pollution, 139, 140
Belle Ile, *180*
Bellerive Foundation, 207
Bhopal, *270*
bioaccumulation, 33
biological control, *30*
birds:
  coasts and estuaries, 166, 167
  extinction, 196
  forest destruction and, 88
  wetlands and, 151, *155*, 163
  see also wildlife
birds of paradise, *38*
bison, 117, *118*
Black Forest, Germany, *63*
black rhinoceros, 36, *37*, 121
Blauloch Spring, *144*
bogs, 151-63
Bombay, 266
boreal forests, composition, 59
Boro river, 158
Botswana:
  pastoral societies, 124, *126*
  seasonal floods, 157
  wetlands, 160
Brazil:
  cattle ranching, 80, *80*
  colonization of border areas, 79
  dams, 84, 129, 132
  land rights, 281
  mining, 82
  and oil in Antarctica, 225
  reconstruction, 273
  rubber-tappers, 86, *86*
  see also Amazon Basin
British Columbia, deforestation, 88, *88*
broadleaved trees, 59, 62
Brokopondo Dam, 136
Brundtland, Gro Harlem, 266, 267
bumblebees, *29*
bureaucracy, 243-4, 269-70
Buryatia, 90
bush-tailed possum, 30, *31*

C

cacti, *210*, 211, *216*
cadmium, 32
  in soil, 21
California, water supplies, 133
Camargue, 154, *154*
Cameroons, deforestation, 85
Canada:
  acid rain, 21, 93
  dams, 234
  deforestation and logging, 88, 93
  fishing industry, 181
  mixed forests, *63*
cane toads, *31*
carbon dioxide:
  atmospheric, 16-21
  greenhouse effect, 46-50
  industrial emissions, 256
  photosynthesis, *18*, 28, 29, 121, 261
  savannas and, 121
carbon monoxide pollution, 246
carcinogens, 32, 113, 143, 250, 251
caribou, 228, 234
Carmanah Valley, 88, *88*
carrs, 151-63
cars, 242, 246-7, 270-1
  recycling, 279
Carter, Jimmy, 12
cassava trade, 79
cash economies, 268-9
Caspian Sea, 135
cats, *29*, 197
cattle, 124, *125*, *126*
  feeding, 127
  ranching, 80, 117-27
centralization, 243-4
cereals, 260-1
Chad, *260*
Chagga, 203
chalk deposits, 20-1, *20*
chamois, 207
charcoal production, 82, *83*
chemical waste and pollution, 14, 30-3, 247
  agricultural, 112-14
  bioaccumulation, 33
  disposal, 273
  dumping, 198
  rivers, *15*, *138*
  toxic chemicals, 31-2, 247
  water supplies, 145-7, *147*, 249
  wetlands, 159
Chernobyl, 14, *14*, *36*, 114, 231, *231*
Chesapeake Bay, 170, *171*
Chile:
  krill catches, 223
  semi-deserts, *119*
China:
  agriculture, *102*, 111, 112
  dams, 136
  industrial pollution, *32*
chinampas, 156
Chipko, 27
chlorinated hydrocarbons, 32
chlorofluorocarbons (CFCs), 32, 41-5, *43*, 48, 255-6
Christmas Island, *198*
Churchill, Canada, *232*
Ciba-Geigy, 140
Cienaga Grande, 160
cities:
  life in, 241
  urban wastelands, 241-2
  urbanization, 241
Citizens Clearing House for Hazardous Wastes (CCHW), 273
climate:
  coral reefs and, 193
  forests and, 74-5
  industry and, 256-7
  mountains, 201
  see also global warming; greenhouse effect
climax vegetation, 58-9, 99
coal, 246, 253
coasts, 165-71
  erosion, 171, 187
  nature of, 165-6
  pollution, 168-71, 179
  tourism, 171
coccolithophorid algae, 20, 21
Colmar, 147
Colorado River, *132*, *133*
Colombia:
  forests, *87*
  mangrove swamps, 160
  mountains, 202
Combined Heat and Power (CHP), 276
companion planting, 104
conifers, 59, 62-3
  plantations, 94-5, *95*
conservation of resources, 280-1
consumption, 241, 267
Convention on International Trade in Endangered Species (CITES), 36, 120
coral reefs, 187-93, *188*
  building, 189
  coastal erosion and, 171, 187
  destruction, 189-93
  protection, 193
Cordillera, 205
cornet moths, *71*
cotton crops, 97, *99*, 101
cougars, *202*
cranes, 151
Cree Indians, 234
crop rotation, 106
crown-of-thorns starfish, 191, *191*
ctenophore, *174*
cyanobacteria, 18, *18*

D

dams, 84, 130-6
  adverse effects of, 134-6
  Arctic, 232-4
  taming floods, 136-9
date palms, *149*
DDT, 27, 177, 251
decomposers, 39, *39*, 68, 110
deer, damage by, 30, *31*
deforestation, *51*, 273
  climate and, 74-5
  coral reefs and, 191
  erosion, 70-3
  islands, 198
  losses through, 70-5
  mountain areas, *206*
  rainforests, 76-87
  see also forests
Denmark, waste materials, 279
desertification, 127, 265
deserts, 118, 209-17
development, ideal of, 268
dieldrin, *138*
Dinka, 157, 163
dioxins, 114
diseases:
  industrial pollution, 247
  malaria, 27
  reservoirs and, 135
  waterborne, 248-9
DNA, radioactivity and, 34
dodo, 196
Dogon, *105*, *248*
dolphins, *173*
drift nets, 173, *184*
drought, 148, 206
drugs, natural, 70
dry tropical forests, 61, *61*
Duffek Massif, 223

Dumont d'Urville, *220*
Dust Bowl, 122

E

earthworms, 110
East Anglia, 31
economic growth, 266, 268-9
ecosystems, balance of nature, 27-39
Ecuador:
  fishing, *181*, 185
  mangrove swamps, 163
egrets, *153*
Egypt:
  agriculture, 112
  coastal erosion, 171
  population increase, 258
Ehrlich, Anne and Paul, 257
El Nino current, *181*
El Salvador, agriculture, 205
electricity supplies:
  dams, 130, 132-3
  deforestation and, 84
  efficient use, 275-6
  from renewable sources, 276
  generation, 232-4
elephants, 120-1, *120*
Eletronorte, 129
energy supplies, 253-5, *253*
  agriculture, 255
  coal, 253
  efficiency of use, 274, 275-6
  energy conservation, 255
  nuclear fuels, 253-5
  oil, 253
  renewable sources, 253-5
  see also electricity
environment, balance of nature and, 27-39
erosion, *26*, *79*, 81-2, *82*, 102, *111*, 121-2
  coastal, 171, 187
  coral reefs and, 191
  deforestation and, 70-3
  mountains, 205-6
  rivers, 136
Etosha Pan, *47*
estuaries, 165-71
  nature of, 165
  pollution, 168-71
Ethiopia:
  drought, 206
  famines, *46*
ethylene dibromide (EDB), 113
eucalyptus, 95, 148
  reforestation with, 85
Euregio Rhine-Meuse, 139-40
eutrophication, 31
Everglades, Florida, *155*, *156*, 158
Exxon Valdez, 230, *230*
extinction, 39, 70, *257*
  savanna wildlife, 120-1

F

famines, 260-3
Fangataufa, 190
farming *see* agriculture
Farraka barrage, 163
fens, 151-63
fertility, agriculture, 106-7, 110-11
fertilizers, 21, 31, *107*, 109, 110, *138*, 139, 261
  in coastal waters, 171
  reduction of, 279-80
  wetland pollution, 159
  world consumption, 107
Finland, forests, *58*
fire, forests and, 59, 61
firewood, 76
fish and fishing:
  Antarctica, 223
  Arctic, 230-1
  coral reefs, 189, *189*
  dams and, 134-5
  drift nets and, 173
  inshore fishing, 165, 168, 170
  mangrove swamps, 160, 163
  overfishing, 179-85
  polluted water, 31, 139
  protection, 273
  wetlands and, 155
fjords, 165
flood plains, *100*
floods and flooding, 73, 102
  agriculture, 156-7
  dams, 136-9
  seasonal, 156-7
  wetlands, 158
food:
  contamination, 249-51
  from forests, 74, *74*
  irradiation, 251
  testing, 250-1
  world food crisis, 260-3
  see also agriculture
Food and Agricultural Organization (FAO), 76, 84, 109, 112, 124, 260, 261, 270
food chain, 28-9, *28*
  marine, 174-6
forests, 55-95
  acid rain, 21, 93
  agribusiness and, 79-80
  benefits of, 70-5
  canopies, 67
  clear-cut logging, 90
  climate and, 74-5
  climax vegetation, 58-9, 99
  diversity of, 59-67
  extent, *56-7*, *90-1*
  floods and, 136, *136*
  fragmentation, 37
  as fuelwood, 76
  harvesting minor products, 86, 273
  historical destruction, 13-14
  inhabitants, 74, *74*
  islands, 198
  managed, 90-3
  mining in, 82
  nitrogen pollution, 21
  nutrients, 68, *68*, 70
  plantation forestry, 94-5
  reforestation, 84-5, 273
  regeneration, 85
  saving, 86-7
  structure, 67-8
  temperate, 88-95
  timber trade, 80-2
  understories, 67
  wildlife, 93
  wise use of, 85-6
  see also deforestation
fossil fuels, 253-5
France:
  Antarctica, 219, *220*, 225
  destruction in the Alps, 207
  farmland, *103*, *107*
  nuclear energy, 253-4
  nuclear weapons, 190, 198, *198*
  waste disposal, 146-7
  wetlands, 154
  woodcutting, *92*
Fraser Island, *60*
Fulani, 124
fungi, 29, *39*, 68, 110, 152

G

Gaia theory:
  balance of nature, 27-39
  critical cycles, 20-1
  hypothesis, 16-17
  overloading the system, 21
  in practice, 17-20
Galapagos Islands, *195*, 197
Gambia, agriculture, 109
Ganges, 130, 203
gas deposits, Antarctica, 223-5
gazelles, 211-13
geckos, *70*
General Agreement on Tariffs and Trade (GATT), 271
genetic engineering, 12, 261
George Washington National Forest, 90
Ghana, sewage systems, 249
ghost orchid, *95*
giant otters, *135*
giant tortoises, *195*, 197
glaciers, 203
Global Biodiversity Strategy, 257
global warming, 45-51
  causes, 46
  chain reactions, 50-1
  sea levels, *48*, 49
  see also greenhouse effect
Goa, coastal fishing, *169*
goats, *125*, 196
gorse, 92, 197
grain harvests, 260-1
Grande Carajas programme, 82
grasslands:
  savannas, 117, 120-1
  temperate, 121-2
great auk, 196
Great Britain:
  beechwoods, *62*
  birds, 166
  chemical pollution, 31
  deforestation, 58
  food safety, 250
  industrial pollution, 265
  ivory trade, 120-1
  peat exploitation, *159*
  river pollution, 141
  waste materials, 278, 279
  wildlife, 197
  woodland, 90, 94
Great Plains, 117
Great Rann of Kutch Desert, *211*
Great Whale River, 232
Green Revolution, 109, 261
greenhouse effect, 45-51
  agriculture, 261
  Arctic, 234
  carbon dioxide, 17-21
  cattle feedlots and, 127
  chain reactions, 50-1
  deforestation and, 75
  energy supplies and, 255
  industry and, 256
  peat extraction and, 159
  reducing, 275
  savannas and, 121
  see also global warming
Greenhouse Tax, 280
Greenland, Arctic parks, 230
Grimsby, 165
groundwater, 143-9
  deserts, 217
  pollution, 145
  see also water supplies
Groupement International des Associations Nationales de Pesticides (GIFAP), 270
Guadeloupe, *196*
guanacos, *119*

H

habitat:
  destruction, 36
  fragmentation, 36-9, *37*
  nature reserves, 36-9
Haiti, 55
  deforestation, 55, *83*
  mangrove swamps, *162*
Hanunoo, 85
hares, *104*
Hawaii, *10*
  agriculture, *109*
  deforestation, 198
  wildlife, 196, 197
health, pollution and, 245-6
heavy metals, 32
  coasts and estuaries, 168
  pollution, 140
  river discharges, *140*
  wetlands pollution, 156
Hekstra, Gjerrit, 49
herbicides, 97, 107
  river pollution, *138*
herrings, 155
Himalayas:
  deforestation, 76
  vegetation, *206*
Hodel, Donald, 230
Hoechst, 140
honey, *74*
Hoover Dam, 133
hormones, foodstuff, 251
Humphrey, Hubert, 12
hunter-gatherers, 118

I

Incas, agriculture, 205
India:
  agriculture, 104, 109, 111, 262
  carbon dioxide emissions, 256
  dams, 133, 134
  fishing industry, 183
  flooding, 73, 136
  population levels, 258, 259
  shanty towns, 266, 267
  social forestry, 95
  water supplies, 148
  wetlands, 154
  wind energy supply, 278
Indonesia:
  agriculture, *105*, *109*
  community breakdown, 244
  coral reefs, *190*, *191*
  dams, 134
  fishing industry, 184
  mangrove swamps, 163
  transmigrants, 78-9, *78*
  wetlands, 160
industrial pollution, *32*, *83*, 247
  Arctic, 232
  coasts and estuaries, 168-71, *169*
  reduction, 278
  rivers, *15*, 82, *138*, 139-41
  water supplies, 249
industrialization, 244
industry:
  energy efficiency, 276
  and the environment, 256-7
  lobbyists, 270
  population and, 258
  waste disposal, 139-41
Industry Cooperative Programme, (ICP), 270
INFORM, 278
Inle Lake, *157*
Inner Mongolia, nomads, *122*
insecticides, foodstuffs and, 249-50
  see also pesticides
insects, managed woodland, 93
international agencies, 270
International Monetary Fund (IMF), 259, 268
International Tropical Timber Organization (ITTO), 82
interplanting, 104-5
introduced plants, and the balance of nature, 30
Inuit, 228, 230, *233*
Ireland, peat extraction, *158*, *159*
irradiation of food, 251
irrigation, 97, 102, 111-12, 148, *148*, *149*, 157-8
  mountain areas, 203
islands:
  effects of isolation, *197*
  wildlife, 195-200
Israel, food contamination, 249-50
Itaipu dam, 132
Italy, farming, *107*
Ivory Coast, soil erosion, 73
ivory trade, 120-1, *120*

J

James Bay, 232-4, *234*
James River, 170
Japan:
  farmland, *100*
  fishing industry, 173
  krill catches, 223
  and oil in Antarctica, 225
  whaling, 222, *223*
Java:
  agriculture, 104
  mangrove swamps, 163
Jencoacoara Ecological Reserve, *168*
Johnson Island, 198
Jonglei Canal, 157-8

K

kakapo, 197
Kansas, farmland, *99*
Kariba dam, 134
Kashmir, *100*
Kayapo Indians, 85
Kedung Ombo, 134
Kenya:

ivory trade, 120
water supplies, 124, *125*
kepone, 170
Kerala, 259
Kesterton National Wildlife Refuge, 159
kingfishers, *131*
krill, 221, 223
!Kung, 118, 124, 258-9

L

La Grande River, 232, 234
La Plagne, 207
Lacondan Mayas, 85
Ladakh, *100*
land ownership, 268
land rights, 281-2
landfill sites, 146-7
Lantau Island, *156*
Laoying Dam, 136
Lapps, 231, *231*
laterization, 70-3
Laysan, 196
lead poisoning, 32, 246-7
lemurs, 197
Leningrad, *252*
Leningradskaya, 219
limestone deposits, 20-1
Lindane, 251
Liverpool Bay, 170
livestock, hormone treatment, 251
lobbyists, 270
loggerhead turtles, 171
London:
pollution, 245
traffic, 242
Long Island, New York, 156
Love Canal, 14
Lovelock, James, 16, 20, 189

M

Maasai, *125*
MacMillan Bloedel, 88
McMurdo, 219
Madagascar:
forest destruction, *54*, 55, *72*, *78*, 198
moths, *71*
natural drugs from, 70
timber trade, *81*
wildlife, *196*, 197, 198
malaria, 27
Malaysia, *13*
deforestation, *82*
timber trade, 81
Maldives, 198
Mali, 125, *262*
agriculture, *112*
Manaus, *86*
mangrove swamps, 152, 160-3, *160*, 162, *162*, 176
maquis, 59
marine life, and atmospheric carbon dioxide, 20-1
Mars, atmosphere, 16, *17*
Marsh, George Perkins, 29, 30
Martinique, coral reefs, *190*
material world, 243
Matto Grosso, deforestaion, 77
Maun, 159
Mauritius, dodos, 196
Mediterranean:
pollution, *170*, 177
tourism, 171
Mekong River and Delta, *141*, 166
Melville Bay, *226*
Mendes, Chico, 80, 85, 281
Mercedes Benz, 279
mercury, 32, 232
in the soil, 21
Mersey River, 170
methane, 16, 127
Meuse, River, 139, 140
Mexico:
savannas, 121
swidden agriculture, 85
wetlands, 156, 158
Michigan:
forests, *64-5*
water supplies, 146
Minamata, 14
minerals:
Antarctica, 223, 225
Arctic, 230
mining:
coral reefs, 187
deforestation, 82
mountains and, 201
mink, 197
mires, 151
moas, 196
monkeys, *162*
monocultures, 93, 113, 279
monsoon forests, 61
Mont Blanc, *202*
Mont Etale, *201*
Montana, agriculture, 97
Montreal Protocol, 256-7
Moorea, *193*
mosquitoes, 27
motor vehicles, 242, 246-7, 270-1
Mount Anne, Tasmania, *66*
Mount Bolivar, 202
Mount Colombus, 202
Mount Fubilan, 201
Mount Kilimanjaro, 203
mountains, 201-7
agriculture, 203-5
erosion, 205-6
natural history, 201-3
people, 203-6
wildlife, 202
Mozambique, *61*
multinational companies, 270, 271
Mururoa, 190
mutagens, 32
Muthare Valley, *258*
mycorrhizae, 68

N

Nabateans, 214
Nagami, 273
Namib desert, *216*
Narmada River, 133, 134, 136
National Institute of Amazonian Research, 38
national parks, 257
coral reefs, 193
Nations, James, 76
natural world, balance of nature, 27-39
nature reserves, 36-9
Negev Desert, 209, 217
Negros, Philippines, 79
Nepal:
deforestation, 206
mountains, *204*
Netherlands:
agriculture, 114
birds, 166
fertilizer pollution, 21
population levels, 259
river pollution, 139-40
New Britain, *244*
New Caledonia, *198*
New Guinea:
birds of paradise, *38*
mountains, 201
swidden agriculture, 85
New Zealand:
deforestation, 58, 92, *94*
ecological invasions, 30, *31*
forests, 67
introduction of gorse, 92, 197
wildlife, 196, *196*, 197
Nicaragua, food safety, 251
Niger, agriculture, *115*, 118, *118*
Nile River, 135, 156-7, 171
nitrates, water contamination, 145
nitrogen pollution, 21
nomads, 122-4, *214*
Norfolk Broads, 31
North Sea:
coastal wetlands and, 155
fishing, 182
pollution, 140-1, 171, 177-9
Norway:
coastal pollution, 171
fishing industry, 180-1, 231
krill catches, 223
Lapps, 231, *231*
waste materials, 279
nuclear contamination, 14, *14*, 33-6, *34*, 114, 140, 231, *231*
nuclear energy, 253
power stations, *34*, 35-6
waste disposal, 35-6, 217
nuclear weapons, 35, 263
testing, 190
nudibranch, *20*
Nuer, 157, 163
nutrients:
agriculture, 102
forests, 68, *68*, 70
rain forests, 68
wetlands and, 155

O

oceans, 173-85
coral reefs, 187-93
marine food chain, 174-6
pollution, *176*, 177-9
rising sea levels, *48*, 49, 198, 263
seabed, 174, 176
seashores, 165-71
wave power, 255, 276
see also fish and fishing
Odra River, *140*
Ogallala Aquifer, *144*, 145, 148
oil:
Antarctica, 223-5
Arctic and, 228-30
pollution, *177*, 230
reserves, 253
Ok, 201
Okavango Delta, 157, 158, 159
Okefenokee Swamp, 154, *155*
Olympic National Park, Washington, *67*
Orang Asli, *13*
orchids, *59*
organic chemicals, 32
organic farming, 114-15, 279
Orme River, *131*
overpopulation, 258-9
oxygen:
atmospheric, 16-21
photosynthesis, 28, 29
in polluted water, 31
ozone, *42*, 246
forest damage, 94
ozone layer, 41-5, 256

P

Palawan, 82
Pantanal, *153*
paper, recycled, *278*
pastoralism, 122-4
peasants, deforestation and, 76-9
peat, 152
extraction, *158*, *159*
penguins, *222*
persistent chemicals, 32-3, 113
Peru, agriculture, 205, *205*
pesticides, 97, 107, 109, 112-14
carcinogenic, 245
coasts and estuaries, 168
DDT, 27, 117, 251
foodstuffs and, 249-50
health risks, 32-3
natural, 70
persistent, 32-3, 113
river pollution, *138*
spraying, *250*
testing, 250-1
wetlands pollution, 156, 159
woodland pests, 93
pests:
crop rotation, 106
forestry, 93
modern agriculture, 107
see also pesticides
petrochemical revolution, 12
Pfizer, 250
Philippines, 79
agriculture, 262
coral reefs, 191
mangrove swamps, 163
mountain agriculture, 203, 205
river pollution, *138*
slums, *242*
swidden agriculture, 85
photosynthesis, *18*, 28, 29, 121, 261
phytoplankton, 174-5
Pinnacle Desert, *213*
pioneer ecosystems, 98-101
plants:
deforestation and, 70
interrelationship with animals, 28-9
islands, 195-8
loss of, 36
mountains, 202
nutrients, 31
Socotra, 195
wetlands, 152-4
see also agriculture; forests
plutonium, 35
Point Geologie, 219, *220*
Poland, industrial pollution, 177, 247
polar bears, *232*
political change, 282
Pollagh Bog, *158*
pollution:
acid rain, 21, 93-4, 206, 247
agriculture, 112-14, *138*, 139
Antarctica, 219, *220*
Arctic, 231-2
coal burning, 246
coasts and estuaries, 168-71
coral reefs and, 191
fertilizers, 21, 31, 112-14, *138*, 159, 171
foodstuffs, 245-6, 249-51
health and, 245-6
industrial, *83*
inshore fishing, 165
mangrove swamps, 163
marine, *176*, 177
reduction, 278-9
rivers, *15*, 82, *138*, 139-41, *140*
seas, 177-9
sewage, 21
smogs, *245*, 246
water supplies, 245, 246, 248-9
wetlands, 156, 159
see also chemical pollution; radioactive contamination
polychlorinated biphenyls (PCBs), 32, 33, 113, 170, 177, 179, 232
polyps, 189, 191
Pongsak, Ajaan, 85
population, 258-9
poverty, 258, 266-7, 268
Powell, John Wesley, 127
prairies, 117-27
prickly pear, 30, *30*
Prince William Sound, 230
Prudoe Bay, 228
puffins, *181*

R

radioactive contamination, 14, *14*, 33-6, *34*, 114, 254
Arctic, 231-2
radioactive waste, 14, *14*, 33-6, *34*, 114, 140, 231, *231*
rafflesias, *71*
rainfall:
acid rain, 21, 93-4, 206, 246
deforestation and, 75
erosion, 70-3
rainforests:
commercial value, *87*
destruction, 76-87, 266
fragmentation, 36-9
natural drugs from, 70
nutrient recycling, 68
temperate, 67, *67*
Yacuna Indians' view of, 12
see also forests
Ramsar Convention, 159-60
rangelands, 117-27
recycling materials, 275, 278
reefs *see* coral reefs
Regional Mangrove Project, 163
reindeer, 231-2, *231*
reservoirs *see* dams
Rhine, 140, 249
Rhone, 203
rice crops, 104
rivers, 129-41
dams, 130-6
flooding, 136

pollution, *15*, 82, *138*, 139-41, *140*, 249
Rondonia, 77
deforestation, 77, *79*
Ross Sea, 225
rubber-tappers, 86, *86*, 281
Ruhr Valley, 139-40
rural economies, 268

S

St Gotthard Pass, 207
salinization, 97, 102, 111-12, 132, 157
salmon, 230-1
Salmon River, 93
saltmarshes, 152
San Francisco, *241*
sand dunes, 167, *168*
sand eels, 231
Sandoz, 140
Sarawak:
forests, 270
timber trade, 81-2
savannas, 117, 120-1
Scheld, River, 141
sclerophyllous forests, 61
Scotland:
acid rain, 21
conifer plantations, *95*
seabed, 174, 176
seabirds, 166, *166*
seals, 33, *51*, *218*, 231, *233*
seas, 173-85
Antarctica, 221
coral reefs, 187-93
marine food chain, 174-6
pollution, *176*, 177-9
rising levels, 49, 198, 263
wave power, 255, 276
see also fish and fishing
seashores, 165-71
erosion, 171
pollution, 168-71
wildlife, 166-8
seaweeds, *167*
semi-arid scrubland, 117, 118, *118*
Senegal, agriculture, 105, 267
Senegal River Valley Scheme, 262
Serengeti, 120
Seveso, 114
sewage:
Antarctica, 219
coastal waters, 171
coral reefs and, 191
disease and, 248
freshwater pollution, 31, 139
marine pollution, 21, 170, *176*
Shark Bay, Australia, *18*
sheep, *203*
Siberia, forests, 59
Serra dos Carajas, 82, *83*
sifakas, *196*
silt, 136
coastal erosion and, 171
estuarine, 166, 168
skiing, *206*, 207
slums, 241, *242*
slurry, *147*
smog, 245, *245*, 246
snowy owls, *233*
Socotra, 195
soft drinks, *244*
soil:
drainage, 111-12
fertility, 106-7, 110-11
organic farming, 279
salinization, 97, 102, 111-12, 132, 157
soil erosion, *26*, *79*, 81-2, *82*, 102, 111, 121-2
forest protection, 70-3
solar power, 255, 277
solar system, *16*
Sonoran Desert, *210*
South Africa, ranching, 127
South Georgia, *222*
South Korea, fishing industry, 173
South Sumatra, wetlands, 158
Soviet Union:
agriculture, 97-8, 112, 261
dams, 135
deforestation, 88-9
fishing industry, 231
krill catches, 223
nuclear contamination, 14, *14*, 36, 114
species, extinction, 39, 70, 120-1, 257
Spitzbergen, *228*
Sri Lanka, coral reefs, 187
Stein Valley, 88
steppes, 121-2
Stockholm Conference (1972), 265, 266
storks, *104*
Straits of Malacca, 163
Strathcona Park, 88
stromatolites, 18, *18*
succession, 59
Sudd, 157-8
sulphur, acid rain, 21
Sumatra:
deforestation, *78*
mangrove swamps, 163
Sundarbans, 163
Surinam, dams, 136
swamp cypresses, *152*
swamps, 151-63
Sweden:
acid rain, 21
coastal pollution, 171
plantation forestry, 94
swidden agriculture, 85
Switzerland, destruction in the Alps, 207
Sydney, *245*

T

Taiwan, fishing industry, 173
Tanzania, local fishing industry, 273
Tasmania, forests, *66*, 67
taxation, 281
Tees River, 170
temperate forests, 62-7
acid rain, 93-4
destruction, 88-95
rain forests, 67, *67*
temperate grasslands, 121-2
teratogens, 32
Texas, uncultivated land, *98*
Thailand:
agriculture, *113*
dams, *135*
destruction of the forests, 79, *79*
reforestation, 85, 95
savannas, 121
Thames River, 141
estuary, 168
Third World:
agriculture, 109, 111
coasts and estuaries, 168, 170
community breakdown, 244
dams, 133
deforestation, 58, 74
economic growth, 266
energy supplies, 255
fishing industries, 183
food exports, 262
industrial pollution, 247
local community power, 281
organic farming, 280
overpopulation, 258
pesticide contamination, 251
poverty, 266-7
river pollution, 139, 141
solar power, 277
tensions with industrialized nations, 263
urbanization, 241-2
waste disposal, 147
water supplies, 148, 245, 248
wetlands, 155
Tibet, mountains, *204*
Tierra del Fuego, *62*
timber trade, 80-2
Tonga, 198
toucans, *71*
tourism:
Antarctica, *224*
and coastal ecology, 171
wildlife and, 197
toxic chemicals and waste, 31-2,
agriculture, 112-14
marine pollution, 177-9
water supplies and, 145-7
trade agreements, 271
traffic pollution, 246-7
Trans-Alaska Pipeline, 228, *229*
transhumance, 203, *203*
transmigration, 78-9
trawlers, *182*
tree kangaroos, *70*
trees *see* forests; timber trade
Tropical Rainforestry Action Plan (TFAP), 84-5
tropical rainforests:
climate and, 74-5
composition, 59
destruction, 76-87
nutrients, 68
wildlife, 257
see also forests
Tsembaga, 85
Tuaregs, 118, *118*, 124
tuatara, 197
tundra, *228*, *234*
Tunisia, water supplies, 217
Turkmenia, 97-8
turtles, *153*, *183*
Tuvalu, 198

U

Uganda, wetlands, 154
UN Development Programme, 84, 163
UN Environmental Programme (UNEP), 257, 265
UNESCO, 163
United States:
acid rain, 93
agriculture, 114-15, 261
Clean Air Act, 247
dams, 130-2, 133
deforestation and logging, 58, 88, 90, 94
desertification, 127
energy efficiency, 276
estuarine pollution, 170
Food and Drug Administration, 270
food imports for Third World, 262
food safety, 250
forests, *64-5*, 67, *67*
hazardous waste, 32
irrigation, 112, *112*, 133
lead poisoning, 247
oil supplies, 253
organic farming, 279
ozone damage, 94
ozone pollution, 246
population levels, 259
renewable energy sources, 277
river pollution, 141
soil erosion, 122
toxic waste, 273
water supplies, 132, 143, 145, 146, 148, 249
wetlands, 156, 159
uranium mines, *34*, 35
urban sprawl, 241-2
Uzbekistan, 97-8

V

Val d'Isère, *206*
Valdez, *229*
Vancouver Island, deforestation, 88
varzeas, 152
vegetation:
deserts, 209-11
global warming and, 50
Venezuela, cloud forests, *75*
Venus, atmosphere, 16, *17*
Vicecomodoro Marambio, *220*
Vistula River, *140*, 177
volcanic rocks, *34*

W

waldsterben, 93-4
Wales:
farmland, *100*
water supplies, 95
waste, 241
recycling, 275, 278
reducing, 278-9
Third World, 147
toxic, 31-2, 247, 273
water supplies and, 145-7
water:
coasts and estuaries, 165-71
effect of conifer plantations on, 95
erosion, 70-3
flooding, 73, 102, 136-9, 156-7, 158
rivers, 129-41
saltwater intrusion, 166
wetlands, 151-63
water supplies, 143-9, 245, 246, 248-9
dams, 130-2
drought, 148, 206
pastoral societies, 124
pollution, 145
wetlands and, 158-9
wave-power, 255, 276
Weddell Sea, *224*, 225
wells, 143-9
West Germany:
deforestation, 58
river pollution, 139-40, 141
West Papua, 79
community breakdown, 244
wetlands, 151-63
conserving, 159-60
draining, 158-9
living with, 156-8
mangrove swamps, 160-3, *162*
pollution, 159
value of, 154-6
wildlife, 152-4
whales, *175*, 221, 222-3
wildebeest, 124
wildlife:
Alps, 207
Antarctica, 221
aquifers, 145
Arctic, 227, 234
coasts and estuaries, 166-8
coral reefs, 189
dams and, 134-5
deserts, 209-14
extinction, 39, 70, 120-1, 257
forests, 93
global warming and, 49-50
islands, 195-8
loss of, 36
marine pollution, 179
mountains, 202
rivers, 130
savannas, 120-1
temperate grasslands, 121
wetland pollution, 159
wetlands and, 152-4, 160
wind-power, 255, 277-8, *277-8*
Woburn, Mass., 143
World Bank, 84, 124, 129, 133, 158, 159, 257, 265, 268, 269
World Commission on Environment and Development (WCED), 266
World Health Organization (WHO), 26
World Wide Fund for Nature (WWF), 38, 193
Wright Valley, *220*
Wyoming:
bighorn sheep, *203*
prairies, *117*

X

Xavante Indians, 273
Xochimilco, Lake, 156

Y

Yacuna Indians, 12
yaks, *204*
Yanomani Indians, 82
Yekuana Indians, *84*

Z

Zaire, forests, *69*
Zambezi River, 134
zooxanthellae, 189, 193

FURTHER READING

It is impossible to list all the sources of information used for this book. However, readers who wish to find out more may find the following publications of particular interest:

A Blueprint for Survival, Edward Goldsmith, Robert Allen, Michael Allaby, John Davoll and Sam Lawrence, Tom Stacey, London, 1972.
Africa in Crisis, Lloyd Timberlake, Earthscan, London, 1988.
Altered Harvest: Agriculture, Genetics and the Fate of the World's Food Supply, Jack Doyle, Viking/Penguin, New York, 1985.
Alternative Agriculture, National Research Council, National Academy Press, Washington, DC, 1989.
Ecoscience: Population, Resources, Environment, Paul R. Ehrlich, Anne Ehrlich, John P. Holdren, W. H. Freeman, San Fransisco, 1977.
Food First: The Myth of Scarcity, Frances Moore Lappé and J. Collins, Souvenir Press, London, 1977.
Food, Climate and Man, Asit K. Biswas and Margaret R. Biswas, John Wiley and Sons, London, 1979.
Gaia: A New Look at Life on Earth, J. E. Lovelock, Oxford University Press, 1989.
Gaia: The Thesis, the Mechanisms and the Implications, Peter Bunyard and Edward Goldsmith (eds.), Wadebridge Ecological Centre, Camelford, 1988.
Green Britain or Industrial Wasteland?, Edward Goldsmith and Nicholas Hildyard (eds.), Polity, Cambridge, 1986.
Hazardous Waste in America, Samuel S. Epstein, Lester O. Brown and Carl Pope, Sierra Club, San Francisco, 1982.
How the Other Half Dies, Susan George, Penguin, London, 1976.
Nature's Economy, Donald Worster, Cambridge University Press, 1972.
Poisoners of the Seas, K. A. Gourlay, Zed, London, 1988.
Reaping the Whirlwind: Some Third World Perspectives on the Green Revolution and the Seed Revolution, Clarence Dias, Council on International and Public Affairs, New York, 1986.
State of the World, Worldwatch Institute, W.W. Norton, New York and London, published annually.
Taking Population Seriously, Francis Moore Lappé and Rachel Schurman, Earthscan, London, 1989.
The Ages of Gaia, J. E. Lovelock, Oxford, 1988.
The Fragile Environment, Laurie Friday and Ronald Laskey (eds.), Cambridge University Press, 1989.
The Health Guide to the Nuclear Age, Peter Bunyard, Macmillan, London, 1988.
The Pesticide Conspiracy, Robert van den Bosch, Doubleday, New York, 1978.
The Politics of Cancer, Samuel S. Epstein, Anchor, New York, 1979.
The Population Explosion, Paul and Anne Ehrlich, Hutchinson, London, 1990.
Winds of Change, John Gribbin and Mick Kelly, Headway, London, 1989.
Papers published by the Worldwatch Institute, Washington, are also highly informative. The following titles are particularly recommended:
Building on Success: The Age of Energy Efficiency
Clearing the Air: A Global Agenda
Environmental Refugees: A Yardstick of Habitability
Protecting Life on Earth: Steps to Save the Ozone Layer
Renewable Energy: Today's Contribution, Tomorrow's Promise
Reversing Africa's Decline
Slowing Global Warming: A Worldwide Strategy
The Changing World Food Prospect: The Nineties and Beyond
Water for Agriculture: Facing the Limits
Water: Rethinking Management in an Age of Scarcity
The following magazines and journals are valuable sources of information: *Ambio, The Ecologist, Environment Digest, Environment, ENDS (Environmental Data Services), New Scientist, Nature, Marine Pollution Bulletin, Science, Scientific American, Technology Review.*

QUOTATION SOURCES

14 *Silent Spring,* Rachel Carson, Houghton Mifflin, 1962 **19** *Five Kingdoms,* Lynn Margulis, W. H. Freeman and Company, 1982; *Touch the Earth,* T. C. McCluhan, Outerbridge & Dienstfrey, 1971 **21** *The Guardian,* 19 January 1990 **34** *Silent Spring,* Rachel Carson, Houghton Mifflin, 1962 **68** Letter to Professor J. Henslow, 1832 **72** *Index on Censorship,* Vol. 18, nos 6 & 7, July/August 1989; *Costa Rican Natural History,* Professor Daniel H. Janzen, University of Chicago Press, 1983 **73** *The World's Landscapes: China,* Yi-Fu Tan, Longman 1970 **74** *Amazonia - Oxfam's work in the Amazon basin,* Oxfam **75** Oxfam **77** *Introduction to World Forestry,* Jack Westoby, Blackwell, 1989 **80** *Fight for the Forest, Chico Mendes in His Own Words,* Latin America Bureau, 1989 **81** *Amazonia - Oxfam's work in the Amazon basin,* Oxfam **84** *In the Rainforest,* Catherine Caufield, Heinemann, 1985 **85** *Amazonia - Oxfam's work in the Amazon basin,* Oxfam **85** Dhammanaat Foundation leaflet **85** *Fight for the Forest, Chico Mendes in His Own Words,* Latin America Bureau, 1989 **92** *Touch the Earth,* T. C. McLuhan, Outerbridge & Dienstfrey, 1971 **93** Introduction to *Trees Be Company,* Angela King & Susan Clifford (eds.), The Bristol Press, 1989 **110** *Brother, Can You Spare a Dime?,* Susan Winslow, Webb & Bower, 1976 **111** *Geographical Magazine,* March 1990; *National Geographic,* September 1984 **113** *Silent Spring,* Rachel Carson, Houghton Mifflin, 1962 **114** *National Geographic,* February 1980 **124** *Human Ecology,* Professor Bernard Campbell, Heinemann, 1983 **126** *Far Away and Long Ago,* W. H. Hudson, Everyman, 1939 **127** *National Geographic,* September 1984; *National Geographic,* September 1984; *Rural Rides,* William Cobbett, 1832; *The Creation of World Poverty,* Teresa Hayter, Pluto Press, 1981 **136** *In the Rainforest,* Catherine Caufield, Heinnemann, 1984; *BBC Wildlife,* December 1989 **140** *The Sunday Times,* 22 April 1990; *Water Bulletin,* 27 April 1990 **141** *The Natural History of Selborne,* Gilbert White, 1788 **144** *Silent Spring,* Rachel Carson, Houghton Mifflin, 1962 **146** *New Scientist,* 23 February 1984; *The Listener,* 21 January 1982 **159** *Waterlogged Wealth,* Dr Edward Maltby, Earthscan, 1986 **163** *Island Africa,* Jonathan Kingdon, Collins, 1990 **174** *The Voyage of the Beagle,* Charles Darwin, 1845 **177** *The Sunday Times,* 22 April 1990 **180** *World Conservation,* Charlie Pye-Smith, Macdonald, 1984 and *The Real Cost,* Richard North, Chatto & Windus, 1986 **183** *Close to the Earth, Living Social History of the British Isles,* Judith Cook, Routledge and Kegan Paul, 1984 **184** Greenpeace Campaign Report, May 1990 **185** *The Guardian,* 25 April 1989 **189** *The Silent World,* Jacques Cousteau, Hamish Hamilton, 1953 **197** *Island Years,* Frank Fraser Darling, Readers Union, G. Bell and Sons, 1952 **216** *Sahara Desert,* Professor John Cloudesley Thompson (ed.), Pergamon Press 1984 **221** *Alone,* Admiral Richard E. Byrd, 1938, republished by Queen Anne Press/Macdonald & Co., 1987, quoted in *The Greenpeace Book of Antarctica,* by John May, Dorling Kindersley, 1988 **224** Quoted in *Whale Nation,* Heathcote Williams, Jonathan Cape, 1988; *The 1826 Journal of John James Audubon,* Abbeville Press, 1987 **229** *National Geographic,* December 1988 **240** *Silent Spring,* Rachel Carson, Houghton Mifflin, 1962; *New Statesman,* 21 April 1989 **241** *Small is Beautiful,* E. F. Schumacher, Blond & Briggs, 1973 **243** *Bring Me My Bow,* John Seymour, Turnstone Books, 1977; *Small is Beautiful,* E. F. Schumacher, Blond & Briggs, 1973 **244** *The New Environmental Age,* Max Nicholson, Cambridge University Press, 1987 **246** *The Guardian,* 6 October 1989; *National Geographic,* April 1987 **247** *The Guardian,* 19 January 1990 **249** *The Sunday Times,* 24 September 1989; *Silent Spring,* Rachel Carson, Houghton Mifflin, 1962; letter to *Farmers' Weekly,* 30 June 1989 **251** *The Guardian,* 17 July 1989 **252** *The Abolition of Man,* C. S. Lewis, Macmillan, 1947; Introduction to *Trees Be Company,* Angela King & Susan Clifford (eds.), The Bristol Press, 1989 **254** *The Guardian,* 25 April 1989 **255** *The Guardian,* 15 September 1989 **257** (both) *The Guardian,* 5 July 1988 **259** *Introduction to World Forestry,* Jack Westoby, Blackwell, 1989; *National Geographic,* December 1988 **261** *Human Ecology,* Professor Bernard Campbell, Heinemann Educational, 1983; *Bring Me My Bow,* John Seymour, Turnstone Books, 1977 **262** *National Geographic,* December 1988 **264** *The Guardian,* 25 April 1989; *The Guardian,* 15 August 1989 **265** *The Guardian,* 8 September 1989; *On The Duty of Civil Disobedience,* Henry David Thoreau **267** *Small is Beautiful,* E. F. Schumacher, Blond & Briggs, 1973 **268** *Human Ecology,* Professor Bernard Campbell, Heinemann Educational, 1983 **271** *Developed to Death,* Green Print/Merlin Press, 1989, quoted in *Winds of Change,* John Gribbin and Mick Kelly, Hodder & Stoughton, 1989 **272** *Small is Beautiful,* E. F. Schumacher, Blond & Briggs, 1973; *Fight for the Forest - Chico Mendes in His Own Words,* Latin America Bureau, 1989 **274** *Close to the Earth - Living Social History of the British Isles,* Judith Cook, Routledge and Kegan Paul, 1984 **275** *BBC Wildlife,* February 1990 **276** *The Guardian* 29 September 1989 **278** Introduction to *The Real Cost,* Richard North, Chatto & Windus, 1986 **279** *BBC Wildlife,* December 1989 **280** *When the Bough Breaks - Our Children, Our Environment,* Lloyd Timberlake and Laura Thomas, Earthscan Publications, 1990 **282** *Human Ecology,* Professor Bernard Campbell, Heinemann, 1983

PHOTOGRAPHIC CREDITS

BC Bruce Coleman FL Frank Lane NHPA Natural History Photographic Agency OSF Oxford Scientific Films
PE Planet Earth Pictures SPL Science Photo Library

6 J. A. L. Cooke/OSF **8-9** J. A. L. Cooke/OSF **10** NASA/SPL **13** Derek Hall **14** Bill O'Neill/New Scientist **15** N. A. Callow/NHPA **18t** John Reader/SPL **18b** Prof. David Hall/SPL **18-19** Mark Newman/Frank Lane **20t** Jacques Guillard/Scope **20b** Jeff Foot/BC **20-21** Karl-Heinz Jorgens/NHPA **22-23** Peter Dombrovskis/Envision **24-25** Peter Dombrovskis/Envision **29t** Jane Burton/BC **29b** Jane Burton/BC **30** Frans Lanting/BC **31t** R. van Nostrand/FL **31c** J. Cancalosi/BC **31b** Jan Taylor/BC **32-33** Alain le Garsmeur/Panos **33** Mark Edwards/Still Pictures **34** Penny Tweedie/Impact **34-35** Joe Cornish/Landscape Only **35** Y. Arthus Bertrand/Explorer **37** Hans Reinhard/BC **38** Alain Compost/BC **39** David Scharf/SPL **40** NASA/SPL **43** P. Evans/BC **44-45** F. Jalain/Explorer **45** M. Walker/NHPA **46** Tim Gibson/Envision **47** Carol Hughes/BC **49** Robert Harding Picture Library **51t** John Hartley/NHPA **51b** David Rootes/PE **52-53** François Gohier/Explorer **54-55** Frans Lanting/Minden Pictures **58-59** Eero Murtomaki/NHPA **59** Frans Lanting/Minden Pictures **60** G. E. Schimda/NHPA **61** Peter Johnson/NHPA **62** Geoff Doré/BC **62** Gunter Ziesler/BC **62-63** Daniel Fauré/Scope **63** John & Gillian Lythgoe/PE **64-65** L. West/FL **66** Peter Dombrovskis/Envision **67** Michael Fogden/OSF **69** Frans Lanting/Minden Pictures **70t** Jean-Phillippe Varin/Jacana **70b** Kathie Atkinson/OSF **71tl** Norman Owen Tomalin/BC **71tr** Frans Lanting/Minden Pictures **71b** Erwin & Peggy Bauer/BC **72-73** Frans Lanting/Minden Pictures **73** David Houston/BC **74** Frans Lanting/Minden Pictures **74-75** D. M. Moisnard/Explorer **75** G. I. Bernard/OSF **76** Mark Edwards/Still Pictures **78t** J. Hartley/Panos **78b** Frans Lanting/Minden Pictures **79t** Derek Hall **79b** F. Gohier/Explorer **80-81** Robert Harding Picture Library **81t** Frans Lanting/Minden Pictures **81b** Mark Edwards/Still Pictures **82** Derek Hall **82-83** Luiz Claudio Marigo/BC **83** Mark Edwards/Still Pictures **86t** Luiz Claudio Marigo **86c** Luiz Claudio Marigo **86b** Luiz Claudio Marigo **88** Rolf Bettner/Beautiful British Columbia **89** Gary Fiegehen/Beautiful British Columbia **92t** Eric Schings/Explorer **92b** Eric Crichton/BC **93** Eric Schings/Explorer **94-95** Frances Furlong/BC **95t** John Lythgoe/PE **95b** Rolf Lundqvist **96-97** Nicholas Devore/BC **98-99** John Shaw/BC **99t** Fred Mayer/Magnum **99b** Image Bank **100t** Gerald Cubitt/BC **100b** John Heseltine/SPL **101** Robert Harding Picture Library **102-103** D. Barrett/PE **103t** Jean-Daniel Sudres/Scope **103b** Daniel Faure/Scope **104t** Eric Hosking **104b** Leonard Lee Rue/BC **105t** Bryan & Cherry Alexander **105b** J. Hartley/Panos **106** Daniel Fauré/Scope **106-7** Michael Newton **108** John Lewis Stage/Image Bank **109** Panos **110t** Ernest Hershberger/Envision **110b** Eric Crichton/BC **112** Ken Preston-Mafham/Premaphotos **112-113** J. Hartley/Panos **113** Derek Hall **114** M. J. Thomas/FL **115t** Kit Houghton **115b** J. Hartley/Panos **116-117** Franz Camezind/PE **118-119** Luiz Claudio Marigo **118** J. Hartley/Panos **119** Penny Tweedie/Impact **120** Jonathan Scott/PE **121** R. F. Coomber/PE **122-123** Alain le Garsmeur/Panos **125t** Jonathan Scott/PE **125b** Bryan & Cherry Alexander **126** David Reed/Panos **128-129** M. Moisnard/Explorer **130-131** John Heseltine/SPL **131t** John Shaw/NHPA **131b** Stephen Dalton/NHPA **132** Ronald Toms/OSF **133** Michael Freeman/BC **134** Domenico Ruzza/Envision **135tl** Luiz Claudio Marigo **135tr** Steve McCutcheon/FL **135b** Derek Hall **136-137** Trygve Bolstad/Panos **138** Jean-Luc Barde/Scope **139t** Stephen J. Krasemann/NHPA **139b** Jacques Brun/Explorer **142-143** Penny Tweedie/Impact **144** Dr. Eckhart Pott/BC **147t** A. J. Roberts/FL **147b** Martin Bond/SPL **148-149tc** Jacques Guillard/Scope **149t** Adrian Deere-Jones/BC **149b** Mark Boulton/BC **150** Anthony Bannister/NHPA **152** Mike Holley/Envision **153t** David Maitland/PE **153b** Luiz Claudio Marigo **154** Jacques Guillard/Scope **154-155** M. L. van Nostrand/FL **155** Fritz Polking/FL **156t** John Lythgoe/PE **156b** Derek Hall **157** D. Barrett/PE **158** David Woodfall/NHPA **161** Peter Dombrovskis/Envision **162l** Alain Compost/BC **162r** Mark Edwards/Still Pictures **164** Michael Newton **166-167** Arthur Butler/OSF **167** Frans Lanting/Minden Pictures **168** Luiz Claudio Marigo **169t** Neil Cooper/Remote Source **169b** Walter Deas/PE **170** Frans Lanting/Minden Pictures **171** Mike Price/BC **172-173** Luiz Claudio Marigo **174** Larry Madin/PE **175** Steve McCutcheon/FL **176t** Robert Hessler/PE **176b** Greenpeace/Morgan **178** Fred Mayer/Magnum **181-181** Gunter Ziesler/BC **180** Jean Daniel Sudres/Scope **181** Kenneth Day **182t** J. Duncan/PE **182b** Robert Harding Picture Library **183** Greenpeace/Morgan **184** Greenpeace/Grace **184-185** Phillippa Scott/NHPA **186** Nancy Sefton/PE **188** Bill Wood/NHPA **190** Mike Coltman/PE **190-191** Rod Salm/PE **191t** Bill Wood/BC **191b** Martin Coleman/PE **192-193** Nicholas Devore/BC **194-195** Frans Lanting/Minden Pictures **196r** Frans Lanting/Minden Pictures **196bl** John McCammon/OSF **196tl** Michel Guillard/Scope **198-199** Marcel Isy-Schwart/Image Bank **199tl** Richard W. Beales/PE **199tr** Eric Pacaud/Scope **200** C. Somner/Explorer **202t** Andre Fournier/Scope **202bl** Judd Cooney/OSF **202br** David E. Rowley/Envision **203** Daniel Fauré/Scope **204t** Brian J. Coates/BC **204b** Dieter & Mary Plage/BC **205** Walter Rawlings/Robert Harding Picture Library **206t** Robert Harding Picture Library **206b** Louis Audobert/Scope **207** Jacques Serpinski/Scope **207r** David Tomlinson/NHPA **208-209** James Carmichael/NHPA **210t** Ken Preston-Mafham/Premaphotos **210b** John Shaw/BC **210-211** Arup Shah/PE **212-213** Anthony Bannister/NHPA **213t** Ken Preston-Mafham/Premaphotos **213b** G. Deichmann/PE **214** Hans Christian Heap/PE **215** Hans Christian Heap/PE **216** Ken Preston-Mafham/Premaphotos **217** W. Wisniewski/FL **218** P. V. Tearle/PE **220t** Greenpeace/Morgan **220bl** Greenpeace/Culley **220r** Kim Westerskov/OSF **221** Frans Lanting/Minden Pictures **221-222** Frans Lanting/Minden Pictures **223** Greenpeace/Morgan **224** Frans Lanting/Minden Pictures **226** Bryan & Cherry Alexander **228t** M. Ogilvie/PE **228b** Gary Crandall/Envision **229t** Bryan & Cherry Alexander **229b** Steve McCutcheon/FL **230** Greenpeace/Merjenburgh **231t** Bryan & Cherry Alexander **231b** Duncan Murrell/PE **232** Stephen Krasemann/Jacana **232-233** Bryan & Cherry Alexander **233** Brian Hawkes/NHPA **234** Bryan & Cherry Alexander **235** Mark Newman/FL **238-239** Philip Quirk/Wildlight **240** Derek Hall **242t** David E. Rowley/Envision **242b** Oliver Strewe/Wildlight **244** Oliver Strewe/Wildlight **245** Philip Quirk/Wildlight **247** Derek Hall **248** Bryan & Cherry Alexander **250** Philip Quirk/Wildlight **251** J. P. Ferrero/Jacana **252** Sandra Buchanan **254** James Manson/Colorific **255** Mark Edwards/Still Pictures **258** Susanna Pashko/Envision **260** J. Hartley/Panos **262** J. Hartley/Panos **264** Bart Barlow/Envision **266** Ken Preston-Mafham/Premaphotos **267** Grace Davies/Envision **269** Oliver Strewe/Wildlight **270** Bartholomew/Frank Spooner Pictures **271** Patrick Walmsley/Envision **272** David Reed/Panos **277** Michael J. Howell/Envision **278** James Holmes/SPL

ILLUSTRATION SOURCES

All artwork by Oxford Illustrators, except pp.57-95: Vanessa Luff

28 After *The Nature of the Environment*, Andrew Goudie, Basil Blackwell 1989 **36** Compiled after computer simulations by the Lawrence Livermore National Laboratory, California and US Air Force weather data **37** After papers in *Nature Conservation: The Role of Remnants of Native Vegetation*, Surrey Beatty & Sons Pty, Chipping Norton, Australia, 1987 **48** After a paper by S. Jelgersma in the *Workshop on the Impact of Sea Level Rise*, Delft Hydraulics Laboratory, Holland, 1986, and on *Winds of Change*, John Gribbin and Mick Kelly, Hodder & Stoughton 1989 **79** After map prepared by Technischen Fachhochschule, Berlin **84** Adapted from *Annals of Carnegie Museum* 52, September 1983: Parker et al., *Resource Exploitation in Amazonia: Ethnological Examples from Four Populations* **87** After map in *The Ecologist*, November-December 1989 **118** After *The Nature of the Environment*, Andrew Goudie, Basil Blackwell 1989 **120** After *BBC Wildlife*, September 1989 **159** After *Green Magazine*, January 1990 **170** After *Marine Pollution*, R. B. Clarke, Oxford Science Publications 1989 **179** After map in *The State of the Environment*, OECD 1985 **253** After *The Times Atlas*

Unless otherwise indicated, the illustrations were compiled from a number of sources, including data published by government agencies.

The authors would like to acknowledge the special debt they owe to the following individuals and groups, whose work has proved invaluable in the writing of this book:

Anil Agarawal, Miguel Altieri, Jayanta Bandyopadhyay, Sunderlal Bahuguna, Bank Information Centre, Stephen Boyden, David Brower, Brent Blackwelder, Derek Bryce-Smith, Catherine Caufield, Marcus Colchester, Joseph Collins, Joseph E. Cummins, Heramn Daly, Bharat Dogra, Anne Ehrlich, Paul Ehrlich, Paul Ekins, Samuel Epstein, Friends of the Earth, Robert Goodland, Susan George, Greenpeace, Charles A. S. Hall, Ross Hume Hall, Gjerrit Hekstra, David Hyndman, Hugh Iltis, Institute for Food and Development Policy, International Institute for Environment and Development, International Rivers Network, Sandy Irvine, N.D. Jayal, Alwyn K. Jones, Greg Katz, Mick Kelly, Martin Khor, Smitu Kothari, Rajni Kothari, Francis Moore Lappé, Larry Lohmann, Lokayan Institute, London Food Commission, James Lovelock, Amory Lovins, Hunter Lovins, José Lutzenberger, John Madeley, Edward Maltby, Robert Mann, R.D Mann, Lynn Margulis, Zhores A. Medvedev, Norman Myers, Thomas Outerbridge, John Papworth, David Pearce, Fred Pearce, Jonathon Porritt, Karl Polanyi, Darrell Posey, Probe International, William E. Rees, Bruce Rich, Jeremy Rifkin, Wolfgang Sachs, John Seymour, Vandana Shiva, Steven Shrybman, Peter Snell, Survival International, Third World Network, Lloyd Timberlake, Richard Webb, B. B. Vohra, World Bank, World Resources Institute, Worldwatch Institute, World Rainforest Movement, Donald Worster.

The authors would also like to thank Pete Wilkinson who contributed the chapter on Antarctica; Brian Alexander and Cherry Alexander who contributed the chapter on the Arctic; Linda Gamlin who contributed the chapter on Islands and Judith Perera for the maps and other material on the Aral Sea. Particular thanks are due to David Campbell, who originated the idea for the book; to our editors at Editions du Chêne, David Burnie and Linda Gamlin; to Ruth Prentice who designed the book, and to Bénédicte Servignat who co-ordinated the French edition.